bues

Agrochemical Fate and Movement

About the Cover

Steeply ridged bluffs of western Iowa's Loess Hills rise abruptly from the cultivated Missouri River floodplain in Harrison County. The hills are characterized by narrow ridge crests, steep side slopes, and well-defined alluvial valleys often with incised channels that terminate at an active gulley head.

ACS SYMPOSIUM SERIES **751**

Agrochemical Fate and Movement

Perspective and Scale of Study

Thomas R. Steinheimer, EDITOR
U.S. Department of Agriculture

Lisa J. Ross, EDITOR
California Environmental Protection Agency

Terry D. Spittler, EDITOR
Cornell University

American Chemical Society, Washington, DC

Library of Congress Cataloging-in-Publication Data

Agrochemical fate and movement : perspective and scale of study / Thomas R. Steinheimer, Lisa J. Ross, Terry D. Spittler.

    p. cm.—(ACS symposium series , ISSN 0097–6156 ; 751)

    Includes bibliographical references.

    ISBN 0–8412–3608–9

    1. Agricultural chemicals—Environmental aspects—United States. 2. Pesticides—Environmental aspects—United States.

    I. Steinheimer, Thomas R., 1938– . II. Ross, Lisa J., 1957– . III. Spittler, Terry D., 1953– . IV. American Chemical Society. Meeting (215th : 1998 : Dallas, Tex.) V. Series.

TD196.A34 A373   2000
628.5′29—dc21   99–53365
                                                                                                             CIP

The paper used in this publication meets the minimum requirements of American National Standard for Information Sciences—Permanence of Paper for Printed Library Materials, ANSI Z39.48–1984.

Copyright © 2000 American Chemical Society

Distributed by Oxford University Press

All Rights Reserved. Reprographic copying beyond that permitted by Sections 107 or 108 of the U.S. Copyright Act is allowed for internal use only, provided that a per-chapter fee of $20.00 plus $0.75 per page is paid to the Copyright Clearance Center, Inc., 222 Rosewood Drive, Danvers, MA 01923, USA. Republication or reproduction for sale of pages in this book is permitted only under license from ACS. Direct these and other permissions requests to ACS Copyright Office, Publications Division, 1155 16th Street, N.W., Washington, DC 20036.

The citation of trade names and/or names of manufacturers in this publication is not to be construed as an endorsement or as approval by ACS of the commercial products or services referenced herein; nor should the mere reference herein to any drawing, specification, chemical process, or other data be regarded as a license or as a conveyance of any right or permission to the holder, reader, or any other person or corporation, to manufacture, reproduce, use, or sell any patented invention or copyrighted work that may in any way be related thereto. Registered names, trademarks, etc., used in this publication, even without specific indication thereof, are not to be considered unprotected by law.

PRINTED IN THE UNITED STATES OF AMERICA

Foreword

THE ACS SYMPOSIUM SERIES was first published in 1974 to provide a mechanism for publishing symposia quickly in book form. The purpose of the series is to publish timely, comprehensive books developed from ACS sponsored symposia based on current scientific research. Occasionally, books are developed from symposia sponsored by other organizations when the topic is of keen interest to the chemistry audience.

Before agreeing to publish a book, the proposed table of contents is reviewed for appropriate and comprehensive coverage and for interest to the audience. Some papers may be excluded in order to better focus the book; others may be added to provide comprehensiveness. When appropriate, overview or introductory chapters are added. Drafts of chapters are peer-reviewed prior to final acceptance or rejection, and manuscripts are prepared in camera-ready format.

As a rule, only original research papers and original review papers are included in the volumes. Verbatim reproductions of previously published papers are not accepted.

ACS BOOKS DEPARTMENT

Contents

Preface   xi

OVERVIEW

1. **Agrochemical Movement: Perspective and Scale-of-Study Overview**   2
   Thomas R. Steinheimer, Lisa J. Ross, and Terry D. Spittler

THE KEYNOTES

2. **Monitoring Pesticide Runoff and Leaching from Four Farming Systems on Field-Scale Coastal Plain Watersheds in Maryland**   20
   D. R. Forney, J. Strahan, C. Rankin, D. Steffin, C. J. Peter, T. D. Spittler, and J. L. Baker

3. **Effects of Watershed Scale on Agrochemical Concentration Patterns in Midwestern Streams**   46
   David B. Baker and R. Peter Richards

4. **Use of Laboratory, Field, and Watershed Data to Regulate Rice Herbicide Discharges**   65
   Lisa J. Ross, D. G. Crosby, and J. M. Lee

THE CHESAPEAKE BAY WATERSHED

5. **Potential for Herbicide Contamination of Groundwater on Sandy Soils of the Delmarva Peninsula**   80
   W. F. Ritter, A. E. M. Chirnside, R. W. Scarborough, and T. S. Steenhuis

6. **A Small Agricultural Watershed Study on Maryland's Outer Atlantic Coastal Plain**   95
   W. E. Johnson, L. W. Hall, Jr., R. D. Anderson, and C. P. Rice

7. **Watershed Fluxes of Pesticides to Chesapeake Bay**   115
   Gregory D. Foster and Katrice A. Lippa

8. Watershed Monitoring in Sustainable Agriculture Studies   126
   T. D. Spittler, S. K. Brightman, M. C. Humiston,
   and D. R. Forney

9. FILIA Determination of Imazethapyr Herbicide in Water   135
   M. Lee, R. A. Durst, T. D. Spittler, and D. R. Forney

THE MIDWESTERN PLAINS

10. Spring Season Pattern of Nitrate-N and Herbicide Movement
    in Snowmelt Runoff from a Loess Soil in Southwestern Iowa   146
    T. R. Steinheimer and K. D. Scoggin

11. Runoff Losses of Suspended Sediment and Herbicides:
    Comparison of Results from 0.2- and 4-ha Plots   159
    L. M. Southwick, D. W. Meek, R. L. Bengtson, J. L. Fouss,
    and G. H. Willis

12. Estimation of Potential Loss of Two Pesticides in Runoff
    in Fillmore County, Minnesota Using a Field-Scale
    Process-Based Model and a Geographic Information System   172
    Paul D. Capel and Hua Zhang

13. Herbicide Transport in Subsurface Drainage Water Leaving
    Corn and Soybean Production Systems   185
    T. B. Moorman, R. S. Kanwar, and D. L. Karlen

14. Metolachlor Volatilization Estimates in Central Iowa   201
    J. H. Prueger, J. L. Hatfield, and T. J. Sauer

15. Pesticides in Ambient Air and Precipitation in Rural, Urban,
    and Isolated Areas of Eastern Iowa   217
    M. E. Hochstedler, D. Larabee-Zierath, and G. R. Hallberg

16. The Midwest Water Quality Initiative: Research Experiences
    at Multiple Scales   232
    J. L. Hatfield, D. A. Bucks, and M. L. Horton

17. Reconnaissance Survey of Sulfonamide, Sulfonylurea,
    and Imidazolinone Herbicides in Surface Streams
    and Groundwater of the Midwestern United States   248
    T. R. Steinheimer, R. L. Pfeiffer, K. D. Scoggin,
    and W. A. Battaglin

18. **The Potential of Vegetated Filter Strips to Reduce Pesticide Transport** 272
    J. L. Baker, S. K. Mickelson, K. Arora, and A. K. Misra

CALIFORNIA'S SAN JOAQUIN VALLEY

19. **Organophosphorous Insecticide Concentration Patterns in an Agriculturally Dominated Tributary of the San Joaquin River** 288
    N. N. Poletika, P. L. Havens, C. K. Robb, and R. D. Smith

20. **Pesticide Transport in the San Joaquin River Basin** 306
    Neil M. Dubrovsky, Charles R. Kratzer, Sandra Y. Panshin, Jo Ann M. Gronberg, and Kathryn M. Kuivila

21. **Transport and Fate of Pesticides in Fog in California's Central Valley** 323
    James N. Seiber and James E. Woodrow

22. **The Role of Dissolved Organic Matter in Pesticide Transport through Soil** 347
    J. Letey, C. F. Williams, W. J. Farmer, S. D. Nelson, M. Agassi, and M. Ben-Hur

INDEXES

Author Index 363

Subject Index 367

Preface

In North America we take for granted a high quality, diverse, and inexpensive supply of food. This abundance is the benefit of using human-made chemicals to increase farmland productivity. Enhancing soil fertility by biological or chemical amendment is a long-standing practice in production systems. What has been added in the post-World War II era is the collateral use of chemical pesticides as the first line of defense against predatory and competitive pests that threaten high productivity. This use of synthetic pesticides to mitigate the effects of weeds, insects, fungi, rodents, nematodes, and pathogenic microorganisms in many types of production systems has grown steadily during the past fifty plus years. Today we acknowledge and accept reliance upon chemical agents for pest control. However, in many production systems, pesticides do not always remain on the fields where they are applied. Parent compounds and breakdown products are randomly dissipated and selectively transported to areas where their impacts on non-target natural resources and biota are not always understood. Transport occurs via volatilization or wind-borne particles, surface water flow, or groundwater flow. Their occurrence offsite is a concern to both the public and to regulatory agencies. This is true for herbicides, insecticides, and fungicides, although the herbicides probably represent the greatest mass of pesticide transported because they constitute more than 80% of all pesticides applied.

Understanding the fate of organic pesticides in soil, water, and the atmosphere is complex for several reasons. First, many different types of processes, some biochemical and some geochemical, play a major role in determining fate by controlling dissipation and degradation. Second, the identification of important intermediate species often tests the limits of analytical chemistry, requiring sophisticated and costly techniques for support of monitoring studies. Confounding the complexity issue is the scale at which research should be conducted. For example, when assessing microbial degradation of a pesticide, the scale of research may be conducted at the Petri dish, small plot, or commercial field level. Do these experiments yield the same results in terms of our understanding of the degradation process? Under what circumstances is the Petri dish level preferable over the field level? Is a small field plot adequate? Do researchers need one or more commercial-sized field plots to account for the potential variability in degradation and dissipation processes and rates of reactions? What are the best study designs for pesticide transport in watersheds? Are small tributaries representative of a watershed in general or does each stream and river have its own characteristic pattern of pesticide fate and movement? These are the types of questions researchers face when deciding how to examine the environmental fate of pesticides.

Various stakeholders are conducting additional studies addressing environmental fate and movement questions. Recognizing that degradates may be more important in water quality assessments than earlier believed, stakeholders are also seeking more information on the dissipation of daughter compounds. The need for a detailed and comprehensive understanding of the fate and movement from production areas accompanies the discovery and development of each new pesticide. Furthermore, new factors that have entered the "risk–benefit analysis" dialogue include increased risks to children and at-risk populations and the potential impact of chronic exposure to low levels of pesticides acting as environmental endocrine disruptors. Passage in 1996 of the Food Quality Protection Act requires that cumulative risks, both dietary and environmental, be calculated for individual chemicals as well as for their chemical class together.

These questions, which we have faced on numerous occasions in our own research, prompted us to organize the symposium, *Agrochemical Movement: Perspective and Scale*, sponsored by the Division of Agrochemicals during the 215th American Chemical Society (ACS) National Meeting in Dallas, Texas, March 29–April 2, 1998. We hoped this forum would promote discussion among scientists who conduct studies at various scales to help us better comprehend the role that each level plays in our understanding of the fate of pesticides in the environment. We organized the symposium with two paradigms in mind. First, that the type of information generated concerning the fate and movement of pesticides is scale-of-study dependent. Studies designed to delineate degradation and dissipation processes are different than those designed to assess movement and often require a different scale for investigation. In addition, studies of either processes or movement will be enhanced by use of more than one level of scale. Second, both similarities and differences exist among farming systems across the United States with geographic features, climate patterns, and agronomic practices that influence both pesticide movement and the types of pesticide issues of greatest concern. These differences may lead to a different understanding of the behavior of pesticides and their potential impact on natural resources. The book identifies three agroecoregions, each distinct in its prevailing agronomic practice and scale of operation: the Chesapeake Bay and Eastern Shore, the Upper Midwestern Plains, and the San Joaquin Valley of California. Although pesticide use issues are dominated by herbicides along the east coast and in the midwest, they are dominated by insecticides in the west. The book is intended for practicing professionals in the environmental sciences; natural resource managers; agricultural scientists; farm management advisors and consultants; pesticide research specialists; and pesticide registration specialists and other regulatory professionals. Readers of this volume will find a single resource that provides a national scope and context for comparing agrochemical movement issues across the United States.

The editors thank all chapter authors for their excellent contributions. Each brings valuable insight to scale-of-study issues and to the differences in agrochemical movement issues across the United States. We hope this collective work will serve as a focal point for continued discussion in both the scientific and public realms. The editors express their appreciation to the ACS Division of Agrochemicals for providing the forum for this work. Thomas Steinheimer thanks Jerry Hatfield for his encouraging support of this project and Kenwood Scoggin for assistance in the preparation of several chapters. Terry Spittler thanks Bernice Andersen for invaluable assistance with manuscripts and communications.

THOMAS R. STEINHEIMER
National Soil Tilth Laboratory
Agricultural Research Service
U.S. Department of Agriculture
2150 Pammel Drive
Ames, IA 50011

LISA J. ROSS
California Environmental Protection Agency
California Department of Pesticide Regulation
830 K. Street
Sacramento, CA 95814

TERRY D. SPITTLER
Cornell Analytical Laboratories
New York Agricultural Experiment Station
Cornell University
Geneva, NY 14456

OVERVIEW

Chapter 1

Agrochemical Movement: Perspective and Scale-of-Study Overview

Thomas R. Steinheimer[1], Lisa J. Ross[2], and Terry D. Spittler[3]

[1]National Soil Tilth Laboratory, Agricultural Research Service, U.S. Department of Agriculture, 2150 Pammel Drive, Ames, IA 50011
[2]California Environmental Protection Agency, California Department of Pesticide Regulation, 830 K Street, Sacramento, CA 95814
[3]Cornell Analytical Laboratories, New York State Agricultural Experiment Station, Cornell University, Geneva, NY 14456

Scientists today acknowledge the important role played by agrochemicals in maintaining agricultural production, yet issues remain concerning the movement of these crop protection chemicals away from their points of application and into nontarget natural resources. A wide variety of chemicals are used to address different pest control needs. The choices made by the producer often are determined by the nature of the landscape, pest cycles, cropping systems, weather patterns, and the economics of production. Hence, they differ across the U.S. We present an overview of incidental movement studies of pesticides used in production agriculture and their potential nontarget impacts within three distinct geophysical regions of the United States. Examples from laboratory, test plot, and field investigations of agrochemical movement will illustrate the importance of scale in the study design in the Chesapeake Bay Watershed, on the Midwestern Plains, and in California's San Joaquin Valley.

The abundant, high quality, diverse, and relatively inexpensive food supply that most Americans take for granted is a direct result, for many commodities, of the use of pesticides in production agriculture. Delivery of other nonfood goods and services, often of higher quality and at lower cost to the consumer, also result from the use of pesticides. These economic and quality of life benefits are not achieved without some potential risk to human health and the environment as a consequence of the application of pesticide chemicals. Therefore, laws which regulate the distribution of pesticides and programs which monitor their constantly changing occurrence patterns are necessary to assure the public that unacceptable risks are avoided.

Two central issues regarding the consequences of pesticide usage in production agriculture relate to their potential impacts on nontarget natural resources. The first issue recognizes that quantitative assessments of impact require a detailed understanding of the environmental fate and movement aspects for each chemical when introduced into its intended agricultural setting. In this context, four questions should be raised regarding the fate of an applied pesticide; what happens to it and how much of it moves ?, where does it go ?, how does it get there ?, and, how far offsite from its point of introduction does it move and what does it do ? The first question addresses the efficacy of placement and utilization or degradation reactions, both in the soil and in plants through which it may be translocated. These are chemistry issues addressed by soil chemists, soil microbiologists, and geochemists. The second question acknowledges the importance of the partitioning of a pesticide among the various hydrogeologic compartments which are identified as components of a cropped field. These could include precipitation, surface runoff, unsaturated-zone water and well water. The third question addresses the notion that partitioning through the soil volume within a field is governed by the physical processes of advection and dispersion, and seeks to explain how these processes result in the partitioning. The fourth, and, perhaps, most important question for the general public, addresses edge-of-field issues, and the displacement offsite and entry into nontarget natural resources. For many fields, in every geophysical region of the U.S., drainage from agricultural landscapes often is hydrologically connected to larger basins or drainage conduits, eventually entering either the Nation's largest rivers and/or coastal waters.

The second issue, which has only emerged within the past 15 years, revolves around the recognition that most agricultural pesticides are neither biorefractory nor completely mineralized in the soil environment to which they are applied. This leaves researchers with the challenge of determining dissipation properties, specific breakdown pathways and mechanisms, and overall degradation rates (*1*). The importance of knowing the environmental-fate, pathways, rates of overall dissipation, and potential for environmental risk associated with soil metabolites has only recently been recognized for registration. It is therefore important to establish which metabolites are relevant from an ecotoxicology standpoint. Recent studies suggest that when degradates are not taken into account for groundwater quality studies, as much as 50% of the herbicide occurrence legacy may be overlooked (*2*). A pesticide metabolite may have increased relevance when *i*) it is a pesticide itself, *ii*) its molecular structure is only slightly altered in comparison to the structure of the parent, *iii*) it is known to be toxic or its structure contains moieties which make a substance toxic, *iv*) its field half-life is >3 months, and *v*) it follows from its structure that its soil adsorption is low and persistence relatively high. A metabolite may have diminished relevance when it is a compound which occurs naturally as a product of microbial metabolism, or is retarded as a bound residue. For most pesticides currently in use, many of these questions remain unanswered.

On farm pesticide usage increased from about 182 million kg. in the mid-1960's to nearly 386 million kg. by 1980, due to increased usage in all types of chemical control agents. By far, the greatest growth resulted from the widespread adoption of herbicides for weed control in crop production (*3*). Since the mid-1980's total pesticide consumption has increased only modestly. Table 1 compares recent production statistics for herbicides and insecticides, the two dominant chemical types. Nematocides, fungicides, rodenticides, acaricides, miticides, and plant growth regulators together constitute much smaller percentages. 1987-1995 data are production estimates; 1996-97 are consumption estimates, and do not include production intended for export. Herbicide production peaked in 1994 while insecticide production peaked in 1993. During the past ten years, the net increase in herbicide production and insecticide production are 1% and 4%, respectively; while the net increase for fungicides is >20%. Wide variation in yearly pesticide use patterns result from weather, unexpected pest outbreaks, cropping choices and acreage, and economic factors.

Table 1. Pesticide Production and Consumption in the U.S., 1986-1997.[a]

	Herbicides	Insecticides	Nematocides	Fungicides
1986	625	204	66	70
1987	623	185	76	62
1988	676	203	85	193
1989	676	208	112	207
1990	643	264	67	85
1991	657	181	72	na
1992	696	194	71	85
1993	749	308	82	77
1994	868	253	100	72
1995	819	174	97	143
1996	568	185	69	85
1997	575	192	na	103

[a] Chemical and Engineering News, June 23, 1997 and June 29, 1998. All values are millions of lbs.; does not include production of nematocides and plant growth regulators intended for sale outside the U.S.

Over the past three decades, the variety and chemical complexity of organic pesticides registered and introduced into commerce has steadily increased. Moreover, other recent developments have further complicated pesticide risk debates. First, some existing and many recently registered pesticides actually are mixtures of similar but non-identical chemical components. Their relationship to each other is that they are non-superimposable on their mirror images, a consequence of their molecular assymetry. Known as enantiomers, they exhibit different chemical and biological properties, some of which are useful in enhancing herbicidal activity. Examples of such formulations which are enantiomeric with respect to active ingredients are Dual Magnum[R], Pursuit[R], and Balance[R]. As a consequence, an issue which is gaining attention within the regulatory community is whether or not it is necessary or appropriate to conduct pesticide exposure assessments through monitoring activities for specific enantiomers. Analytically, separations of individual enantiomers of a host of biologically active chemicals is one of the hottest research areas in all of environmental analytical chemistry. Second, those concerned with the long-term legacy of pesticide usage, primarily in agriculture, must acknowledge that dissipation in the field of the registered agent over time does not equate to zero impact on non-target environments. If assessments of impacts are to be supported by the best science, then degradate chemicals must be included in analytical schemes for all samples. Third, the possibility that long-term exposure to low-levels of anthropogenic contaminants may interfere with endocrine system function in many species has been raised, largely by population biologists. As a class of commercially important synthetic chemicals, pesticides have come under scrutiny due to their wide-spread distribution throughout our terrestrial environment. Today more than 30 classes of chemicals with pesticidal properties are now widely acknowledged by most scientists and regulators, alike. Illustrating this diversity are the variety of chemicals registered for a broad spectrum of problems in weed, insect, and fungal control.

Herbicides	*Insecticides*	*Fungicides*
Arylanilines	Carbamtes	Azoles
Benzoic Acids	Organochlorines	Benzimidazoles
Bipyridyliums	Organophosphates	Carboxamides
alpha-Chloroacetamides	Organotins	Dithiocarbamates
Cyclohexadione Oximes	Oximinocarbamate	Morpholines
Dinitroanilines	Pyrethroids	Organophosphates
Diphenyl Ethers & Esters		Phenylamides
Hydroxybenzonitriles		Strobilurine Analogs
Imidazolinones		
Organophosphates		
Phenoxyacetic Acids		
Sulfonylureas		
Thiocarbamates		
sym-Triazines		
unsym-triazinones		
Uracil		
Ureas		

Water Quality

In an agricultural cropping system, there are three compartments to which pesticides are applied which become the sources for subsequent loss to water. These are the vegetation, the soil surface, and the soil below the surface. Leaching, the process which may contaminate groundwater directly, occurs as water percolates down through the soil carrying with it a load of pesticide whose properties facilitate vertical movement through the soil volume. Runoff, the process which may contaminate surface water directly, occurs whenever precipitation rate exceeds infiltration rate, thereby generating flow across the soil surface. This flow solubilizes pesticides directly and mobilizes soil-particle attached pesticides. Both leaching and runoff may be enhanced by washoff from plant surfaces.

In a 1986 report to Congress the USEPA stated that the nation's remaining water quality problems were largely attributable to pollution from nonpoint sources (NPS). About 50-70% of the assessed surface waters were adversely affected by agricultural nonpoint source pollution. A major component of this NPS pollution is pesticide losses in runoff together with leaching to groundwater. NPS pollution issues are further complicated by several uniquely defining characteristics (4):

- NPS pollution is not easily associated with a defined process producing a specific discharge.

- NPS pollution often is intermittent and related to the intensity of an intermittent event, making it difficult to quantify and even more difficult to control.

- NPS pollution originates over a broad area making identification and source assessment difficult.

- NPS pollution issues often transcend political jurisdictions.

- NPS pollution sources often resist regulatory-based controls.

- NPS pollution sources often resist regulatory-based controls.

Runoff and leachate from agricultural production activities are the primary NPS processes responsible for impairment of our nation's water resources. In addition to direct effects on surface and groundwater quality, water leaving farmed and/or husbandry landscapes carrying agricultural chemicals can also cause physical habitat alterations as well as flow or riparian-zone functional modifications. The extent of these impacts depends on several factors. Application factors include site, type of applicator equipment, the chemical nature of the formulation, and the application amount/volume and timing. All impact runoff and leaching behavior of pesticides. Site of application could be directly on the growing crop or undesirable weed plants, the soil surface, or by incorporation into the subsurface soil. Other application considerations are the properties of the active ingredient chemical. Another factor which impacts NPS processes is persistence and mobility. Some pesticide-soil combinations result in such strong binding to soil particles that the pesticide moves only if the soil moves, as is the case with water or wind erosion. Many pesticides now in use degrade so rapidly on soil and crop surfaces that rainfall must occur within a short time following application for significant movement to occur. Pesticides must be relatively persistent and mobile to leach to ground water because the time required for water to percolate to deep aquifers can range from months to years. However, once a pesticide has leached into subsurface soil horizons, biological activity and binding capacity there are often diminished in comparison to surface soil. Thus, in general, as the pesticide moves down through the soil, it becomes more persistent and more mobile. Persistent and mobile pesticides represent a greater threat for runoff. However, that portion of the pesticide load which is most available for movement as runoff, that at the topmost surface of the soil, is also the portion most rapidly dissipated by evaporation and photodegradation. Furthermore, runoff events are often complete in hours, and the accompanying erosion can move immobile pesticides attached to soil. Thus, for many widely used chemicals, pesticide runoff is less dependent on pesticide properties than pesticide leaching, and much more dependent on the timing of runoff events in the hydrologic regime relative to application.

Textural composition of soil and landscape topography are another pair of related factors affecting NPS processes. Soils differ widely in their capacity to absorb water. The slope and drainage pattern of a field or a watershed greatly affect its potential to generate runoff water. Well-drained soils such as sands and sandy loams have the greatest leaching potential, and poorly-drained clays and silty-clays have the greatest runoff potential. In addition, watershed drainage area has an important effect on pesticide concentrations in runoff. Small streams adjacent to pesticide treated fields may carry very high peak concentrations, even exceeding 1 ppm, but with a trend showing a rapid decrease over the time of the runoff event. Whereas, in large river basin sized areas, peak concentrations are likely to be much lower but with a trend showing an elevated load remaining over a longer time period. Weather and climate also impact NPS processes because both determine the types of crops grown, the intensity of pest-control problems, and the persistence of the chosen pesticides. The intensity, amount, and timing of precipitation with respect to application determines how much pesticide movement occurs. While not controllable by the farmer, probabilities for both runoff and leaching can be minimized. Careful scheduling of pesticide application on dry low-surface-wind days and around imminent precipitation is possible. Today farmers are provided with information from regulators and manufacturer's necessary to prevent pesticide contamination. Knowledge of erosion patterns and control issues on each field, best application techniques, and diligent monitoring of weather patterns, are the best approaches to minimizing pesticide losses.

The issues underlying the potential for agrochemicals to damage nontarget environmental settings transcend every region of the U.S., and, at the same time, are of concern to both public and private sectors. Thus, the co-organizers determined at the outset of the planning that the Symposium should bring together three elements. First, it should have as much of a national scope as is practical while recognizing that practices in production agriculture are vastly different across the U.S. Second, the dialogue should be fully open to all and welcome the active participation of all agrochemical interests; academia, producers, manufacturers, and government agencies; those with a regulatory mission, those with a commercial mission, and those with a research mission. Third, all presentations be given the opportunity to be expanded into a chapter for inclusion in the ACS Symposium Series volume. Following are synopses selected from contributed talks. We begin with highlights from the Keynote Session and follow through with discussions of water quality issues representative of each of the three major regions of the continental U.S. where intensive agricultural production is the dominant land use.

The Keynote Session

As are most of the agricultural operations along the east coast of the U.S., those of the Delmarva Peninsula are characteristically small 10-100 ha. fields divided by numerous natural barriers such as streams, inlets and woods, plus the anthropogenic boundaries of roads, structures and property lines. The eastern shore of the Chesapeake Bay is highly agricultural and intimately associated with coastal streams, rivers and inlets discharging directly into the Bay. There is relatively little agriculture along the Atlantic Coast because of the unsuitability of shore and dune growing conditions, and the other uses which occupy most of the ocean shoreline. Field applied pesticides need to be transported only short distances before they are either added to the ebb and flood of the Chesapeake Bay's tidal movements or have reached the generally shallow 1-5 m. water tables typical of these near-sea level regions. R. A. Forney and others describe field studies of four grain cropping and tillage schemes set out on four discrete watersheds, each 2-10 ha. in area. Leaching of applied pesticides to the water table by infiltration (wells) and runoff (through autosampler-equipped flumes) could be tracked for these defined, confined test systems. Vadose zone infiltration in the absence of runoff was simultaneously measured in nearby replicated level plots equipped with pan and suction lysimeters. The scale of the watershed trials at Chesapeake Farms is within an order of magnitude or less of many of the region's production operations, and those few that are significantly larger would have broad level centers whose major avenue of agrochemical movement would be infiltration downward, a situation covered by the companion replicated plots at Chesapeake Farms. In this region, studies are focused on approximately the same scale as the geographical units. This program is focused on minimizing both production costs and environmental burdens by optimizing tillage practice, cultivar rotation, application rates and timing, and desirable chemical properties. The complicating factor is the chronology and intensity of rainfall events relative to applications. However, the relevance of each of these controllable variables is demonstrated in the absence of extreme meteorological circumstances.

The concept of the "watershed" is applicable at spatial scales ranging from 1 m^2 boxes set in environmental chambers in laboratory settings to replicated 1 a. plots on research farms to major continental river basins, such as the Mississippi, with a drainage area in excess of 3 million km^2. Studies of runoff displacement of water, suspended sediments, nutrients, and pesticides are reported on areas across these watershed sizes. Often these data reveal systematic shifts in a variety of hydrologic, concentration, and loading characteristics that are related to watershed size. These shifts can be considered scale effects. D. B. Baker and R. P. Richards, have been examining scale effects in assessment of agrochemical runoff patterns and suspended sediment displacement within tributaries of the Lake Erie Basin. Their sampling locations subtend watersheds ranging from 10-16000 km^2 in size across similar land use patterns and soil characteristics. As watershed size increases, peak storm event concentrations decrease while the durations of mid-range concentrations seem to persist for longer periods. The extent of such scale effects is parameter specific, but is most evident for suspended solids. Data are presented to support the hypothesis that these scale effects are attributed to the pathways and timing of contaminant movement from fields and into streams. Furthermore, dilution associated with the routing of water into and through the stream system from differing positions within the watershed also influences observed concentrations and loads. Implications from their results for comparing water quality data and for designing sampling programs for risk assessments are addressed as well.

California's Central Valley is one of the most diverse and intensively farmed areas in the U.S. The valley is hydrologically divided into two regions: the southern San Joaquin-Tulare watershed and the northern Sacramento River watershed. Fertility of this region depends, to a large degree, on a sufficient, high-quality water supply. However, due to the Mediterranean climate of this region the growing season is, for the most part, dry, with a majority of the rainfall occurring in winter months. Most of the water that supports this agricultural region is through a system of dams and manmade conveyances designed to supply water during the warm, dry growing season. Field applied pesticides are transported offsite both during the growing season as well as during the rainy winter season through surface water runoff, the atmosphere, and the soil profile. Many of the offsite transport issues in the Sacramento River are associated with rice culture. In the early 1980s, rice herbicides were responsible for fish kills that occurred in agricultural drains of the Sacramento River watershed. In addition, the presence of certain herbicides in the city of Sacramento drinking water supply generated taste complaints from the public. Health limits for the public, as well as water quality goals to protect fish and wildlife were developed. L.J. Ross and others describe the development of use restrictions based on the physical and chemical properties as well as field data on environmental fate and behavior of these herbicides. In addition to the need for laboratory and small field plot information, data from the actual watershed was essential to refine use restrictions and meet water quality goals. The authors demonstrate the need for information at all scales of research: laboratory, small field plot, and watershed, in the success of a regulatory program to control herbicide discharge from rice fields. Issues in the San Joaquin Valley are more frequently related to pesticide applications to orchards during the dormant season.

Chesapeake Bay and the Eastern Shore

An assessment of agrochemical movement in the Chesapeake Bay watershed region must take several factors into account. First, it must be realized that this fertile region, blessed with ample rainfall and a favorable temperate climate, is transected by the heavily populated and industrialized Northeast corridor—variously described as a vital artery or urban sprawl. Residential, urban and commercial land uses preempt much agricultural production, while at the same time cause their own nonpoint source contributions of nutrients, lawn use and other urban pesticides, plus nonhorticultural runoff and infiltration. Protection of local water supplies is always a consideration, while the sporting and recreational qualities of local lakes, rivers and streams also have strong popular and legal advocacy. However, it is the Chesapeake Bay itself that receives not only the chemical influx from this region, but its close attention as well.

Historically, the Bay has been a sheltered sea of massive shellfish resources, near shore fisheries and an extended transportation system, as well as a major breeding ground for migratory marine species and waterfowl. Loss of this habitat and its long producing resources is recognized as being potentially ecologically and economically devastating. Second, agricultural operations are of relatively small acreage—originally limited by what could be cleared and maintained by nonmechanized labor, and now, realistically, by natural and jurisdictional dividers that would be uneconomical or unacceptable to remove. It does, however, contain some of the country's most intensive agricultural operations, to wit, the poultry farms and processing facilities that drive the surrounding grain production and frequently overpower the local ecosystems' capabilities to absorb massive amounts of natural runoff from manure and processing wastes. The sum total of the regions numerous and diverse agricultural operations still makes for a relatively high percentage of agricultural land use in the numerous drainages comprising the Chesapeake watershed. For instance, the southern tier of New York State is in this region: agriculture is still New York's biggest business. Third, the watershed is geologically quite diverse, with low relief coastal plains on the eastern and near western shores, but with major western tributaries reaching back through the piedmont region to the Appalachian foot hills. In fact, the entire Appalachian range is transected by the Susquehanna River's course from New York through Pennsylvania. The result of this is that homogenous geographical systems are small, capable of being accurately approximated by investigations on scales that may not be appreciably smaller than their subjects. Conversely, attempts to include large geographical areas must immediately accept the compromise that they represent the average of many uses, numerous chemistries and multiple, competing geological routes. As with politics, most measurable pesticide movement is local on the eastern shore: field runoff goes quickly into riparian zones (marshland headwaters of near-coastal streams) or runs via branch streams to minor tributaries to coastal river tributaries of the Bay. The additive effect of all these small watersheds continues to the fall-line of the river. There, at the tidal head, further definition is complicated by multidirectional flows from tides and other Bay currents. This is the bottom line for Bay-bound and Bay-borne contaminants: are they of high enough concentration, timing and duration to adversely affect the thousands of marine and other species that rely on this largest of estuarian systems for breeding grounds or life-long habitat? Minimization of the influx of anthropogenic materials, either toxicants or nutrients, will help protect this extremely important ecosystem.

Delaware's level coastal plain is not only relatively far removed from either the Bay's eastern shore or the Delmarva's east coast, its essentially sandy soil also helps make infiltration the almost exclusive transport mechanism for applied agricultural pesticides. With water tables normally five meters or less in this area, the initial response for planning an appropriate research structure might be that representative soil columns could be established in the laboratory and produce percolation data predictive of field conditions. In fact, this approach would be a poor simplification, as applied chemicals can be found at watertable levels within minutes of a rain event because of nonhomogeneous macropore flow through wormholes and root voids. William Ritter reported on multiple tillage, application timing and chemical combinations used on field scale plots and modeling experiments with simulated rainfall events of selected magnitude. While the watertable to which the majority of transportable material infiltrates is not a vast dynamic aquifer flowing down gradient to supply other regions, it can be perturbed by extensive pumping for irrigation or municipal uses. Plus, rural dwellings relying upon wells for water may have their sources' quality affected by local agricultural operations.

Those sections of the coastal plain, the predominately agricultural eastern shore in particular, which are heavily subdivided by bays, inlets, rivers and a maze of tributaries, potentially afford applied agrochemicals rapid transit to the Chesapeake system. While the vertical drop from field to stream level is small, the distance may be very short, thus allowing for development of a reasonable gradient for runoff under certain hydrologic conditions. Several runoff producing events each year are not uncommon with the chemical burden primarily a function of the time between last application and the event. Brinsfield, et al., determined that in moderate to heavy rain events loose, silty-loam soil of conventionally plowed plots rapidly formed a poorly permeable, muddy upper layer that encouraged increased runoff compared to the thatch and organic litter on the surface of the adjacent no-till plot. Upon drying, this surface formed a hard crust that further abetted runoff in subsequent events.

Once agrochemicals have been transported by runoff or lateral groundwater movement into the Chesapeake Bay's tributary systems, the multiplicity of operations contributing to this loading requires that they be considered nonpoint source pollutants (NPS). Occasionally a stream is contained within a defined, homogenous watershed such that knowledge of the relevant operations allows the data gathering to be limited to applied materials. However, most tributaries receive discharge from multiple, uncharacterized sources, thus limiting the study to a few chemicals of interest, or requiring that a screen of all probable or suspected chemicals be conducted on collected samples if a realistic measurement of the anthropogenic burden is desired. Quite logically, as the sampling site(s) move further down the tributary network, study areas become larger and less well defined.

Conducting studies on larger areas requires the compromises of event to sampling time lags, dilution, band broadening, and carryover from prior events. W. E. Johnson and coworkers studied two monitoring sites 2.7 km apart on the German Branch of the Choptank River, an estuary of the coastal plain. In contrast to the western shore tributaries that have extensive networks reaching far inland, eastern coastal plain streams are small basins that flow directly to tidal water. The German Branch drainage is over a quarter million acres, of which 70% is agricultural use. Twelve agrochemical/nutrient parameters were included in the analyses of the weekly samplings, and one major event was intensely followed using four-hour sampling intervals. Coupling this with stream flow measurements, analyses of rainwater-borne materials, and determinations of agrochemical concentrations associated with mobile sediment, the authors provide an in-depth overview of down gradient movement.

Monitoring data for metolachlor and atrazine collected from tributaries of all ranks over a two year period has been collected and compared to toxicology benchmarks for selected estuarian species by Lenwood Hall, Jr., et al. In his presentation, he noted essentially negligible concentrations from late summer through early spring, followed by their rapid appearance after spring application, with subsequent later detection at lower levels in the major tributaries and mainstem region. Overall concentrations were deemed too low in all habitat samples to constitute ecological risks.

The western shore tributaries, which have diverse use drainages extending hundreds of miles westward and northward into the piedmont and Appalachian regions, bring a varied contribution to the Bay's ecosystems. In studies by G. D. Foster and coworkers, streams were sampled at the fall-line, or tidal head, and the samples analyzed for a suite of eighteen pesticides. The analyte list included both herbicides used in coastal plain grain production and many from upland landuse protocols for vegetables, turf control, and ornamentals. Coupled with river flow measurements, these data produced seasonal discharge flux approximations consistent with first-principal expectations. Two eastern shore rivers had the highest pesticide concentrations, which was typical for lower order rivers (less dilution) close to the application areas. More complex river systems introduce more complex variables which influence the seasonal flux of chemicals in the seven western shore rivers. All discharge producing operations in these watersheds are also subject to the same variable that dominates smaller scale studies; rain event timing and intensity.

Throughout the entire eastern geographical realm one type of high intensity horticultural operation has been increasing relentlessly. Over 500 new golf courses are added every year in the U.S., most having zero-tolerance, high standards of performance and maintenance. Following the selection of optimum soil type, grass cultivar and topographic profile, an intense fertilization and pesticide routine is followed to guarantee their integrity. Both runoff and infiltration are controlled to insure consistent soil moisture and the absence of non-intentional standing water. Martin Petrovic has long studied the optimization of these conditions at Cornell University, and presented a full spectrum of experimental scales ranging from greenhouse flower pots, through meter-scale plots with movable covers to give complete control over precipitation, and on to whole golf course monitoring studies. Turf, including lawns, recreational and public areas is the country's largest acreage crop. Its potential for contributing mobile agrochemicals to a region as heavily developed as the eastern United States must be taken very seriously.

The Midwestern Plains

The midcontinent region of the U.S. is the largest and most intensive agricultural region in the country. Consequently, the majority of all agricultural chemicals are applied to field crops in this region. The intense use of soluble and mobile herbicides poses potential problems for NPS pollution of streams and groundwater throughout the midwest. Virtually all of the modern herbicides registered for weed control on cropland across the North American continent can be found in use on a field somewhere within the midwest each growing season. In Iowa, in 1990 alone, herbicide usage on row crops rose to more than 21 million kg of active ingredients applied to some 9 million ha. Similar intensity-of-usage patterns are found from the Dakota's to Pennsylvania and extend as far south as Missouri. While the total poundage may have decreased somewhat since that time, herbicides continue to be the major tool for weed control used by most producers. Furthermore, the chemical choices, both in the number of new products together with reformulating or re-packaging of current products, continues to increase. Consequently, assessments of their occurrence in environmental compartments and mechanisms of their movement continues to be studied. Runoff from fields shortly after application in the Spring can result in high concentrations of herbicides in surface water (*5-9*). Similarly, in many areas of the midwest on which relief across the landscape is relatively low, edge of field movement via runoff is not a major loss route. However, vertical displacement of chemicals often occurs by movement through the root-zone, into the intermediate vadose zone, and ultimately to the water table. Subsurface flow subsequently transports chemicals offsite (*10-14*).

The movement of herbicide through subsurface drainage to surface water is the focus of research conducted on an experimental farm in northeast Iowa by T. B. Moorman and coworkers. These replicated one-acre plot scale studies are carried out on predominantly silt-clay loam soil underlain by seasonally high water tables and benefit from subsurface drainage. The drainage consists of tile lines in place near the base of the root zone down the center and along the borders of each plot. The effects of both herbicide banding and swine manure application on the leaching of atrazine and metolachlor through these soils are described. Herbicide losses are highly variable, ranging from 0.01-10 g/ha, and are determined by precipitation patterns, tillage practices, and soil-herbicide interactions. Observations comparing atrazine and metolachlor loss on manured plots suggests that the largely organic amendment reduces atrazine losses to tile water to a greater degree than it reduces metolachlor losses. Herbicide banding lowers the application rate and reduces herbicide losses proportionately.

For several families of pesticide chemicals, especially organophosphate insecticides and several classes of herbicides, field losses resulting from volatilization from both plant canopy and soil surfaces can be a major loss mechanism. Pesticides in the atmosphere have been studied for some time, but only recently has the emphasis shifted to measuring fluxes and decay curves for selected herbicides in air samples at times which permit direct correlation of concentration with both agronomic practices and weather conditions. J. R. Prueger and coworkers describe their efforts attempting to correlate observed metolachlor volatility profiles with other measurements in agricultural meteorology and develop a predictive capability for assessing herbicide loss to volatilization. In contrast to the water quality losses through runoff and leaching, this loss mechanism dissipates rapidly, and is at its greatest levels in the days immediately following application. Consequently, high-frequency sampling becomes very important. Metolachlor volatility profiles are compared for two different methods of herbicide application.

Atmospheric transport and deposition of pesticides in precipitation are significant questions when addressing risk assessment from a human exposure perspective. In a one-year study conducted at four sites in Iowa chosen to represent both rural and urban environments, M. E. Hochstedler and coworkers identified twenty-eight pesticides in precipitation between October 1996 and September 1997. Their study emphasizes comparisons of sites based upon predominant landuse patterns and discusses implications for both the temporal variability of atmospheric loading of these chemicals as well as possibilities of long-range transport. The discussion addresses the appropriateness of chosen methodology for both field sampling and residue analysis of low-conductivity rainwater samples and the constant dilemma over minimum sample volume for analysis.

Patterns of herbicide usage across the midwest, primarily in corn and soybean, have changed since the mid-1980's. Reduced tillage has grown and with it an increased need for new chemistry compatible with the requirements of post-emergence weed control in conservation tillage operations. In the upper midwest, especially in Iowa, Illinois, and southern Minnesota, among the most widely adopted by farmers of these new chemistries are the sulfonylurea and imidazolinone families. Both are characterized by high efficacy at low application rates, and relatively low toxicity to mammals, arthropods, marine life, and aviary species. Because they are relatively new, all are post-1985 initial registration, their potential impacts on water quality are not well known. In a somewhat unique collaborative effort involving two federal agencies and a producer of herbicides, a reconnaissance study of midwestern water resources which hold the potential to be impacted by farming practices employing these new chemicals was carried out in five states between June and October of 1997. In addition, new residue methodology for water analysis developed jointly by a consortium of herbicide manufacturer's together with USEPA was evaluated in this effort. Sixty-two surface and groundwater samples, collected from areas presumed to be most vulnerable to both runoff and leaching, were analyzed for sixteen of the new herbicides. T. R. Steinheimer and coworkers discuss the new method and its nuances, offer recommendations for improvements to it which should improve its sample-throughput value, and report results from the reconnaissance survey.

Minimizing off-field displacement of agrochemicals from cropped areas via surface runoff is of continuing concern. One approach which is endorsed by regulatory agencies, conservationist interests, and farmers is vegetative filters. Vegetative filters act through a variety of retention and transformation mechanisms which are qualitatively known. For field situations which act on runoff directly, biological assimilation and transformation are not sufficiently rapid to reduce contaminant concentrations without additional mechanisms for increasing the residence time of the runoff within the filtering system. J. L. Baker and coworkers are identifying the physical parameters and variables which are critical in determining the net effectiveness of these filters for mitigating the offsite movement of chemicals. Their field-scale work in filter strip technology has shown that two factors are critical; the cropped-to-filter area ratio, and the properties of the chemical. Their discussion includes recommendations for optimizing the retention efficiency for a given landscape.

The San Joaquin Valley of California

The San Joaquin Valley, approximately 12,000 mi^2, can be divided into two drainage basins, the San Joaquin and Tulare Basins, which comprises the southern two thirds of California's Central Valley (15). The Tulare Basin is a closed basin: water drainage begins and ends within the basin boundaries. In addition, surface water streams are all ephemeral (16). In contrast, the San Joaquin Basin drains into the Sacramento-San Joaquin Bay Estuary, a valuable fishing and wildlife resource. The basin contains surface water streams and rivers, both ephemeral and perennial in nature. The San Joaquin River has three major tributaries on the east side of the valley:

the Merced, Tuolumne, and Stanislaus Rivers, which originate in the Sierra Nevada Mountains. In addition, there are a number of small irrigation district drains which carry excess irrigation water as well as agricultural runoff water from the valley floor to the San Joaquin River and these tributaries. Soils on the east side of the valley, which originate from the Sierra Nevada batholith, are generally coarse textured and well drained (*16*). On the west side of the valley, surface water streams are ephemeral and originate in the Coastal Range. These tributaries frequently carry rain and irrigation runoff from agricultural fields. Soils on the west side, which originate from the marine shales of the Coastal Range, are generally fine textured and highly erodible (*16*).

The San Joaquin River flows through the northern portion of the San Joaquin Valley, an area of diverse and intensive agriculture. Major crop acreage includes alfalfa, almonds, beans, corn, grapes, tomatoes, walnuts, and wheat. Over 300 pesticide active ingredients are used in the San Joaquin-Tulare Basins, with an annual reported usage of over 88 million lbs (*17*). Given the variety and amount of pesticides used in this region, the correlation between toxicity tests and pesticide concentrations in the watershed, and the valuable Sacramento-San Joaquin Bay Estuary natural resource, there has been increasing interest and concern about the transport and fate of agrochemicals in the San Joaquin River watershed.

The Mediterranean climate, combined with the hydrological and geological features of this region, plays a role in the temporal and spatial distribution of pesticides in this watershed. During winter months when the San Joaquin Valley receives almost all of its annual rainfall, pulses of pesticides, particularly those used to control overwintering orchard pests, are found in the San Joaquin River, rising and falling with river discharge. In summer months, irrigation runoff water and drift are factors affecting pesticide occurrence and distribution. In addition, soils of the eastern side of the valley are well drained, with deep groundwater aquifers, while those of the west side more poorly drained with shallow aquifers. Hence, pesticides applied to fields on the west side may be more likely to move to surface water through overland flow, subsurface flow, and attached to eroded soil particles than on the east side. We are only beginning to understand the various mechanisms of entry to surface water and know even less about their relative contribution to stream loads in this watershed. This information will be critical if management practices to control off-site movement are to be developed and effectively implemented.

In a small agriculturally dominated creek on the west side of the San Joaquin Valley, Poletika and coworkers found pesticides in irrigation runoff was an important route of entry into surface water. Daily sampling was conducted for an entire year and measured concentrations were related to application, discharge, and weather patterns. An above average water year, i.e. heavy rainfall during winter months, restricted the number of dormant orchard applications and therefore the contribution of rain runoff may not have been fully addressed. Consequently, pesticide input from irrigation runoff was found to be a more important route of entry to the creek than rain runoff and spray drift. However, pesticide loads were heaviest during winter months, even though applications were fewer than during the drier part of the year. Results indicate farming practices designed to reduce irrigation and rain runoff of pesticides may therefore be more effective for controlling off-site movement to surface water than application management practices.

Results from surface water sampling in the San Joaquin River watershed indicate the spatial and temporal distribution of pesticides are affected by application and cropping patterns, and subbasin hydrology. Dubrovsky and others examined spatial and temporal patterns of pesticide occurrence during the course of one year in three subbasins of the watershed. Both summer irrigation and winter rain-runoff water carried pesticides into the watershed. High rates of application corresponded with high concentrations in surface water during the summer irrigation season. In contrast, winter storm runoff generated maximum concentrations for some pesticides even though the highest use occurred at different times of the year. Other factors of potential importance in transport to surface water include application method and type of irrigation. Due to the complexity of the hydrology of this watershed and the heterogeneous nature of land use in the Valley, other factors influencing transport mechanisms need to be explored. In addition, the relative magnitude of the various routes of entry is important for minimizing off-site transport into surface water.

In addition to transport to surface water, airborne pesticides have been measured in the San Joaquin Valley. In the atmosphere, pesticides exist as vapors or associated with liquid or solid aerosols. The removal of these constituents from air is principally by three processes; degradation, wet deposition, and dry deposition. Pesticides have been detected in air, rainfall, snow, ice, and fog water in the Central Valley and Sierra Nevada mountains. Pesticides in fog water may reach concentrations greater than those predicted by their vapor-water distribution coefficients. Seiber and Woodrow studied the transport and fate of pesticides in fog in the valley and found high enrichment factors for fog water. In addition, fogwater deposition has been implicated in the occurrence of inadvertent residues on non-target crops and high-risk exposures for raptors. Atmospheric transport (air, rain, and snowfall) of organophosphate insecticides to the Sierra Nevada was reported, indicating the potential for long-range transport of these chemicals.

Pesticides may also be transported through the soil profile into aquifers or through subsurface flow to surface waters. Transport into soil layers deeper than would be expected from transport models has been explained by preferential flow pathways, macropore movement where the chemicals bypassed the soil matrix. Co-transport of pesticides complexed with dissolved organic matter is another mechanism investigated by Letey and coworkers. In laboratory studies, dissolved organic matter played a significant role in the rapid and deep transport of napropamide, while preferential flow was found not to be a factor. Their additional investigation of sewage sludge and pesticide movement may influence our thinking on the use of organic matter for retaining chemicals at the soil surface or for bioremediation purposes.

Summary

Pesticides used in production agriculture have been detected in surface water and ground water samples from across the United States, indicating that many major water resources are contaminated (*18*). These conclusions emerge from the results of extensive studies using high sensitivity analytical techniques and data analysis tools. Concentrations of herbicides and insecticides in agricultural streams, and in many rivers in agricultural regions, were highest in those areas of greatest agchemical usage. Herbicide concentrations are greatest in midwestern streams, where use is most extensive. Similar patterns emerge in all region of the U.S. Nationally, the number of detections constitute a very small percentage of the total analyses done and usually are at concentration levels deemed well below those which threaten humans or aquatic life. However, harmful effects have been documented in some areas of high pesticide use. Detections above human-health-based drinking water standards, while rare, do occur.

Toward a New Agricultural Ethic. While agriculture is widely considered to be the nation's biggest nonpoint source of water pollution, it has often been treated separately and less strictly by the regulatory community than other sectors of the national economy. The environment, the farm sector, and the taxpaying public would all benefit from national policies that simultaneously address the economic and the environmental consequences of modern U.S. agriculture, with its evolving technology and ever increasing capacity. Because environmental problems linked to agriculture are so widespread and diverse, the voluntary participation of farmers seems essential if we are to achieve environmental goals. Environmentalists need to recognize that there are limitations to the speed and the degree to which agricultural programs and farming practices can be altered to achieve improved environmental quality endpoints. It is equally important for the agriculturalists to recognize the need for policies which weigh equally production and environmental protection. Pesticide manufacturer's continue to develop new synthetic products with improved environmental-compatibility attributes, while developing others from natural products which express a characteristic pest-control activity. To be sure, the rapidly evolving age of "transgenic agriculture", with its hopes, promises, and fears, will exert its inevitable influence upon food and fiber production early in the new millennium. In the end, all humankind will benefit from a safe and affordable food supply, a prosperous farm sector within our national economy, and a sustainable and productive farming ecosystem.

The Symposium organizers developed the program by invitation to experienced researchers so as to insure the highest quality of both platform and poster presentations. Unfortunately, prior commitments and extremely heavy workloads prevented some from contributing a chapter from their talk to this volume. In Table 2 the editors wish to acknowledge all Symposium presenters, their program title, and Abstract reference.

Table 2. Symposium Presenters, Their Program Title, and Abstract Reference. [Division of Agrochemicals, PICOGRAM and Abstracts, Issue 54, Spring 1998]

PRESENTERS	PROGRAM TITLE	ABSTRACT #
R.D. Wauchope	Pesticide Movement and Fate: Integrating Spatial Scales via Physical Modelling, Simulation Modelling, and Geographic Information Systems	005
D.B. Baker; R.P. Richards	Effects of Watershed Scale on Agrochemical Concentration Patterns in Midwestern Streams	006
D.R. Forney; J. Strahan; C. Rankin; D. Steffin; C.J. Peter; T.D. Spittler; J.L. Baker	Monitoring Pesticide Runoff and Leaching from Farming Systems On Field-Scale Coastal Plain Watersheds in Maryland	007
J.L. Hatfield; D.A. Bucks; M.L. Horton	The Midwest-Water Quality Initiative: Research Experiences at Multiple Scales	008
D. Crosby; L. Ross; M. Lee	Use of Laboratory, Field, and Watershed Data to Regulate Rice Herbicide Discharges	009
L.W. Hall; R.D. Anderson; D.P. Tierney	Exposure Assessments of Atrazine and Metolachlor in the Mainstem, Major Tributaries, and Small Streams of the Chesapeake Bay Watershed: Implications for Ecological Risk	018
A.M. Petrovic; S.Z. Cohen	Fate of Agrochemicals in the Turfgrass Environment: From the Laboratory to the Golf Course	019
	Continued on next page.	

Table 2. *Continued*

PRESENTERS	PROGRAM TITLE	ABSTRACT #
G.D. Foster; K.A. Lippa; C.V. Miller	Watershed Transport and Fluxes of Organic Agrichemicals to Northern Chesapeake Bay	023
R.N. Lerch; E.E. Alberts; F. Ghidey; P.E. Blanchard	Herbicide Transport in Surface Water at Field, Watershed, and Basin Scales	029
R.S. Kanwar; T.B. Moorman; D.L. Karlen	Effect of Banding and Swine Manure Application on Herbicide Transport to Subsurface Drains	030
A.C. Newcombe; N.D. Simmons; P. Hendley; M. Mills; D.I. Gustafson; A.J. Klein	Acetochlor Monitoring and Movement Studies	031
E.M. Thurman; A.G. Bulger; E.A. Scribner; D.A. Goolsby	The Relative Attenuation Factor (Ar): A New Conceptual Technique for Comparing Regional Water Quality Data in the Mississippi River Basin, USA	032
T.R. Steinheimer; R.L. Pfeiffer; C.J. Peter; M.J. Duffy; W.A Battaglin	Reconnaissance Survey of Sulfonylurea Herbicides in Streams and Groundwater of the Midwestern U.S.	033
J.H. Prueger; J.L. Hatfield; T.J. Sauer	Metolachlor Volatilization Estimates From Broadcast and Banded Fields in Central Iowa	034
N.N. Poletika; P.L. Havens; C.K. Robb; R.D. Smith	Organophosphorus Insecticide Concentration Patterns in an Agriculturally Dominated Tributary of the San Joaquin River	041
N.M. Dubrovsky; C.R. Kratzer; S.Y. Panshin; J.M. Gronberg; K.M. Kuivila	Pesticide Transport in the San Joaquin River Watershed	042
J. Seiber; S. Datta; J. Woodrow	Transport and Fate of Pesticides in Fog in California's Central Valley	043
L.S. Aston	Transport of Organophosphates from the San Joaquin Valley to the Sierra Nevada Mountains	044
J. Letey; W.J. Farmer; C.F. Williams; S.D. Nelson; M. Ben-Hur	The Role of Dissolved Organic Matter in Pesticide Transport Through Soil	045

Table 2. *Continued*

PRESENTERS	PROGRAM TITLE	ABSTRACT #
Hochstedler, M.E; D. Larabee-Zierath	Pesticides in Ambient Air and Precipitation in Rural, Urban, and Isolated Areas of Iowa	072
T.R. Steinheimer; K.D. Scoggin	Spring Snowmelt Runoff Losses of Nitrate and Herbicides from Four Small Agricultural Watersheds in Southwestern Iowa	073
M.Y. Lee; R.A. Durst; T.D. Spittler; D.R. Forney	FILIA Determination of Imazethapyr Herbicide in Watershed MonitoringStudies	074
T.D. Spittler; S.K. Brightman; M.C. Humiston; D.R. Forney	Watershed Monitoring in Sustainable Agriculture Studies	076

Disclaimer

Mention of specific products, suppliers, or vendors is for identification purposes only and does not constitute an endorsement by the U. S. Department Of Agriculture to the exclusion of others.

Literature Cited

(1) *Herbicide Metabolites in Surface Water and Groundwater*; Meyer, M.T., Thurman, E.M., Eds., ACS Symposium Series 630: American Chemical Society: Washington, DC, 1996.
(2) Kolpin, D.W.; Thurman, E.M.; Linhart, S. M. *Arch. Environ. Contam. Toxicol.*, **1998**, *35*, 385-390.
(3) *Pesticides Industry Sales and Usage, 1994 and 1995 Market Estimates*; Aspelin, A.L.: U. S. Environmental Protection Agency, Washington, DC, 1997, 35 pp.
(4) Vigon, B.W., *Water Resour. Bull.*, **1985**, *21(2)*, 179-184.
(5) Wauchope, R.D.; *J. Environ. Qual.*, **1978**, *7*, 330-336.
(6) Frank, R.; Braun, H.E.; Holdrinet, M.V.; Sirons, G.J.; Ripley, B.D., *J. Environ. Qual.*, **1982**, *11*, 497-505.
(7) Leonard, R.A. In *Herbicides in Surface Waters;* Grover, R., Ed.; Environmental Chemistry of Herbicides; CRC Press, Inc., Boca Raton, FL, 1989, Volume I; pp 45-87.
(8) Wu, T.L.; Corell, D.L.; Remenapp, H.E.H., *J. Environ. Qual.*, **1983**, *12*, 330-336.
(9) *Pesticides in Surface Waters - Distribution, Trends, and Governing Factors;* Larson, S.J.; Capel, P.D.; Majewski, M.S., Eds; Pesticides in the Hydrologic System: Ann Arbor Press, Inc., Chelsea, MI., 1997, Volume III, 373 pp.
(10) Hallberg, G.R., *Agric. Ecosys. Environ.*, **1989**, 26, 299-367.
(11) Smith, C.J. In *Hydrogeology with Respect to Underground Contamination;* Hutson, D.H.; Roberts, T.R., Eds., Environmental Fate of Pesticides, John Wiley & Sons, Chichester, U. K., 1990, pp 47-99.
(12) Rao, P.S.C.; Alley, W.M. In *Pesticides;* Alley, W.M., Ed.; Regional Groundwater-Quality; Van Nostrand Reinhold, New York, NY., 1993, pp 345-382.
(13) Pesticides in Ground Water - Distribution, Trends, and Governing Factors; Barbash, J.E.; Resek, E.A., Eds.; Pesticides in the Hydrologic System; Ann Arbor Press, Inc., Chelsea, MI.,1996, Volume II, 588 pp.
(14) Steinheimer, T.R.; Scoggin, K.D., *J. Environ. Monit.* **1999**, *in press*.

(15) Gronberg, J.A.M.; Dubrovsky, N.M.; Kratzer, C.R.; Domgalski, J.L.; Brown, L.R.; Burow, K.R. Environmental Setting of the San Joaquin-Tulare Basins, California. U.S. Geological Survey, Water Resources Investigations Report 97-4205. 45 pp.
(16) Domagalski, J.L. Nonpoint Sources of Pesticides in the San Joaquin River, California: Input from Winter Storms, 1992-93. U.S. Geological Survey, Open-File Report 95-165. 15 pp.
(17) California Environmental Protection Agency, Dept. of Pesticide Regulation. 1991 Pesticide Use Report Database. Sacramento, CA.
(18) Gilliom, R.J.; Barbash, J.E.; Kolpin, D.W.; Larson, S.E. *Environ. Sci. Technol.*, **1999**, *33*, 164A-169A.

THE KEYNOTES

Chapter 2

Monitoring Pesticide Runoff and Leaching from Four Farming Systems on Field-Scale Coastal Plain Watersheds in Maryland

D. R. Forney[1], J. Strahan[1], C. Rankin[1], D. Steffin[1], C. J. Peter[1], T. D. Spittler[2], and J. L. Baker[3]

[1]DuPont Agricultural Products, Chesapeake Farms, Chestertown, MD 21620
[2]Cornell Analytical Laboratories, New York State Agricultural Experiment Station, Cornell University, Geneva, NY 14456
[3]Iowa State University, Ames, IA 50011

> Losses of herbicides in runoff and leachate were measured in four different production systems. These production systems varied in the amounts and types of herbicides used, from preemergence applications of products used in the kg/ha, to selective use of postemergence herbicides used in g/ha. The highest herbicide concentrations in runoff samples were found in the first runoff event after application. Herbicides used at higher use rates were found in higher concentrations in both runoff and leachate samples. Incorporation of herbicides greatly reduced runoff losses. Herbicides used at similar rates but with shorter half-lives were found at lower concentrations in leachate samples. No-till treatments resulted in the highest concentrations in the leachate.

Crop protection from insects, diseases, nematodes, and weeds is a critical component of agricultural sustainability (*1*). Synthetic pesticides are among a suite of technologies that have afforded tremendous gains in agricultural productivity, profitability, and efficiency in the last 50 years. While consumers are increasingly concerned about environmental consequences of production methods, pesticides remain the primary means of controlling pests. Off-site movement of agricultural chemicals into surface and groundwaters is a common consequence of their use (*2*). Perfecting the use of pesticides is an ongoing need in the development of environmentally sound, economically viable, and socially acceptable production systems (*1*).

Concerns about pesticides in surface and groundwater have spawned numerous studies over the years to quantify the amounts of different materials leaving treated fields and to characterize the pesticide concentrations of receiving waters (*3-8*). Studies have focused on the more widely used compounds, and those that appear to represent a threat to human or aquatic ecosystem health based on their physical, chemical, and biological properties. Herbicides are the most widely used class of pesticides, and have received considerable attention. Their losses in surface runoff water have been observed to be highly dependent on their properties, notably soil adsorption and persistence, together with rate, timing, and method of application (9). Also, the timing of runoff producing rainfall events, and factors such as tillage and buffer or filter strips that influence the movement of water and sediments from fields, influence pesticide movement.

Herbicides Used

Among the herbicides used in U.S. crop production, triazines have long been a focus of concern due to their widespread use, relatively high rates of application, extended persistence, and moderate soil adsorption characteristics. Among these, atrazine is the pesticide most often detected in surface and groundwater resources today (6). With application rates in the range of 0.5 to 2.0 kg/ha, it was used on 71% of surveyed U.S. corn acres in 1996 at an average rate of 1.2 kg/ha, with corn applications totaling 24.3 million kg (10,11). It is the second most widely used pesticide in the Chesapeake Bay watershed (12). It is sold in a number of formulations, all of which are classified as "Restricted Use Pesticide" due to ground and surface water quality concerns.

Like atrazine, but less persistent, cyanazine is a widely used corn herbicide that also has been the focus of water quality concerns. Uncertainty as to its regulatory fate in the U.S. in light of water quality and toxicological concerns led to an agreement between DuPont and the U.S. EPA for a gradual withdrawal of cyanazine from the marketplace, with all uses to be canceled by 2003. It is sold in a number of formulations, all of which are classified as "Restricted Use Pesticide" due to toxicological and ground and surface water concerns. In 1996, it was used on 13% of surveyed U.S. corn acreage at an average use rate of 2.6 kg/ha, with corn applications totaling 9.4 million kg (11).

Metribuzin, a triazine herbicide used on soybeans, has declined in importance due to the introduction of other new herbicides, but was still used on 9% of surveyed soybean acreage in 1996, at an average use rate of 0.34 kg/ha, with soybean applications totaling 1.9 million kg.

The acetanilide, or chloroacetamide herbicides, including alachlor and metolachlor, are frequently cited among compounds of concern for aquatic or toxicological risk (8). They are widely used due to their utility in both corn and soybean production. While alachlor is being phased out in recent years in favor of acetochlor, it was still used in 1996 on 9% of surveyed corn acreage and 5% of surveyed soybean acreage. Average use rates were 2.4 kg/ha in corn and 1.8 kg/ha in soybeans, with U.S. corn and soybean applications totaling 6.9 million kg (11). It is sold in a number of formulations, all of which are classified as "Restricted Use Pesticide" due to oncogenicity. Metolachlor was used in 1996 on 30% of surveyed corn acres and 5% of surveyed soybean acres, at average use rates of 2.2 and 2.0 kg/ha, respectively, with corn and soybean applications totaling 20.5 million kg (11). Together with the triazines, atrazine and cyanazine, the acetanilides alachlor and metolachlor are routinely detected in surface and subsurface waters draining from agricultural lands (13). In addition, they are routinely detected at sub-ppb levels in precipitation. Precipitation may deposit more alachlor in the Mississippi River Basin than is transported out of the basin in stream flow (6). These herbicides are most often applied to the soil at or right before planting time to provide residual control of weeds.

Triazines and acetanilides were among the earliest truly selective and widely effective herbicides available to U.S. corn and soybean farmers, and have been in use in U.S. crop production for approximately 20 to nearly 40 years. During this period, pesticide discovery and development efforts have brought new chemistries with new characteristics to the market. At least two new classes of herbicides, imidazolinones and sulfonylureas, have been widely adopted. In soybeans, imidazolinones, imazethapyr and imazaquin widely replaced metribuzin and the acetanilides in the late 1980s and 1990s. With far lower use rates typical of newer pesticides, these two materials were used on 43% and 15% of surveyed U.S. soybean acres in 1996 at use rates of 0.07 and 0.10 kg/ha, respectively, with applications totaling 0.54 and 0.32 million kg (11).

Along with the adoption of imidazolinones, sulfonylureas have also had a tremendous impact on the quantities of pesticides used because of their low use rates. Worldwide, substitution of sulfonylureas for older, higher-use-rate materials in cereals has afforded an estimated reduction in herbicide use of approximately 99 million kg, with an associated reduction of perhaps 3 billion kg of manufacturing waste (14).

While much is known about the physical, chemical, and biological properties of sulfonylureas and their degradation, mobility, and sorption processes, little water quality monitoring data have appeared in the literature for these compounds (15, 16). Losses of 1.1 to 2.3% of applied nicosulfuron and chlorimuron were reported for a rainfall simulated runoff study, similar to the percentage losses typically reported for other higher use rate herbicides (17).

In corn, nicosulfuron is used as a postemergence (foliar) spray for annual and perennial grasses and select broadleaf weeds, at rates of 35 to 70 g/ha (18). Following first commercial use in 1990, its use has grown to 12% of surveyed corn acres, with a total of 111 thousand kg applied, in 1996 (11).

Chlorimuron ethyl has both preemergence and postemergence uses, with rates of approximately 35 and 10 g/ha respectively. It was used on 14% of surveyed U.S. soybean acres, with applications totaling 65 thousand kg (11).

Thifensulfuron methyl is used in soybeans, corn and winter cereals, and tribenuron methyl is used in winter cereals (among other uses) at rates of approximately 5 to 35 g/ha. They would appear to pose minimal threat to surface and ground water quality due to their very short persistence (15,16).

Many approaches are used to estimate or determine the contamination of surface and groundwater by pesticides. In the U.S., EPA registration of a pesticide under FIFRA requires the demonstration that the active ingredient is not likely to accumulate in water at levels that pose unreasonable risk to human health or aquatic ecosystems. Methods employed include model predictions based on known physical and chemical properties, laboratory soil-column leaching studies, small-scale prospective ground water monitoring studies, natural or artificial rainfall field-edge runoff monitoring, whole watershed studies, and monitoring of receiving waters. Each approach offers its own benefits and limitations. Modeling efforts offer the opportunity to simulate wide ranges of conditions but can be limited in accuracy based on the assumptions that must be made when knowledge of the precise impact of variables is not known. Field studies are site-specific with regard to soils, hydrology, agronomy, and prevailing weather.

Evaluating environmental impacts of agricultural production practices, including pesticides, has evolved toward a more systems-oriented holistic approach in recent years, in accord with recommendations put forth by the National Research Council (*19*). In the Midwest, the Management Systems Evaluation Areas (MSEA) program represents an ambitious regional approach that includes the goals of identifying and evaluating agricultural management systems that can protect water quality, and assess the impact of agricultural chemicals and practices on ecosystems associated with agriculture (*20*). This program recognizes that field studies are necessarily site specific and that a diversity of conditions are encountered from site to site, but also that results of multiple projects sharing common objectives will form a powerful tool in solving regional problems.

Materials and Methods

"A Sustainable Agriculture Project at Chesapeake Farms" is a watershed-focused long-term comparison of four cash grain farming systems relevant to the mid-Atlantic region of the U.S. The project arose out of discussions among representatives of DuPont Agricultural Products and a number of progressive thinkers from diverse perspectives concerned about agricultural sustainability. Collaborators include environmental and agricultural non-profit organizations, government agencies, academia, agri-business, and a group of concerned farmers. Each contributes specific talents, diverse thinking, and resources needed to ensure the project's success and give the project the depth of thought needed to help meet the groups' common challenges. A Farmer Advisory Board guides all decisions, assists with outreach, and ensures compliance with local concerns.

Chesapeake Farms encompasses 1335 ha of agricultural and wildlife habitat lands on the Eastern Shore of the Chesapeake Bay near Chestertown, MD. Soils range from gravelly upland soils to low-lying hydric soils. Predominant agricultural soils are silt loams, with management concerns including crusting and drainage. The predominant soil type in the area under investigation is Mattapex Variant silt loam (fine-loamy, mixed, mesic Aquic Hapludult), considered deep and moderately well-drained, formed in silty loess deposited by wind on older, clayey soil. The AP horizon (approximately

28 cm) has about 2% organic matter, CEC of around 8.0 meq/100 g, and pH is maintained in the 6.0 to 7.0 range by liming.

The four farming systems at Chesapeake Farms are subject to distinct rotation schemes, tillage practices, fertility programs, and methods of pest control, all options currently in use on farms in the mid-Atlantic region of the U.S. Each system covers approximately 4 ha and occupies a discreet watershed so that runoff water, soil, nutrients, and pesticide losses can be monitored. Grass waterways and berms manage surface runoff for each watershed, and a flume with automated monitoring equipment is used to measure and samples runoff water (Figure 1). The systems demonstrate varying reliance on rotational diversity and in-season management and labor; make optimal use of fertilizers, crop protection chemicals, and other purchased input; and employ Best Management Practices to keep sediments, nutrients, and pesticides on the field and out of the water. A replicated, small-plot experiment based on the four cropping systems in the watershed study is being conducted in an adjacent field of similar soil type. Comparisons are made on yield, soil tilth, fertility, and pest populations, as well as nutrient and pesticide movement through the soil profile via both passive and suction soil-water samplers.

Cropping Systems. Each cropping system is managed using commercial farming equipment, with all procedures being performed as close to the ideal time, considering prevailing conditions, as possible for optimum crop production. Pesticide applications were made with commercially formulated products, in water or liquid fertilizer, by either custom applicators or on-farm equipment.

System A. System A consists of continuous no-till corn, with rye cover crop (A2) and without rye cover crop (A1). Systems A1 and A2 employ a "programmed" approach to pest and nutrient management, in which crop protection is accomplished largely through pre-planned applications of products known to control the anticipated pests. System A employs approximately 3600 g/ha of pesticides annually, including atrazine and cyanazine for burndown of existing vegetation at planting time (together with 2,4-D, as well as paraquat for added burndown of the rye cover crop). Insecticide is included in furrow at planting time (tefluthrin in 1994 and 1995, chlorethoxyfos in subsequent years), and es-fenvalerate is included in the "burndown" herbicide application (for which the carrier is a nitrogen fertilizer solution). A combination of nicosulfuron and dicamba is applied postemergence for perennial weeds. Within this program approach, the triazines were applied in 1994 and 1995 at rates designed to provide season-long annual weed control, the program then being completed with a spot application of the nicosulfuron/dicamba mixture, which was applied to 30% of the field in 1994. By 1995, however, due to a long history of triazine use on the farm, escaping triazine-resistant weeds required treatment on 80% of the field. For subsequent years, a planned approach was employed with a reduced rate of the triazines and a postemergence broadcast application of the nicosulfuron/dicamba mixture. Pesticide runoff monitoring efforts for this system have focused primarily on the triazines and nicosulfuron (Tables I and II).

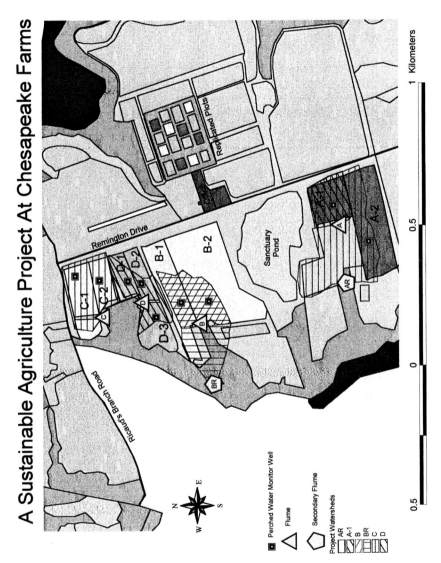

Figure 1. A Sustainable Agriculture Project at Chesapeake Farms.

Table I. Annual Herbicide Concentrations and Losses for Watershed A

Herbicide	Avg. Conc. (µg/L)	Max. Conc. (µg/L)	DAA days (event)	Loss (%)	Loss (g/ha)	Amount applied (g/ha)
1994						
Atrazine	5.16	240.00	2(1)	1.25	8.79	701.8
Cyanazine	25.54	>1000	2(1)	1.96	43.46	2213.5
Nicosulfuron	0.002	0.07	20(1)	0.04	0.003	8.8
1995						
Atrazine	19.62	327.00	1(1)	14.29	100.26	701.8
Cyanazine	20.04	212.00	11(4)	4.67	103.45	2213.5
Nicosulfuron	0.672	10.55	5(1)	10.40	2.45	23.6
1996*						
Atrazine	120.97	280.00	8(2)	15.43	77.33	501.3
Cyanazine	53.93	550.00	7(1)	4.31	68.10	1581.1
Nicosulfuron	2.03	3.85	3(1)	4.67	1.37	29.4

* Includes estimated flow and loss for 2nd event after application, a major event when sampler failed. Flow and loss were estimated based on measured flow for this event in other watersheds and their known relationships to this watershed, and pesticide concentrations in similar events the same week in this watershed.

Table II. Annual Herbicide Concentrations and Losses for Watershed A2

Herbicide	Avg. Conc. (µg/L)	Max. Conc. (µg/L)	DAA days (event)	Loss (%)	Loss (g/ha)	Amount applied (g/ha)
1994						
Atrazine	43.22	680.00	1(1)	3.64	15.94	438.4
Cyanazine	84.79	1035.00	1(1)	2.26	31.28	1382.8
Nicosulfuron	0.01	0.13	34(2)	0.12	0.003	2.6
1995						
Atrazine	7.86	114.00	11(2)	3.87	16.04	414.6
Cyanazine	8.48	162.00	9(1)	1.32	17.31	1307.9
Nicosulfuron	0.18	1.28	19(1)	3.36	0.43	12.9
1996						
Atrazine	50.18	150.00	7(1)	11.60	31.91	275
Cyanazine	43.48	290.00	7(1)	2.97	25.72	867.3
Nicosulfuron	1.17	7.37	3(1)	4.97	0.81	16.2

System B. This system employs a two-year corn and soybean rotation with deep-tillage prior to corn, and full-season no-till soybeans. System B also relies on programmed pest and nutrient management. System B employs approximately 3400 g/ha of pesticides annually. For corn, these include atrazine, cyanazine, and metolachlor applied and incorporated into the top 5 cm of the soil prior to planting corn. Spot applications of a nicosulfuron/dicamba mixture are made to control perennial weeds. The rotation with soybeans, with the herbicides used there, has prevented any buildup of triazine-resistant weeds. The programmed approach for soybeans includes a planting-time application of a mixture of metribuzin, chlorimuron ethyl, and alachlor, together with glyphosate if needed for additional "burndown",

Table III. Annual Herbicide Concentrations and Losses for Watershed B

Herbicide	Avg. Conc. (µg/L)	Max. Conc. (µg/L)	DAA Days (event)	Loss (%)	Loss (g/ha)	Amount applied (g/ha)
1994						
Alachlor	4.23	36	0(1)	0.15	2.60	1733.1
Metolachlor	2.35	11	3(1)	0.25	1.44	578.9
Atrazine	2.85	8	8(2)	0.83	1.75	210.3
Cyanazine	1.39	12	3(1)	0.13	0.85	663.3
Metribuzin	0.01	0.12	54(4)	0	0.01	133.9
Chlorimuron ethyl	1.08	12	0(1)	2.61	0.62	23.9
Nicosulfuron	0.02	0.08	21(3)	0.33	0.01	4.2
1995						
Alachlor	0.72	2.97	54(4)	0.11	1.23	1161.3
Metolachlor	0.41	8.6	29(4)	0.08	0.70	863.9
Atrazine	0.7	8.08	29(4)	0.38	1.20	313.8
Cyanazine	0.3	11.7	29(4)	0.05	0.51	989.9
Metribuzin	0.01	0.52	25(3)	0.03	0.02	89.8
Chlorimuron ethyl	0.18	5	12(2)	1.76	0.26	14.9
Nicosulfuron	0.07	0.53	21(2)	1.72	0.11	6.3
1996						
Alachlor	2.89	210	4(1)	0.77	13.33	1733.1
Metolachlor	2.5	9.6	28(2)	2.03	11.73	578.9
Atrazine	2.08	8.2	28(2)	4.57	9.61	210.3
Cyanazine	0.85	5	26(1)	0.59	3.93	663.3
Metribuzin	0.62	37	4(1)	2.14	2.87	133.9
Chlorimuron ethyl	0.41	25.3	4(1)	1.17	0.26	22.6
Thifensulfuron methyl	<0.05	<0.05	NA	<LOD	<LOD	0.1

Table IV. Annual Herbicide Concentrations and Losses for Watershed B-R

Herbicide	Avg. Conc. (µg/L)	Max. Conc. (µg/L)	DAA days (event)	Loss (%)	Loss (g/ha)	Amount applied (g/ha)
1995						
Alachlor	0.341	1.82	25(2)	0.07	0.55	781.6
Metolachlor	0.021	8.16	29(4)	0.07	0.32	468
Atrazine	0.89	8.8	148(11)	0.84	1.43	170
Cyanazine	0.174	10.82	29(4)	0.05	0.28	536.2
Metribuzin	0.016	0.43	25(2)	0.04	0.03	60.4
Chlorimuron ethyl	0.025	0.34	25(2)	0.3	0.03	10.1
Nicosulfuron	0.011	0.15	21(2)	0.39	0.01	3.4
1996						
Alachlor	0.69	22.94	4(1)	1	9.34	938.93
Metolachlor	0.0779	3.07	28(2)	2.706	10.54	389.61
Atrazine	1.31	2.24	28(2)	12.54	17.74	141.53
Cyanazine	0.281	1.49	28(2)	0.853	3.81	663.27
Metribuzin	0.197	3.69	4(1)	1.38	2.66	89.75
Chlorimuron ethyl	0.17	2.96	4(1)	1.34	0.18	13.1

followed by a postemergence spot application of a mixture of chlorimuron ethyl plus thifensulfuron methyl plus quizalofop for perennial weeds. Monitoring efforts in system B focused on the triazines, acetanilides, and sulfonylureas (Tables III and IV).

System C. System C is a two-year corn, wheat, and soybean rotation with no-till corn, conventionally tilled high-management wheat, and no-till double-crop soybeans. System C uses prescription pest and nutrient management, in which crop protection practices and fertility applications are based on scouting and diagnostics. System C employs approximately 200 g/ha of pesticides annually. In corn, glyphosate with or without 2,4-D is applied at planting time to destroy existing vegetation. A combination of nicosulfuron and dicamba was applied postemergence for annual weed control in 1994 and 1995, atrazine was added to this mixture in 1996. In wheat, a combination of thifensulfuron and tribenuron was applied each year for control of winter-annual weeds, wild garlic, and Canada thistle. For soybeans, glyphosate was applied at planting time if needed (1996 only), followed by a mixture of chlorimuron ethyl, thifensulfuron methyl, and quizalofop applied postemergence. This postemergence treatment was made to both the replicated plots and the watershed field in 1994 and 1995, but only to the plots in 1996 after scouting revealed no weeds needing treatment in the watershed field. Monitoring efforts for system C focused on the sulfonylureas (Table V).

System D. This production system is a three-year corn, soybean, and wheat rotation with rye and hairy vetch cover crops. It has relied whenever possible on cultural and biological pest and nutrient management. System D employs

Table V. Annual Herbicide Concentrations and Losses for Watershed C

Herbicide	Avg. Conc. (µg/L)	Max. Conc. (µg/L)	DAA days (event)	Loss (%)	Loss (g/ha)	Amount applied (g/ha)
1994						
Chlorimuron ethyl	0.07	0.47	10(2)	0.9	0.06	6.80
Nicosulfuron	0.03	0.45	14(3)	0.24	0.03	10.90
Thifensulfuron methyl	<0.05	0.24	8(1)	0.02	0.002	11.00
Tribenuron methyl	<0.05	<0.05	NA	<LOD	<LOD	5.50
1995						
Chlorimuron ethyl	0.18	0.29	10(1)	8.14	0.30	3.60
Nicosulfuron	0.79	7.23	12(1)	7.04	1.29	18.40
Tribenuron methyl	0.01	1.6	4(1)	0.32	0.02	6.50
Thifensulfuron methyl	0.01	0.83	4(1)	0.5	0.02	3.30
1996						
Atrazine	14.66	128	3(1)	7.17	22.48	313.54
Chlorimuron ethyl*						
Nicosulfuron	0.53	7.55	3(1)	9.5	1.04	10.90
Tribenuron methyl	0.02	0.47	6(2)	0.39	0.04	11.00
Thifensulfuron methyl	0.06	0.34	6(2)	2.22	0.12	5.50

*not applied in 1996, not needed based on IPM scouting

Table VI. Annual Herbicide Concentrations and Losses for Watershed D

Herbicide	Avg. Conc. (µg/L)	Max. Conc. (µg/L)	DAA days (event)	Loss (%)	Loss (g/ha)	Amount Applied (g/ha)
1994						
Nicosulfuron	<0.05	<0.05	NA	<LOD	0	4.4
Thifensulfuron methyl	0.04	2.54	8(1)	0.41	0.024	5.8
Tribenuron methyl	<0.05	0.69	8(1)	0.05	0.001	2.9
1995						
Nicosulfuron	0.34	1.4	1(1)	7.3	0.358	4.9
Thifensulfuron methyl	<0.05	0.12	4(1)	0.04	0.002	3.9
Tribenuron methyl	<0.05	0.09	4(1)	0.07	0.001	2
1996						
Thifensulfuron methyl	<0.05	0.07	12(1)	0.12	0.006	5.3
Tribenuron methyl	0.06	0.23	14(2)	3.1	0.082	2.7

approximately 22 g/ha of pesticides annually. For corn, a flail mower kills the hairy vetch cover crop and many of the existing weeds prior to corn planting. A postemergence application of a nicosulfuron plus dicamba mixture is made to a band 38 cm wide as needed to control weeds in the corn row (mechanical cultivation being used between the rows). In 1995 and 1996, when the initial corn planting failed, glyphosate was applied when corn was replanted, and in 1996 this eliminated the need for the postemergence herbicides (Table I). In system D soybeans, glyphosate is applied to destroy the rye cover crops and any existing weeds prior to planting. A postemergence application of imazethapyr is made to a band 38 cm wide as needed to control weeds in the corn row (mechanical cultivation being used between the rows). In system D wheat, a combination of thifensulfuron methyl and tribenuron methyl was applied each year for control of winter-annual weeds, wild garlic, and Canada thistle. Monitoring efforts in system D have focused on the sulfonylureas (Table VI).

Watersheds. The small watersheds on which these systems are evaluated range in size from 2.08 to 9.00 ha. They have been surveyed to the nearest 3-cm elevation, and the surveyed boundaries were georeferenced using differential GPS technology (footnote, Starlink, Inc.). Grassed (tall fescue) waterways were installed in the watersheds according to NRCS design as Best Management Practices for controlling soil erosion. The watersheds were divided into fields as needed to accommodate the different rotational crops for each system, with an objective of having the different fields of a watershed contribute approximately equally to the runoff from that watershed. In addition to the primary field-edge flumes for the watersheds of systems A1, B, C, and D (these watersheds and flumes being designated A, B, C, and D, respectively), two secondary watersheds were characterized. The first, AR, encompasses the A1 watershed plus a portion of field A2 plus a substantial additional acreage of perennial vegetative cover, including both grassed waterways and grass-legume mixtures maintained for wildlife use. Flumes A, B, C, D, and AR are served directly by grassed waterways. The other secondary watershed (BR) encompasses the system B watershed plus a substantial area of mixed vegetation, including perennial grasses, shrubs, and trees. Flume BR is served primarily by an intermittent woodland stream that is subject to substantial bank erosion during strong flows, despite the presence of an apparently well-functioning riparian buffer area between the cropped fields and the woodland.

Watershed A includes 1.74 ha of cropped land and 0.34 ha of grassed waterway. Watershed AR includes the 2.08 ha of watershed A, plus 2.23 ha of field A2, plus 3.58 additional acres of grassed waterway and perennial grass/legume mixture.

Watershed B includes 3.29 ha of cropped land and 0.02 ha of grassed waterway. Watershed BR includes the 3.31 ha of watershed B plus and additional 0.56 ha of cropped land plus 2.24 ha of grassed an forested riparian buffer.

Watershed C includes 3.24 ha of cropped land and 0.66 ha of grassed area, most of which is in the waterway.

Watershed D includes 2.07 ha of cropped land, plus approximately 0.66 ha consisting mostly of grassed waterway, with small portion of a graveled service road.

Sampling equipment. A 75-cm or 90-cm "H" flume (Plasti-Fab, Inc., Tualatin, Oregon) was installed in a soil berm defining the outlet point of each watershed during the project implementation phase in the spring of 1993. These flumes were outfitted with flow monitors and autosamplers (American Sigma, Inc., Medina, NY). Autosamplers were programmed to collect samples on both a time basis and on a flow basis to ensure representative samples from all runoff events producing significant flow. A goal was to collect at least two samples from each event throughout the year, except when fluctuating temperatures near the freezing point threatened to damage the flow monitors (several events each winter were not monitored nor sampled). For events occurring soon after pesticide applications, we attempted to obtain more frequent samples through the time course of the event.

Small plots. The five farming systems were also managed on a set of replicated plots, each measuring 27x18 meters of similar soil type, with little to no slope. The experiment was in Latin square design with four replications. The main plots were the systems, with subplots for each crop in rotation in each system.

One replication of each system was equipped with networks of soil-water samplers. One type of sampler consisted of a length of PVC pipe 2 m long, with "well screening" (0.1 mm slots) on 0.5 m of one end, installed 0.76 m deep, horizontally, the screened end extending into the plot and the other end terminating in a pit at the edge of the plot. When rainfall created conditions of soil saturation, water flowed passively to a collection jug in the pit. Each subplot was equipped with three of these "passive lysimeters", which were installed during the winter of 1993-1994. Samples were removed for analyses following each sample-generating rainfall event.

Another set of soil water samplers were installed in the spring of 1995, of the pressure-vacuum type with porous ceramic cup for allowing the entry of water under vacuum (Soil Moisture Equipment Co., Santa Barbara, CA). Five of these samplers were installed in each subplot of one replication of each system, 1.22 m deep. These were sampled monthly, with varying success.

Following installation, both sets of samplers remained in place with no further soil disturbance nor interference with farming operations. For reporting purposes, results from the two types of samplers were pooled.

Sample handling and processing. Surface runoff samples were handled in the following manner. Following significant rainfall events, samplers at each flume were visited and samples brought into the field laboratory and placed under refrigeration (approx. 4°C). Hydrographs and related data from the autosamplers were examined to determine a reasonable subset of samples to process for analysis. The objective was to send at least two samples from each flume for each event for analysis during most of the year, but to select greater numbers of samples from events occurring immediately after pesticide application. Samples were split into separate containers, then frozen and held at -4°C until shipment on dry ice by overnight carrier to the various laboratories.

During examination of hydrographs and data, an estimation was made of the proportion of the event represented by each selected sample. Later, pesticide concentrations were matched up with flow proportions and total flow measurements to calculate pesticide masses contained in each runoff event.

Soil pore water (leachate). Soil pore water samples were collected from underneath small plots which received the same inputs as did the watersheds. There are a total of nine different cropping sequences practiced due to crop rotations. Because of crop rotations there is one lysimeter set in systems A1 and A2, two lysimeter sets in systems B and C, and three lysimeter sets in system D. As a result, more samples were collected in systems with the higher number of lysimeter sets. The soil pore water was collected by both passive and suction lysimeters. The passive lysimeters were located 70 cm below the soil surface and the suction lysimeters were 120 cm below the soil surface. Sampling of the passive lysimeters started in March, 1995, while sampling of the suction lysimeters commenced in April 1996. Most samples were collected during the late fall, winter, and spring of the year. This is when the recharge is occurring. Following significant rainfall events, samplers at each monitoring site were examined for flow from the passive lysimeters. Whenever a sufficient quantity was present, it was collected, split into separate containers for the different labs, then frozen and held at -4°C until shipment on dry ice by overnight carrier to the various laboratories. Suction lysimeters were visited once per month and sampling attempted. Vacuum was applied to each sampler per manufacturer instructions, and samples collected 24 h later. When no sample was present, the procedure was repeated at least twice more in attempt to obtain sample. Sampling success varied widely between and among samplers, influenced by animal activity (mice chewing on suction lines), soil water content, and other undetermined factors. Whenever a sufficient sample was collected, it was split into separate containers for the different labs, then frozen and held at -4°C until shipment on dry ice by overnight carrier to the various laboratories.

Sample analysis. Water samples were sent to the Cornell Analytical Lab for analysis of alachlor, metolachlor, atrazine, cyanazine, metribuzin, and imazethapyr. Details of the sample handling and analytical methodology used are contained in a separate chapter in this publication (*21*).

Imazethapyr water samples where were analyzed by an immunoassay method at the Cornell Analytical Lab (*22*). The results of these analyses were questioned when the calculated amount of imazethapyr was in some cases equal to or greater than the amount applied. Forty-one water samples including some of those which had had the highest reported imazethapyr concentrations were reanalyzed using a capillary electrophoresis (CE) method. There was no correlation between the immunoassay and the CE method. In a number of samples where the immunoassay method indicated imazethapyr concentrations of greater than 10 ppb, no imazethapyr was detected by the CE method. The highest concentration of imazethapyr reported where there was good agreement between the two methods was 8.5 ppb. Because of the general lack of agreement between the two methods, it was decided to only report this value as the highest concentration detected.

Analysis of water samples for sulfonylureas was conducted using immunoassays. Water samples were shipped frozen to the DuPont Corporate Center for Analytical Sciences in Ponca City, Oklahoma for the analysis of nicosulfuron, chlorimuron ethyl, tribenuron methyl, thifensulfuron methyl and rimsulfuron. The samples remained frozen and were allowed to thaw in the refrigerator overnight just prior to analysis. Samples were analyzed by proprietary enzyme-linked immunosorbent assays (ELISA). The limit of quantitation for each analyte was 0.10 ppb.

Each assay was developed to operate using the same basic principle. Details of the assay development and validation, as well as of specific applications, can be found in publications authored by Strahan (23,24). In general, polyclonal antibodies are added to samples containing unknown amounts of antigen and are incubated. Aliquots of the samples are then transferred to an antigen-coated 96-well microtiter plate where antibody not bound to antigen in the sample may bind to antigen immobilized in the microwells. A second antibody-enzyme conjugate is added which then binds the immobilized first antibody. Color is generated following addition of the substrate; the intensity of the color is measured at 405 nm and is inversely proportional to the amount of antigen present in the sample.

Once thawed, samples were removed from the refrigerator and allowed to reach room temperature. They were filtered, diluted 1:4 with Phosphate Buffered Saline (PBS, pH 7.4) and assayed (n=3) in batches of 15. Each plate also included a seven-point standard curve and two controls - an unfortified control water and the control water fortified with 0.10 ppb of the analyte of interest.

Data was accepted if the standard curve met established performance criteria, the precision of replicate analyses had a relative standard deviation (RSD) of 10% or less and the mean recovery of the controls fell within 70-120% of the level of fortification.

No confirmatory analyses were run.

Results and Discussion

Runoff Losses. The average total amount of herbicides applied to each watershed between 1994 and 1996 is presented in Table VII. The average annual herbicide use ranged from 3384.75 g/ha on watershed B to 12 g/ha on watershed D.

Table VII. Average Annual Herbicide Applications and Surface Water Runoff Losses from Four Watersheds, 1994-1996

Watershed System	Average Annual Application (g/ha)	Average Annual Loss (g/ha)	Average Annual Loss (%)
A	2658.24	135.07	5.08
AR	1440.33	46.48	2.82
B	3384.74	7.07	0.39
C	34.05	0.97	3.13
D	12	0.16	0.82

This demonstrates the vast differences in herbicide inputs among the production systems. The watersheds where planned applications of preemergence herbicides were used (A, A2, and B) received amounts of herbicides at least 40 times greater than the watersheds (C and D) where postemergence herbicides were used. The maximum rate of each herbicide applied is given in Table VIII.

Table VIII. Maximum Concentrations of Various Herbicides Detected at Chesapeake Farms

Compound	Max. Use Rate (treated area) (g/ha)	Maximum Detected Conc. Surface Runoff (ppb)	Soil-Water (ppb)	HAL (ppb)
Cyanazine	3326	1035.00	71.00	1
Atrazine	1055	680.00	75.00	3
Alachlor	2912	210.00	3.30	2
Metolachlor	1452	11.00	2.60	70
Metribuzin	225	0.52	<0.1	100
Chlorimuron ethyl	38	25.00	0.26	140
Nicosulfuron	35	10.55	4.38	8750
Thifensulfuron methyl	21	2.54	<0.05	90
Tribenuron methyl	11	0.69	<0.05	6
Imazethapyr	68	8.50	NS	?

The use rates ranged from 3326 g/ha for cyanazine to 11 g/ha for tribenuron methyl. The average annual herbicide loss for each watershed and the percent average annual loss in surface water run-off are also presented. The percent average annual run-off loss ranged from 0.82 to 5.08. These values are within the range of losses reported in the literature (25, 26, 27). The maximum concentration of each herbicide in runoff and soil pore water collected by the suction lysimeters is also given. There is a significant positive relationship between the maximum use rate of each herbicide and the maximum concentration detected in surface water runoff and soil pore water. Given that all the herbicides tested are moderately water soluble, this finding is not surprising. This observation also supports the Best Management Practice (BMP) of reducing pesticide rates to improve water quality. The current models used to predict the amount of surface water runoff, GENEEC (GENeric Estimated Environmental Concentration) (*28*) and leaching, SCI-GROW (Screening Concentration In GROund Water) (*29*) of pesticides take into account the use rate of the pesticide. While use rate is one important factor controlling the amount of herbicide that runs off a treated field or leaches there are other factors that are also important. Other important factors which influence the amount of herbicide which runs off a treated field include: the degree of adsorption to the soil, the type of application, the type of tillage, the time interval between application, and the first runoff event. In predicting the concentration of a pesticide which leaches, other important factors would include; the half-life in soil, the amount of herbicide adsorption to soil, the soil type and organic matter content, and the rainfall timing, amount and intensity (*8*). While there are many other factors which need to be considered if one is to accurately predict the amount of runoff or leaching, this study clearly shows that use rate accounts for the majority of the difference among the herbicides tested. The data also clearly shows that for the herbicides studied that the concentrations lost in surface water runoff are much higher than the concentrations that were found in soil pore water. Maximum concentrations found in surface water runoff were between 2.4 and 96 times higher than the maximum concentrations found in soil pore water. This information strongly suggests for the products in this study that the risk of an adverse effect is much higher from surface water runoff than from soil pore water.

One risk endpoint which is often used in discussing the risk of herbicides in water is the drinking water standard. Drinking water standards are either the lifetime health advisory level (HAL) or the lifetime maximum contaminant level (MCL) (*30*). These values which are set by US EPA are based upon the toxicity of the individual product. While the methodology to calculate the MCL and HAL is the same, only MCLs are legal drinking water standards. These lifetime MCL and HAL values are based upon the premise that an individual will drink two liters of water every day for 70 years. If they do so, and the water contains a concentration of a particular herbicide at the MCL, no adverse health effect is expected. Comparing HAL and MCL values which are drinking water standards to concentrations in surface water runoff at the edge of field or soil pore water is inappropriate because this water would not or could not be used as a drinking water source. While being an inappropriate comparison this information does serve to put the level of human safety regarding drinking water into perspective.

An analysis of the data by watershed and compound revealed several important conclusions. Information in Tables 1-6 show that the maximum concentration of every herbicide tested generally occurred in the first runoff event after application. This observation supports the work of other researchers (*3, 8, 31*). A dramatic reduction in the amount of herbicide in surface water runoff can be achieved by incorporating the herbicide into the soil compared to a surface application. By incorporating the herbicides, far less of the herbicide is on the soil surface that is subject to loss by surface water runoff. The reduction in average percent loss with incorporation was not as great, 59 percent for atrazine and 89 percent for cyanazine. The reason for the lower reduction in the percent loss was a lower volume of runoff in system A. No-till has been shown to decrease the volume of surface water runoff.

While this study clearly demonstrates the effectiveness of soil incorporation of herbicides in reducing herbicide runoff in surface water, a couple of other factors must also be considered. No-till farming had not been practiced in watershed A prior to the start of this experiment. It has been shown that the effectiveness of no-till on reducing herbicides in surface water runoff increases with time. It is therefore likely that the differences in herbicide runoff between the two farming systems will decrease over time. The other factor that should be considered is the soil that is lost from the two systems. Sediment is the number one surface water problem from agriculture. Sediment losses in 1995 were 72% greater from watershed B than watershed A on a per acre basis *(data not given)*. Society must decide if reductions in herbicide concentrations are more or less valued than reductions in erosion.

A number of observations were made about surface water runoff losses for specific herbicides. These observations generally confirmed the results of other researches and added new information about the runoff losses of sulfonylureas. The patterns of surface water runoff losses for sulfonylureas were similar to those of other moderately water-soluble herbicides with similar half-lives but were much lower in concentrations. This point is exemplified by Figure 2 which shows the surface water runoff concentrations for atrazine and nicosulfuron from system A during 1995. The ratio of the application rate, in g ai/ ha, to the initial concentration, in ppb, in runoff was nearly identical. The rate of decrease in runoff concentration over time was also very similar. One difference was that atrazine was detected in surface water runoff for most of the year while nicosulfuron was not detected in the later runoff events. This could be the result of a couple of factors. Atrazine is adsorbed to soil more than nicosulfuron. This would keep more of the atrazine near the soil surface where it would be subject to loss in surface water runoff. The half-life of nicosulfuron (26 days) is shorter than that of atrazine (60 days). Nicosulfuron's shorter half-life would mean there would be less compound available for loss by surface water runoff as time between application and runoff events increased.

The surface water runoff results also demonstrate the water quality attributes of the shorter residual sulfonylureas such as thifensulfuron methyl and tribenuron methyl. Field half-life for thifensulfuron methyl and tribenuron methyl ranges from 2 to 14 days and 1 to 23 days, respectively *(32)*. The rapid degradation of thifensulfuron methyl is mainly due to rapid metabolism by soil microbes *(33)* while rapid hydrolysis accounts for the short half-life of tribenuron methyl *(34)*. In 1995, all detected loss of thifensulfuron methyl and tribenuron methyl occurred in one event 4 days after application (Tables 5 and 6), and performance of the methods are exemplified with this event (Figure 3). The materials were applied to 31% and 19% of watersheds C and D, respectively. Rainfall of 2.39 cm occurred within 3.5 hours of application, resulting in runoff flow from watersheds C and D. Concentrations of the two compounds in runoff water from C ranged from 0.56 to 1.86 µg/L for thifensulfuron methyl, and from 0.28 to 0.94 µg/L for tribenuron methyl, resulting in losses of 20.71 and 10.82 mg/ha of the two compounds, respectively, approximating the 2:1 application rate ratio of the compounds. For watershed D, concentrations were uniformly lower, and far more so than expected based on watershed application rates

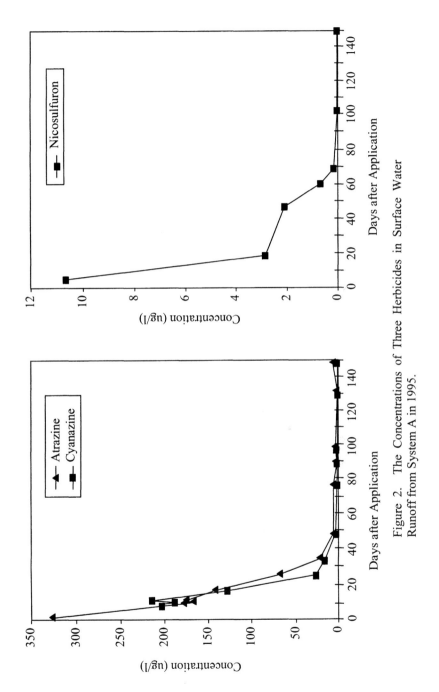

Figure 2. The Concentrations of Three Herbicides in Surface Water Runoff from System A in 1995.

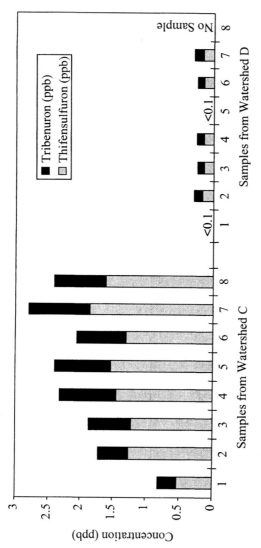

Figure 3. Concentrations of Thifensulfuron Methyl and Tribenuron Methyl in Surface Water Runoff on April 25, 1995.

(watershed D receiving 60% the application rate of watershed C). Furthermore, the concentrations of thifensulfuron methyl declined more than those of tribenuron methyl, resulting in losses of 1.58 and 1.33 mg of the two compounds, respectively. Both compounds apparently decomposed more rapidly in watershed D than in watershed C, and thifensulfuron methyl more rapidly than tribenuron methyl in watershed D. This may be due to greater microbial activity in watershed D, which received an application of poultry litter one month prior to the herbicide application, compared to watershed C which received commercial fertilizer. When runoff events occurred later, after application in 1996, the ratio of losses of the two compounds again indicated greater longevity for tribenuron methyl, and again system watershed D appeared to have less thifensulfuron methyl lost in surface runoff as compared to watershed C than would be expected from the application rates. When runoff began with an event one day after application in 1994, loss of thifensulfuron was inexplicably higher for both compounds for watershed D than C, and higher for thifensulfuron methyl than tribenuron methyl.

Chlorimuron ethyl was applied to watershed B as a preemergence and postemergence treatment to soybeans and to watershed C as a postemergence treatment to soybeans. The amount of chlorimuron ethyl applied to watershed B ranged from 14.9 to 23.9 g/yr. In watershed C, the amount of chlorimuron ethyl used ranged from 3.6 to 6.8 g/yr. The surface water runoff concentrations reflected the higher use rates used in watershed B. Maximum concentrations of chlorimuron ethyl detected from watershed B ranged from 5-25 ppb while maximum chlorimuron ethyl concentrations in surface water runoff from watershed C were 0.29 to 0.47 ppb. Pantone *et al.* (*35*) compared the runoff losses from equal rate preemergence and postemergence applications of atrazine under simulated rainfall conditions and found much greater losses from the preemergence treatments. This information shows there are additional water quality benefits to applying herbicides postemergence compared to preemergence since the ratio of maximum chlorimuron ethyl concentrations in the runoff water from the postemergence compared to the preemergence treatments was much less than the amounts applied. Several factors might be responsible for this additional benefit. One is that some of the postemergence application would be intercepted by the crop and weeds. The amount that is taken up by these plants or adsorbed into the plant surfaces is not available for loss by runoff. Another possible factor is lower soil moisture. There is an actively growing crop when postemergence applications are made. It is likely that the ancedent soil moisture is much lower at the time postemergence applications are made than at the time preemergence applications are made. It would therefore take more rainfall to produce a runoff event following a postemergence application than a preemergence application. The greater rainfall will dilute the herbicide running off more resulting in a lower maximum concentration.

Alachlor and metolachlor were only applied to watershed B. Alachlor applications were made in association with a combination of chlorimuron ethyl and metribuzin. This combination of herbicides was applied as a burn down treatment for the production of soybeans. In this production system, there is a significant amount of weeds present at the time of application. So, this treatment could be viewed as a postemergence treatment. The metolachlor treatments were made in combination with cyanazine and atrazine and incorporated into the soil just prior to corn planting. The maximum concentration of alachlor in the in the runoff ranged from about 3 to 210 ppb. While the absolute values were different, the pattern of maximum runoff concentrations was roughly the same for alachlor, chlorimuron ethyl and metribuzin. This observation suggests that the factors determining the maximum runoff concentrations were more environmental than compound related. The maximum alachlor concentration in runoff for 1994 was 36 ppb which was produced by a 0.75-inch rainfall event on the day of application; the resulting runoff volume was 44,526 gal/ha. The maximum alachlor in runoff in 1996 was 210 ppb which was produced by a 0.5-inch rainfall event four days after application with a resulting runoff volume of 920 gal/ha. This information suggests that the highest herbicide runoff concentrations are likely to be produced by rainfall events that occur shortly after application but produce a limited amount of runoff. Maximum runoff concentrations is the parameter which should be of greater concern since short-term acute toxicity is likely to be the endpoint of greatest concern. The metolachlor maximum runoff concentrations from year to year were much more consistent and generally lower than the alachlor concentrations ranging from 8.6 to 11 ppb. These values were similar to those for cyanazine and atrazine which were applied at the same time as the metolachlor. This information demonstrates the ability for incorporation to decrease herbicide runoff losses. These findings support the work reported by Baker (*36*).

This research has shown there are a number of factors that control the concentrations and amounts of herbicides which move off the field in surface water runoff. Some of these factors can be controlled. The type of tillage is one factor. These results show that incorporation consistently produced low runoff concentrations as compared to losses from no-till. Postemergence applications generally produced lower runoff losses than preemergence applications. Applying compounds at low use rates and compounds which degrade faster produce less runoff losses than high use rate products with longer half-lives. Rainfall is the primary factor influencing runoff losses that cannot be controlled. Generally, the shorter time interval between application and a rainfall event which produces runoff, the greater the runoff losses will be. How much runoff there is will influence the concentration in the runoff.

Leaching Losses. The results of the testing are given in Table IX. While almost all herbicides were tested for in all plots, results are reported for only those herbicides which were applied to individual cropping systems. A number of conclusions can be drawn from the soil pore water data. The maximum concentrations detected in the suction lysimeters, which sampled soil pore water at a greater depth than the passive lysimeters, were significantly lower than the maximum concentrations in the passive lysimeters. Maximum concentrations of atrazine and cyanazine collected in the passive lysimeters were 75 and 71 ppb, respectively. While maximum concentrations for these compounds detected in the suction lysimeters were 4.4 and <LOQ, respectively. This same trend was also noted of the sulfonylurea herbicides. There was only one detection of a sulfonylurea in the suction lysimeters.

Table IX. Soil pore water concentrations found in suction and passive lysimeter located in the various production systems

Active	Lysimeter Type	System A1 # of detects/ # of samples	System A1 Max Conc. (ppb)	System A2 # of detects/ # of samples	System A2 Max Conc. (ppb)	System B # of detects/ # of samples	System B Max Conc. (ppb)	System C # of detects/ # of samples	System C Max Conc. (ppb)	System D # of detects/ # of samples	System D Max Conc. (ppb)
Atrazine	P	31/33	75	14/17	22	11/18	2.9	14/34	24		
	S	1/2	1.5	2/5	3.6	1/20	2.0	13/32	4.4		
Cyanazine	P	14/33	<71	11/17	53	2/18	0.72				
	S	0/2	<LOQ	0/5	<LOQ	0/20	<LOQ				
Nicosulfuron	P	13/33	4.38	9/18	1.65	1/18	0.19	1/36	0.74	6/75	0.85
	S	0/1	<LOQ	0/3	<LOQ	0/28	<LOQ	1/34	0.12	0/22	<LOQ
Chlorimuron ethyl	P					4/18	0.26				
	S					0/20	<LOQ				
Alachlor	P					12/18	4.2				
	S					0/20	<LOQ				
Metolachlor	P					5/18	1.8				
	S					4/20	1.5				
Metribuzin	P					0/18	<LOQ				
	S					0/20	<LOQ				
Thifensulfuron methyl	P							0/37	<LOQ	0/77	<LOQ
	S							0/34	<LOQ	0/15	<LOQ
Tribenuron methyl	P							0/37	<LOQ	0/75	<LOQ
	S							0/34	<LOQ	0/22	<LOQ

Large differences in maximum concentrations were also noted among systems. Systems A1, A2, and B all received applications of cyanazine and atrazine. The maximum concentrations of cyanazine and atrazine in systems A1 and A2 were at least 20 ppb in the passive lysimeters, while in system B the maximum concentration of these herbicides never exceeded three ppb. Systems A1 and A2 were in no-till while system B was conventional tilled. The higher concentrations detected in the passive lysimeters in systems A1 and A2 are likely due to preferential flow due to no-till (27). The highest cyanazine and atrazine concentrations occurred in samples collected in May 1996 following an application of Extrazine II® at the end of March 1996. These higher concentrations found beneath the no-till plots may not translate into increased groundwater concentrations. The herbicide concentrations in the suction lysimeters samples, which were located beneath the corn rooting zone, were much lower. There were no detections of cyanazine in the suction lysimeters and the highest level of atrazine from suction lysimeters in these systems was 3.6 ppb. The decrease in concentrations from the passive lysimeters to the suctions lysimeters is probably due to a combination of factors including crop uptake, degradation, adsorption, and dilution. This same trend was observed for nicosulfuron. Nicosulfuron was applied to all cropping systems. The highest concentrations were again found in the no-till systems A1 and A2. There was only one detection of nicosulfuron in any of the suction lysimeter samples. This would suggest that under these conditions it is unlikely that nicosulfuron would be found in groundwater.

Metribuzin, alachlor, and metolachlor were applied in system B. Metribuzin was not detected in any of the lysimeter samples. Metribuzin has a relatively short half-life and was applied at one-fourth the rate of atrazine. The rates of metolachlor application were twice that of alachlor and the concentration of alachlor detected in the passive lysimeter was approximately twice that of metolachlor. There were no detections of alachlor in the suction lysimeters, but metolachlor was detected. Metolachlor is reported to have a longer half-life than alachlor (37).

Thifensulfuron methyl and tribenuron methyl were not detected in any lysimeter samples. Both of these sulfonylureas are applied at low rates (10-21 g ai/ha) and have short half-lives.

The general trend observed across all the systems was that herbicides applied at lower rates and which degraded faster were detected at lower concentrations or not detected at all. This observation is in alignment with the parameters used in SCI-GROW to predict pesticide concentrations in shallow ground water.

While all herbicides were not applied to all plots in most cases, all herbicides were tested for in all pore water samples (data not shown). This led to the observation that some herbicides were being detected in soil pore water underneath plots which had not received application of certain herbicides since the start of the experiment in 1994. For example, in system D, no applications of atrazine, cyanazine, alachlor, metolachlor or metribuzin were made during the study. Detections of atrazine, cyanazine and metolachlor were made in soil pore water samples taken from plots in system D. Records show that atrazine, cyanazine, and metolachlor had not been applied to this field since 1992. In contrast, concentrations of some herbicides, even in shallow soil pore water, as shown by this study, can last several years.

Conclusion: The water quality data generated by this study show that the concentrations of herbicides which will runoff in surface water or leach in soil pore water vary greatly depending upon which herbicides are used and how they are applied. Discussion of the potential risks associated with these herbicide concentrations is beyond the scope of this paper. Practices such as incorporation, which reduce runoff losses, can lead to higher sediment losses. No-till conditions may also lower runoff losses, but increase concentrations in soil pore water. Herbicides that were applied at lower rates or degraded faster were found in lower concentrations in both runoff and soil pore water.

Literature Cited

1. CAST. 1995. Sustainable Agriculture and the 1995 Farm Bill. Special Pub. No. 18. 4420 West Lincoln Way, Ames, IA 50014-3447.
2. Goolsby, D. A., E. M. Thurman, M. L. Pomes, and Battaglin, W. A. **1993**. Temporal and geographic distribution of herbicides in precipitation in the Midwest and Northeast United States, 1990-1991. Proc. *Fourth Nat. Conf. On Pesticides*. VA Water Resources Research Center, Blacksburg, VA 24060.
3. Wauchope, R. D. The Pesticide Content of Surface Water Draining from Agricultural Fields-A Review. *J. Environ. Qual.* **1978**, 4, 459-472.
4. Fawcett, R. S., Christensen, B. R. and Tierney, D. P. The Impact of Conservation Tillage on Pesticide Runoff into Surface Water. *J. Soil and Water Cons.* **1994**, 126-135.
5. Battaglin, W. A., Goolsby, D. A., and Coupe, R. H. Annual Use and Transport of Agricultural Chemicals in the Mississippi River, 1991-92, **1993**, USGS Open-File Report 93-418, 26-40.
6. Goolsby, D. A., Thurman, E. M., Pomes, M. L., Meyer, M., and Battaglin, W. A. Occurrence, Deposition, and Long Range Transport of Herbicides in Precipitation in the Midwestern and Northeastern United States, **1993**, USGS Open-File Report 93-418, 75-89.
7. Tierney, D. P., Williams, W. M., Hahn, A. L., Holden, P. W., and Newby, L. Surface water monitoring for atrazine in the Chesapeake Watershed (1976-1991). Proc. *Fourth Nat. Conf. On Pesticides*. VA Water Resources Research Center, Blacksburg, VA, **1993**.
8. Johnson, W. E., Plimmer, J. R., Kroll, R. B., and. Pait A. S. The occurrence and distribution of pesticides in Chesapeake Bay. CRC Pub. No. 47, Chesapeake Research Consortium, Inc., Edgewater, MD 21037, 1994.
9. Baker, J. L. and Mickelson, S. K. Application Technology and Best Management Practices for Minimizing Herbicide Runoff. *Weed Technol.* **1994**, 8, 862-869.
10. *Herbicide Handbook*; Ahrens, W. H. Ed. Seventh edition; Weed Science Society of America: Champaign, Illinois, **1994**.
11. USDA/NASS/ERS Agricultural Chemical Usage, 1996 Field Crops Summary. USDA/NASS/ERS, Washington, DC, **1997**.
12. Setting, M. E. Meeting pesticide challenges in the Chesapeake Watershed: Best management practices. Proc. *Fourth Nat. Conf. on Pesticides*. VA Water Resources Research Center, Blacksburg, VA, **1993**.

13. Dietrich, A. M., Gallagher, D. L., Schicho, D. L., Hubbard, T., Reay, W., Simmons, G. M., Hayes, M. C., Jourdan, S. W., and Lawruck, T. L. Groundwater transport of pesticides to surface waters and estuaries. Proc. *Fourth Nat. Conf. On Pesticides.* VA Water Resources Research Center, Blacksburg, VA, 1993.
14. Brown, H. M., Lichtner, F. T., Hutchison, J. M., and Saladini, J. L. The impact of sulfonylurea herbicides in cereal crops. Proc. *Brit. Crop Prot. Conf.* **1995**, 3, 1143-1152.
15. Beyer, E. M., Duffy, M. J., Hay, J. V., and Schlueter, D. D. Sulfonylureas. In *Herbicides- Chemistry, Degradation and Mode of Action;* Kearney, P. C. and Kaufman, D. D., Eds.; Marcel Dekker, Inc., New York, NY, **1988**, Vol. 3; 117-189.
16. Brown, H. M. Mode of action, crop selectivity, and soil relations of sulfonylurea herbicides. *Pestic. Sci.* **1990**, 29, 263-281.
17. Afyuni, M. M., Wagger, M. G., and Leidy, M. G. Runoff of two sulfonylurea herbicides in relation to tillage system and rainfall intensity. *J. Environ. Qual.* **1997**, 26,1318-1326.
18. Anonymous, *Crop Protection Reference 13th ed. 1997*, Chemical and Pharmaceutical Press: New York, NY, **1997**.
19. Hallberg, G. Soil and water quality and agricultural ecosystems. Proc. *National Agricultural Ecosystem Management Conf.*, New Orleans, LA, 12/13-15/1995, Conservation Technology Information Center, West Lafayette, IN, **1995**.
20. Hatfield, J. L., Anderson, J. L., Alberts, E. E., Prato, T., Watts, D. G., Ward, A., Delin, G., and Swank, R. Management Systems Evaluation Areas – An Overview. *Proc. Agricultural Research to Protect Water Quality Conf.*, 2/21-24/1993, Soil and Water Cons. Soc., Ankeny, IA, **1993**.
21. Spittler, T. D., Brightman, S. K., Humiston, M. C. and Forney, D. R., *Watershed Monitoring in Sustainable Agriculture Studies*, In *Pesticide Movement: Perspective and Scale*, Steinheimer, T. R., Ross, L. J. and Spittler, T. D. Eds., *ACS Symposium Series*, in press.
22. Lee, M., Durst, R. A., and Spittler, T. D., *FILIA Determination of Imazethapyr Herbicide in Water*, In *Pesticide Movement: Perspective and Scale*, Steinheimer, T. R., Ross, L. J. and Spittler, T. D. Eds., *ACS Symposium Series*, in press.
23. Strahan, J. *Development and Application of an Enzyme-Linked Immunosorbent Assay Method for the Determination of Multiple Sulfonylurea Herbicides on the Same Microwell Plate*, Chapter 7, ACS Symposium Series 646, **1996**, 66-73.
24. Strahan, J. and Wank, J. *Determination of Sulfonylurea Herbicides in Water by Multianalyte ELISA*, Current Protocols in Field Analytical Chemistry, **1998**, 2C.5, 1-11.
25. Seta, A. K., Blevins, R. L., Frye, W. W. and Barfield, B. L. Reducing Soil Erosion and Agricultural Chemical Losses with Conservation Tillage. *J. Environ. Qual.,* **1993**, 22, 661-665.
26. Webster, E. P., and Shaw, D. R. Impact of Vegetative Filter Strips on Herbicide Loss in Runoff from Soybean *(Glycine max). Weed Sci.* **1993**, 44, 662-671.
27. Hall, J. K., Mumma, R. O., and Watts, D. W. Leaching and runoff losses of herbicides in a tilled and untilled field. *Agriculture, Ecosystems and Environ.* **1991**, 37, 303-314.
28. Parker, R. D., Nelson, H. P. and Jones, R. D. GENEEC: A screening model for pesticide environmental exposure assessment. In *Proceedings of the International Symposium on Water Quality Modeling,* Heatwole, C. Ed., American Society of Agricultural Engineers, **1995**.
29. Barrett, M. R. The Screening Concentration in Ground Water, document circulated to Exposure Modeling Work Group. **1998**.

30. Barrett, M. R., Williams, W. M., and Wells, D. Use of Ground Water Monitoring Data for Pesticide Regulation. *Weed Technol*. **1993**, 7, 238-247.
31. Baker, J. L., and Johnson, H. P. The Effect of Tillage Systems on Pesticides in Runoff from Small Watersheds. *Transactions of the ASAE*, **1979**, 22, 554-559.
32. Hornsby, A. G., Wauchope, R. D., and Herner, A. E. *Pesticide Properties in the Environment*, Springer-Verlag, New York, NY, 1996.
33. Brown, H. M., Joshi, M. M. and Van, A. T. Rapid soil microbial degradation of DPX-M6316 (Harmony®). *Weed Sci. Soc. Am. Abstr*. **1987**, 27, 62.
34. Brown, H. M. Mode of action, selectivity, and soil relations of sulfonylurea herbicides. Pestic. Sci. **1990**, 29, 263-281.
35. Pantone, D. L., Young, R. A., Buhler, D. D., Eberlein, C. V., Koskinen, W. C., and Forcella, F. Water quality impacts associated with pre- and postemergence applications of atrazine in maize. *J. Environ. Qual*. **1992**, 21, 667-573.
36. Baker, J. L. Effects of tillage and crop residue on field losses of soil-applied pesticides. In *Fate of Pesticides and Chemicals in the Environment*, Schnoor, J. L., Ed.; John Wiley and Sons, Inc., New York, NY, 1992.
37. Zimdahl, R. L., and Clark, S. K. Degradation of three acetanilide herbicides in soil. *Weed Sci*., **1982**. 30, 545-548.

Chapter 3

Effects of Watershed Scale on Agrochemical Concentration Patterns in Midwestern Streams

David B. Baker and R. Peter Richards

Water Quality Laboratory, Heidelberg College, Tiffin, OH 44883

For selected Lake Erie Basin tributaries, detailed studies of nutrient and sediment runoff have been underway since 1974 and of pesticide runoff since 1983. The monitoring stations subtend watersheds ranging in size from 11 km^2 to 16,400 km^2 and having similar land use and soils. Examination of the agrochemical concentration and loading patterns at these stations reveals systematic changes related to watershed size (scale). As watershed size increases, peak storm event concentrations decrease while the durations of mid-range concentrations lengthen. The extent of these scale effects is parameter specific, being most evident for suspended solids. We hypothesize that these scale effects are attributable to the pathways and timing of pollutant movement from fields into streams, coupled with dilution associated with routing of runoff water into and through the stream system from differing positions in the watershed. These scale effects need to be considered when comparing concentration and loading data from different watersheds and when designing sampling programs.

The concept of "watersheds" can be applied at spatial scales ranging from less than 1 m^2 boxes in laboratory settings to major continental river basins, such as the Mississippi with a drainage area of 3.2 million km^2. Studies of runoff of water, suspended sediments, nutrients, and pesticides have occurred throughout the above range of watershed sizes. Examination of the resulting data reveals systematic shifts in a variety of hydrologic, concentration, and loading characteristics that are related to watershed size. These shifts can be considered scale effects.

Many factors in addition to watershed size contribute to observed differences in runoff characteristics among watersheds (1). Watershed shape and topography influence hydrology. Watershed land use has major impacts on runoff characteristics. Land use may vary greatly among watersheds, in relation to variations in soil and water resources and other geographic factors. Small agricultural watersheds, such as research plots or individual fields, generally have a single land use, crop, and set of management practices. As the size of agricultural watersheds increases, the diversity of crops and sets of management practices within its boundaries also increases. As watershed size increases still further, tributaries draining watersheds having predominantly urban or forested land uses may join those draining agricultural

© 2000 American Chemical Society

watersheds, resulting in changing hydrologic, concentration and loading characteristics. These and many other factors influence runoff characteristics, and consequently confound efforts to examine the effects of watershed scale (size) per se.

In northwestern Ohio, our laboratory has monitored a set of watersheds ranging in size from 11.3 km^2 to 16,395 km^2 for up to 23 years. These watersheds have very similar land uses (Table 1) and have generally similar soils. Point sources contribute very little to the total watershed outputs, even for the large watersheds. Consequently, they do allow an evaluation of scale effects with minimal confounding by other variables which influence runoff characteristics and stream concentrations. In this paper, we describe scale effects on hydrology, concentration durations, and loading patterns of pollutants derived from agricultural land use. The magnitude of scale effects varies systematically among parameters. We hypothesize that these variations are derived from the differences among parameters in export pathways and timing as pollutants are delivered from the land into stream systems. These effects interact with increasing dilution of local runoff by increases in channel storage as streams become larger. This dilution effect is magnified by the downstream movement of the flood front as a kinematic wave.

Most of the scale effects described below have been noted in various reports and papers published by our laboratory (2-5). These same scale effects, especially as they relate to pesticide concentrations, have been noted in USGS publications (6). This paper focuses on scale effects, expanding our previous analyses and offering hypotheses regarding the generation of these scale effects in river transport.

Methods

All of the sampling stations identified in Table I are located at or near USGS continuous stream gauging stations. Submersible pumps located just above the stream bottom pump water continuously from the streams into the sampling stations. Refrigerated automatic samplers containing 24 polypropylene bottles and located in the gage houses are used to collect three samples per day, with the sampler pumps taking water supplied to the gage house by the submersible pumps. At weekly intervals, sample bases are changed and samples are returned to the laboratory for analyses. During runoff periods, all three samples for each day are analyzed, while during non-runoff periods, a single sample per day is analyzed. During the pesticide runoff season (April 15 - August 15) additional automatic samplers containing glass bottles are used to collect samples for pesticide analyses. During the "non-runoff" season, grab sampling techniques are used to collect pesticide samples twice per month.

The analytical program for each sample includes suspended solids, total phosphorus, soluble reactive phosphorus, nitrate plus nitrite, nitrite, ammonia, total Kjeldahl nitrogen, sulfate, chloride, silica and conductivity. Automated, colorimetric procedures are used for the analyses of the above nutrients (8). From 1983 through 1992, pesticides were analyzed using dual column GC with nitrogen-phosphorus detectors. Beginning in 1993, pesticides were analyzed by GC/MS. In 1993, the laboratory shifted from liquid-liquid extraction to solid phase extraction. Pesticide analytical procedures have generally followed the methods outlined in successive versions of EPA Method 507. More details on the sampling and analytical procedures can be found elsewhere (3, 5).

Results and Discussion

Storm Hydrographs. The effects of watershed size on unit area hydrographs are illustrated in Figure 1. For Rock Creek, the peak runoff rates during storms, in mm per acre, are much higher than for the Maumee River. The duration of the individual storms is also much shorter for Rock Creek than for the Maumee River. In Rock Creek flows return to base flow between successive storms more often than for the Maumee River.

Table I. Monitoring program for agricultural watersheds in the Lake Erie Basin

Tributary	USGS Station Number	Basin Area (km²)	Period of record	Land use (percent)					Nutrient samples analyzed	Pesticide samples analyzed
				C	P	F	W	O		
Maumee R.	04193500	16395	1976-1978 1982-1997	76	3	8	4	9	9,532	1,093
Sandusky R.	04198000	3240	1975-1997	80	2	9	2	7	10,689	1,214
Honey Cr.	04197100	386	1976-1997	83	1	10	1	6	11,393	1,328
Rock Cr.	04197170	88	1983-1997	81	2	12	1	4	8,928	1,186
Lost Cr.	04195440	11.3	1982-1993	83	0	11	1	5	6,636	720

[1] Land use categories indicate percent of basin in : C, cropland; P, pasture; F, forest; W, water/wetland; O, other. Data from (7).

For the Maumee River, the Sandusky River, Honey Creek and Rock Creek, we have analyzed sets of individual storm runoff events through the 1995 water year. The resulting data base included 226 storms for the Maumee River, 237 for the Sandusky River, 261 for Honey Creek, and 158 for Rock Creek. Data for Lost Creek were not analyzed in this manner. In Figure 2, various storm characteristics are compared for the four rivers. Figure 2a illustrates changes in hydrograph shape, as described by the ratio of peak flow to storm volume. As the watershed size decreases, this ratio increases. Figure 2b shows the duration of storm events in relation to watershed size. Storm durations decrease as watershed size decreases. Figure 2c shows the distribution of the ratios of peak flow to basin area (peak unit area discharge). As watershed size decreases the peak unit area discharges increase. The above hydrologic scale effects are well known and described in great detail in the hydrological literature (9-11).

Pesticide Concentration Patterns. Although it has the smallest watershed, Lost Creek was not used for these comparisons because its period of record is shorter than that of the other stations, half the pesticide samples were concentrated in the four-year interval 1984-87, and several major pesticide runoff events have occurred since the station was discontinued.

The effects of watershed size on annual herbicide chemographs are illustrated using 1989 data for atrazine in the Maumee River and Rock Creek (Figure 3). Peak atrazine concentrations were higher in Rock Creek than in the Maumee, with concentrations falling to near baseline between storms for Rock Creek. In contrast, the Maumee had one broad peak of atrazine concentrations.

The distribution of annual maximum concentrations of atrazine and metolachlor for the Maumee River, the Sandusky River, Honey Creek and Rock Creek are shown in Figure 4. The graphs cover the 14 year period from 1983 to 1996. Annual maximum atrazine and metolachlor concentrations increase as watershed size decreases. The exception is that Rock Creek has lower maximum concentrations than the larger Honey Creek. This is because Rock Creek is less suitable for growing corn than Honey Creek, and consequently the amount of atrazine applied in the Rock Creek basin is disproportionately low.

The distribution of atrazine and metolachlor concentrations within the lowest quartile of values during the May through August period of 1983-1996 is shown in Figure 5. As watershed size increases, the concentrations of samples within the lowest quartile of values for the time period increases.

Concentration exceedency curves for atrazine and metolachlor for the Maumee River and Rock Creek are shown in Figure 6. These graphs include the 40% of the time with the highest concentrations for each river and cover the time period from water years 1985 through 1996. The highest concentrations are greater in Rock Creek. However, the concentrations exceedency curves cross, demonstrating that

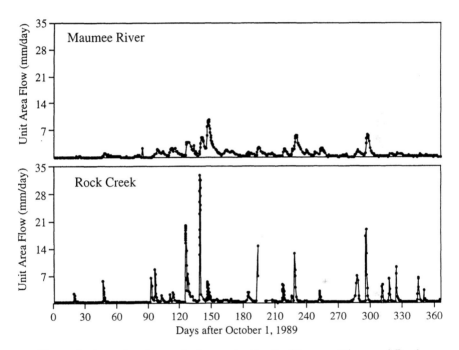

Figure 1. Annual unit area hydrographs for the Maumee River and Rock Creek, 1989 water year.

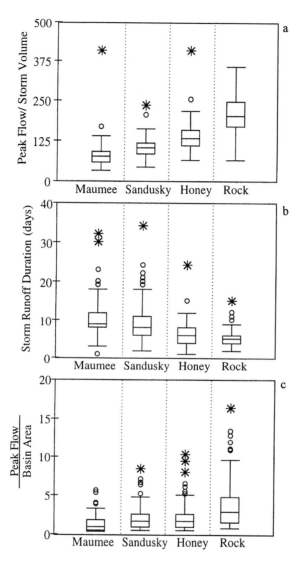

Figure 2. Effects of watershed size on peak flow/storm volume, storm runoff duration, and peak flow to basin area. A dashed lines separating groups of boxplots indicates a statistically significant difference between the populations on either side of the line (Mann-Whitney U, p=.05, ties omitted).

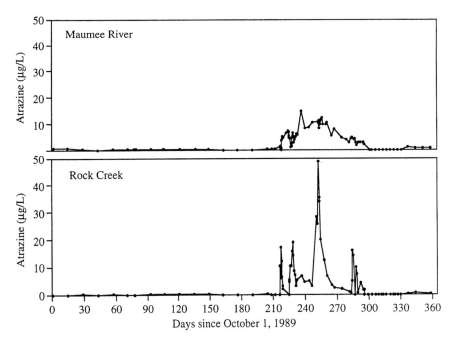

Figure 3. Annual chemographs for atrazine for the Maumee River and Rock Creek, 1989 water year.

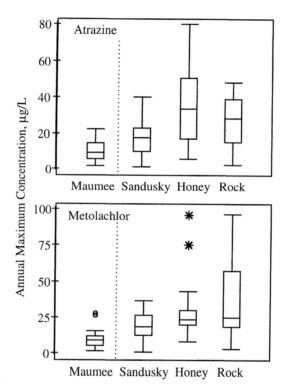

Figure 4. Distribution of annual maximum concentrations of atrazine and metolachlor in relation to watershed size, 1983-1996. A dashed lines separating groups of boxplots indicates a statistically significant difference between the populations on either side of the line (Mann-Whitney U, $p=.05$, ties omitted). Within the atrazine group on the right, Sandusky is significantly different from Honey, but the other pairwise tests are not significant.

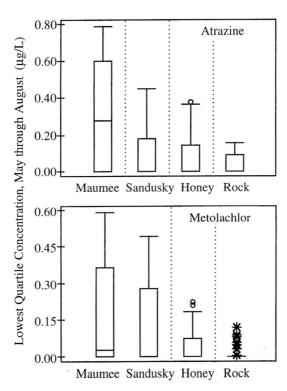

Figure 5. Distribution of atrazine and metolachlor concentrations within the lowest quartile of values for the May through August period, in relation to watershed size, 1983-1996. A dashed lines separating groups of boxplots indicates a statistically significant difference between the populations on either side of the line (Mann-Whitney U, p=.05, ties omitted).

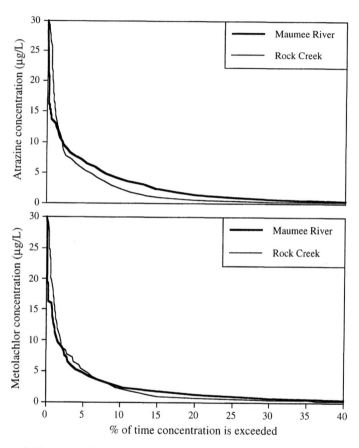

Figure 6. Concentration exceedency curves for atrazine and metolachlor for the Maumee River and Rock Creek, 1985-1996.

intermediate concentrations in the larger Maumee River are greater than those in the smaller rivers. The curves for the Sandusky River and Honey Creek fall between those for the Maumee River and Rock Creek.

Nutrient and Sediment Concentrations Patterns. At all of the sampling stations, the shape of concentration exceedency curves varies among parameters. In Figure 7, concentration exceedency curves are shown for suspended solids, total phosphorus, nitrate + nitrite, chloride and conductivity for the Maumee River and Lost Creek. In order to plot parameters with differing concentrations on the same graph, each parameter was expressed as a percent of the concentration exceeded 0.5% of the time (99.5 percentile concentrations) for the period of record at that station. Suspended solids concentrations drop off much more quickly, relative to their 99.5 percentile concentrations than do conductivity values. The graphs "stack up" by parameter in the same way for both watersheds. The order from the bottom of the figure to the top is suspended solids, total phosphorus, nitrate + nitrite, chloride and conductivity.

For a given parameter, the concentration exceedency curves vary with watershed size (Figure 8). For both suspended solids and nitrate + nitrite, the smallest watershed (Lost Creek) has the steepest drop from its 99.5 percentile concentration and the largest watershed (Maumee River) has the smallest drop. For both parameters the curves stack up in relation to watershed size, with the smallest watershed at the bottom of the stack.

Nutrient and Sediment Loading Patterns. The temporal distribution of loading combines the effects of scale on the hydrologic response of the river to storm events the effects of scale on runoff-related concentrations patterns. Data for a particular station and parameter were ranked by instantaneous loading rate (concentration x flow) from the highest to the lowest rate for the period of record. Loads associated with each sample were calculated (loading rate x time). Cumulative loads and cumulative times were then calculated and expressed as a percentage of the total load and total time for the period of record. The percentage of the total load can then be plotted as a function of the percentage of the total time, as shown in Figure 9. For the Maumee River, the 20% of the time with the highest export rates of suspended solids accounted for about 90% of the total export of suspended solids while the 20% of the time with the highest chloride loads accounted for about 60% of the total chloride export. For both the Maumee River and Lost Creek the parameters stack up in a particular order, with chloride and conductivity at the bottom, followed by discharge, nitrate + nitrite, total phosphorus and suspended solids.

Cumulative loading curves for the five watersheds are shown for suspended solids and nitrate + nitrite in Figure 10. For a particular parameter, the cumulative loading curves stack up in relation to watershed size. The smaller the watershed, the greater the proportion of the total load that is exported in a given percentage of time. Thus, the 5% of the time with the highest suspended solids export rates accounted for 67% of the total export rate from the Maumee River and more than 95% of the total export from Lost Creek.

In Table II, the percentages of the total loads exported during the 1% of the time with the highest loading rates are shown for various parameters and streams. For suspended solids, the 1% of the time with the highest export rates accounted for 62% of the total export from Lost Creek and 32% of the total export from the Maumee River. For nitrate + nitrite, the 1% of the time with the highest export rates accounted for 20% of the total export from Lost Creek and 8% of the total export from the Maumee River.

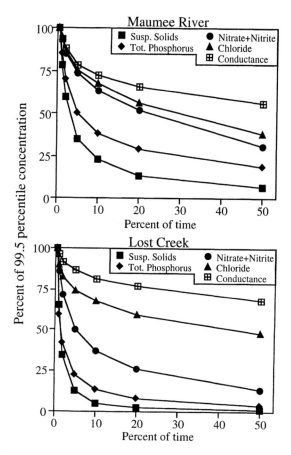

Figure 7. Concentrations exceedency curves for various nutrients, relative of 99.5 percentile concentrations, for the Maumee River and Lost Creek.

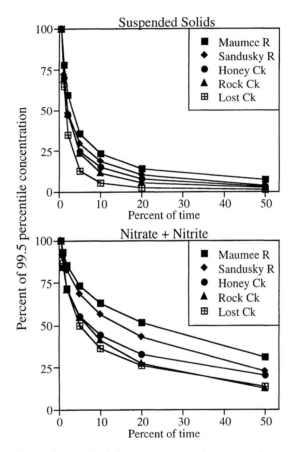

Figure 8. Effects of watershed size on concentrations exceedency curves for suspended solids and nitrate.

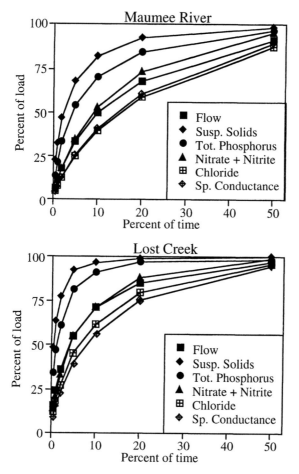

Figure 9. Variations in percent of total load versus percent of time for various parameters for the Maumee River and Lost Creek.

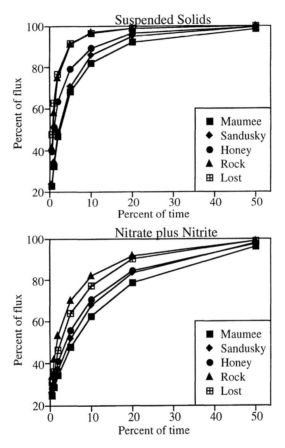

10. Effects of watershed size on percent of loads versus percent of time for suspended solids and nitrate.

Table II. Percent of the total load accounted for by the 1% of the time with the highest loading rates (fluxes).

Watershed	Suspended Sediment	Nitrate	Discharge	Chloride
Maumee	32%	11%	11%	8%
Sandusky	34%	13%	14%	12%
Honey	51%	16%	15%	11%
Rock	58%	27%	25%	16%
Lost	62%	20%	24%	18%

Hypothesized Causes of Scale Effects. We believe that many of the scale effects noted above are a direct consequence of the routing of water and chemicals through drainage networks during storm runoff events, coupled with differing pathways of chemical movement from land to streams. The pattern of concentrations and loading at a specific stream location depends on its position in the drainage network.

Whenever two streams merge with differing concentrations of chemicals, the resulting concentrations of the mixed water will be intermediate to the two parent streams. The resulting mixed concentrations will depend not only on the concentration differences between the streams, but also on the flow of the streams. For mixing in flowing systems the resulting concentration is the sum of the instantaneous fluxes of the two streams divided by the sum of the instantaneous discharges.

Several factors give rise to differences in concentrations and flows in two streams as they mix. Generally where streams join, their individual watersheds will have differing sizes. Consequently, even when a rainfall event occurs simultaneously over the entire watershed, the streams will be in different phases of their runoff response to the storm as they mix. Chemical concentrations vary in characteristic ways during runoff events, and peaks of chemographs may precede, coincide with, or trail the peak of the hydrograph, depending on the chemical (2,12). In Figure 11, chemograph shapes for suspended sediments, atrazine and nitrate + nitrite are compared with the storm hydrograph for a June 1993 storm in Honey Creek. The asynchrony between peak concentration and peak flow contributes to the potential for dilution during mixing.

One way to illustrate the asynchrony between chemical transport (time integrated instantaneous fluxes) and water discharge (time integrated instantaneous flow) during a storm runoff event is to plot double mass curves for individual storm events. For double mass curves, cumulative loads are plotted against cumulative discharges. To compare different parameters, the cumulative loads can be plotted as a percentage of the total storm load and cumulative discharge as a percentage of the total discharge for the storm. In Figure 12 double mass curves are shown for suspended solids, atrazine and nitrate + nitrite for the June 1993 storm event on Honey Creek (Figure 11). The first 50 percent of the storm discharge water accounted for 76% of the total storm load of suspended sediments, 58% of the total storm load of atrazine, and 36% of the total storm load of nitrate + nitrite. For suspended sediments the flow weighted average concentration of suspended solids was three times higher during the first half of the storm discharge than during the second half. The corresponding atrazine concentration was 1.38 times higher, while that for nitrate + nitrite was 1.63 times lower. Based on the distribution of these chemicals during this runoff event, the capacity for dilution during routing is greatest for suspended solids and least for atrazine. Considerable volumes of water with low sediment concentrations are present during the receding portion of the storm runoff event and considerable volumes of water with low nitrate + nitrite concentrations are present early in the runoff event.

We believe that the asymmetry present in chemical transport is due to the different pathways of movement of pollutants from land surfaces to streams. Pollutants associated with surface runoff commonly exhibit a "first flush"

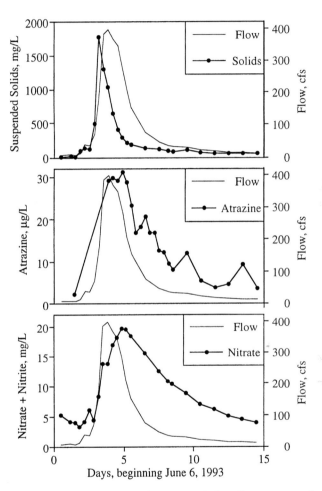

11. Comparison of chemographs for suspended sediments, atrazine and nitrate with the hydrograph for a June 1996 storm in Honey Creek.

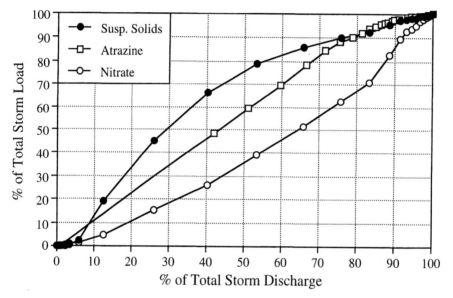

12. Percent of total load versus percent of total storm discharge for suspended sediments, atrazine, and nitrate for a June 1996 storm in Honey Creek.

phenomena. It is well known that in urban runoff studies, pollutant concentrations are particularly high early in the runoff event (first flush) (*13*). In field runoff studies from cropland, sediment and particulate phosphorus concentrations also have higher concentrations early in the runoff period than later. This generates the "asynchrony" in stream sediment concentrations that support scale effects through dilution. The pathways of most of the nitrate + nitrite export in this area involve tile flow and interflow. Since these flows are delayed relative to surface runoff, and persist for longer durations, most of the nitrate + nitrite is delivered to the stream late in the runoff period. The chemographs for soluble herbicides are generally much broader than the chemographs for sediments and particulate phosphorus, with peak concentrations occurring between peak sediment and peak nitrate + nitrite. The on-field kinetics of herbicide dissolution from the soil surface and upper soil layers may account for the extended delivery of pesticides from fields to streams. For some soils, pesticide contributions from tile drainage may also contribute to the broad shape of the pesticide chemograph.

Another factor contributing to the observed scale effects involves increasing amounts of storage of pre-storm water in stream channels in a downstream direction. When a small order stream responds quickly to a rainfall event, and that stream enters a high order stream, the amount of pre-storm water present in the receiving stream channel generally increases in a downstream direction. Thus, there is more water present in the channel to dilute incoming local runoff, with its high peak concentrations, in a downstream direction. In addition, the kinematic wave movement of water at the flood front, as that front moves downstream, increases the volume of pre-storm water available to dilute local runoff.

Conclusions and Observations

1. The patterns of chemical concentrations and transport in streams vary systematically with the position of the sampling station in the drainage network. These variations reflect a watershed scale effect and occur even though the landscape and land use may be nearly uniform throughout the drainage network.
2. Peak concentrations of chemicals derived from land runoff are generally higher in streams draining small watersheds than in streams draining large watersheds.
3. The durations of intermediate concentrations are longer for streams draining large watersheds than for streams draining small watersheds.
4. The export or loading of chemicals derived from land runoff occurs in shorter periods of time for small watersheds than for large watersheds.
5. The extent of these scale effects varies among parameters, being most evident for suspended solids.
6. We hypothesize that scale effects are largely a consequence of dilution accompanying the routing of runoff water through the drainage network. Changing channel storage with stream size and kinematic wave movement of the flood front likely also play a role in generating the observed scale effects.
7. We hypothesize that these differences in scale effects among parameters originate from differences in the pathways of movement of chemicals from land surfaces to streams.
8. Awareness of scale effects is important when comparing concentrations and loading data among streams of different sizes and when designing sampling programs aimed at characterizing either concentrations or loadings in stream systems.
9. Models aimed at predicting the concentrations of nonpoint derived pollutants in steam systems should reflect the same scale effects that are apparent in detailed long term data sets such as those we have accumulated.
10. Because of the complexity and variability of factors that interact to produce runoff within large watersheds, individual runoff events may diverge from the general patterns described above.

Literature Cited

1. Saxton, K.E.; Shiau, S.Y. In *Surface Water Hydrology*; Wolman, M.G.; Riggs, H.C., Eds.; The Geological Society of America: Boulder, CO, 1990; pp 55-80.
2. Baker, D. B.; Krieger, K. A.; Richards, R. P.; Kramer. J. K. In *Perspectives on Nonpoint Source Pollution*; EPA 440/5/85-001; U.S. EPA: Washington, DC, 1985; pp 201-207.
3. Baker, David B. *Sediment, Nutrient and Pesticide Transport in Selected Lower Great Lakes Tributaries*; EPA-905/4-88-001; U.S. EPA, Great Lakes National Program Office: Chicago, IL, 1988; pp 1-225.
4. Baker, D. B. *Agriculture, Ecosystems and Environment* **1993**,*46*;197-215.
5. Richards, R. P.; Baker, D. B. *Environ. Toxicol. Chem.* **1993**, *12*,13-26.
6. Fenelon, J. M. *Water Quality of the White River Basin, Indiana, 1992-96*; U. S. Geological Survey Circular 1150. U. S. Geological Survey, Information Services: Denver, CO, 1998; pp 1-7.
7. Resource Management Associates. *Land Resource Information For the Lake Erie Drainage Basin, Co-occurrence of Land Resource Features*. U. S. Army Corps of Engineers: Buffalo, NY, 1979; Vol. 2, pp 96-111, 125-138; Vol. 3, pp 180-189, 230-247.
8. US EPA. *Methods for Analysis of Water and Wastes*; US EPA: Cincinnati, OH, 1979.
9. Matthai, H.F. In *Surface Water Hydrology*; Wolman, M.G.; Riggs, H.C., Eds.; The Geological Society of America: Boulder, CO, 1990; pp 97-120.
10. Leopold, L.B. *Water, Rivers and Creeks*; University Science Books: Sausalito, CA, 1997; pp 39-57.
11. Huber, W.C. In *Handbook of Hydrology*; Maidment, D.R., Ed.; McGraw-Hill: New York, NY, 1993; pp 14.1-14.50.
12. Baker, D. B. *J. Soil Water Conserv.* **1985**, *40*, 125-132.
13. Livingston, E. H.; Cox, J.H. In *Perspectives on Nonpoint Source Pollution*, EPA 440/5/85-001; US EPA: Washington, DC, 1985; pp 289-291.

Chapter 4

Use of Laboratory, Field, and Watershed Data to Regulate Rice Herbicide Discharges

Lisa J. Ross[1], D. G. Crosby[2], and J. M. Lee[3]

[1]California Environmental Protection Agency, California Department of Pesticide Regulation, 830 K Street, Sacramento, CA 95814
[2]Department of Environmental Toxicology, University of California, Davis, CA 95616
[3]Department of Pesticide Regulation, 1020 N Street, Sacramento, CA 95814

Of approximately 188,000 ha of rice grown annually in California, greater than 90% are located in the Sacramento Valley. An estimated 682,000 kg of molinate and 306,000 kg of thiobencarb have been used for weed control in flooded rice fields of this region during a single rice growing season. In the early 1980s, frequent fish kills in agricultural drains and the presence of these herbicides in the Sacramento River, a source of city drinking water, led to their regulatory control in rice field discharges. Development of use restrictions relied on their physical properties, including aqueous solubilities, soil adsorption coefficients, and Henry's Law constants. However, due to differences in degradation rates observed in laboratory and field studies, field data were ultimately employed to develop control strategies. Subsequent monitoring in the Sacramento River watershed was conducted to assure that water quality objectives were met under various river flow conditions and use patterns.

Hydrology. California's Central Valley is hydrologically divided into two regions: the southern San Joaquin River watershed and the northern Sacramento River watershed. The Sacramento River watershed is over 70,000 km^2 in area, and comprises about 17% of the land mass of California (*1*). The Sacramento River is over 515 km long, and stretches from Mount Shasta and the Modoc Plateau in the north, through the fertile Sacramento Valley, to the San Francisco Bay-Delta to the south (Figure 1). The Sacramento Valley

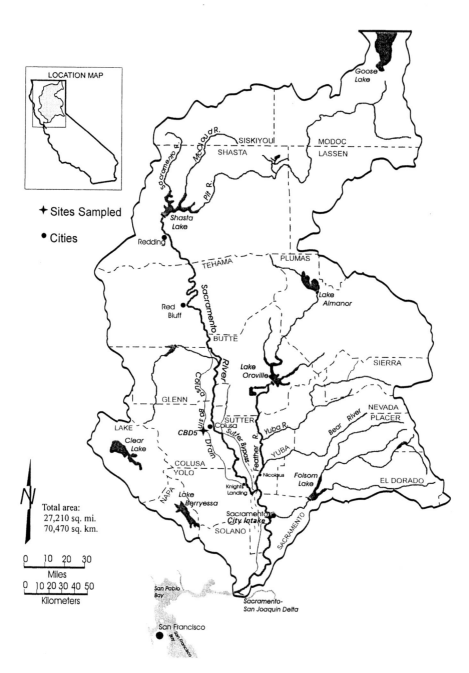

Figure 1. Map of the Sacramento River Hydrologic Basin.

supports a diverse agricultural economy: rice, fruit and nut trees, tomatoes, corn, melons, and sugar beets are some of the major commodities (*2*).

Rice Production and Herbicide Use. The Sacramento Valley is the primary area of rice production in California, where over 90% of approximately 188,000 ha is located (*3*). In this region, rice is grown in flooded, mostly laser-leveled fields. The low percolation rate of clay and hardpan soils in this region make them ideal for rice production. Because downward movement of water is reduced, retention of flood waters on field during the hot, dry growing season is facilitated. Sufficient water supply and good water management practices are critical factors in rice production to maintain weed control, nutrient levels, and crop vigor. The water level within a field is regulated by a series of removable boards placed in weirs built into levees that separate the rice paddies. Excess water is discharged from the fields into a network of agricultural drains, sloughs, and streams which reach the Sacramento River and eventually the San Francisco Bay-Delta.

California, and the United States in general, has the highest yielding rice producing acreage in the world (*4*). This high productivity is due largely to the technology available to the American grower in areas of land preparation, fertilization, irrigation water management, and weed control. Various weed species including sprangletop (*Leptochia fascicularis*), smallflower umbrellaplant (*Cyperus difformis*) and barnyard grass (*Echinochloa crusgalli*) are common problems in the Sacramento Valley. Barnyard grass at an average density of 10.7 plants/m^2 in a rice field may reduce yields by as much as 25% (*5*). Therefore, weed control is economically important to rice growers. Various methods such as cultivation during seedbed preparation, water management strategies, weed-free rice seed, and fertilizer management are all important components of a weed control program. However, because these methods are only partially effective (*6*) for weed control, herbicides, particularly molinate (S-ethyl hexahydro-1 H-azepine-1-carbothioate) and thiobencarb (S-[4-chlorobenzyl] *N,N*-diethylthiocarbamate), are used extensively. An estimated 682,000 kg of molinate and 306,000 kg of thiobencarb have been applied to rice in the Sacramento Valley in a single growing season (Table I, *7*).

Rice Herbicides in the Sacramento River Watershed. As a result of high herbicide use and the discharge of rice field water into surface drains and streams, both molinate and thiobencarb have been detected in various reaches of the Sacramento River Watershed. About 50% of the rice acreage in the Sacramento Valley drains into the Sacramento River along a 105-km stretch between the city of Colusa and the confluence with the Feather River (Figure 1). Rice field drainage may supply between 15 and 30% of the water in the Sacramento River during the rice growing season from May to September (*8*).

The Sacramento River is a natural habitat for many fish species including salmon, trout, striped bass, catfish, and carp. Agricultural drains also provide habitat for fish and wildlife and are an important part of this watershed from a natural resource perspective. Further downstream, the Sacramento River is a source of drinking water for the city of Sacramento. Studies have documented that rice herbicides are present seasonally in water, sediment, and biota of this watershed (*8*). Concentrations as high as 697 µg/L of molinate and 170 µg/L of thiobencarb were found in the Colusa Basin Drain in the early 1980s. In addition, maximum concentrations detected in the Sacramento River reached 27 and

6.0 μg/L, respectively. Biological monitoring during this time period revealed that extensive fish kills observed in the Colusa Basin Drain were due to molinate (*9-10*). In addition, taste complaints received by the city of Sacramento in May and June of 1981 and 1982, were attributed to thiobencarb sulfoxide in drinking water, generated by chlorination of the parent compound (*8,11-12*).

The Need for Regulation. In response to the fish kills and taste complaints, scientists from several state agencies and research institutions participated in a cooperative effort to mitigate these problems and develop a regulatory strategy. A number of tasks had to be accomplished prior to the effective implementation of use restrictions: (1) information about the environmental fate and movement of molinate and thiobencarb under field conditions was needed to develop options for limiting their discharge from the field, (2) action levels associated with the consumption of drinking water that contains molinate and thiobencarb had to be established, (3) water quality goals for the protection of aquatic life had to be determined, and (4) use restrictions had to be developed. Watershed monitoring was also conducted to determine if restrictions on use effectively reduced residues in the watershed to meet the action levels and water quality goals. In this chapter we will focus on the laboratory and field data used to develop use restrictions and on watershed monitoring to emphasize the need for using the appropriate scale of research to resolve issues concerning the movement of agrochemicals from their intended target areas.

Dissipation and Persistence of Rice Herbicides

Environmental dissipation and persistence are a function of a pesticide's physical and chemical properties (*13*). The physical properties, as determined in the laboratory (Table II), showed that molinate has greater water-solubility, lower soil adsorption, and higher vapor pressure and Henry's Law constant than does thiobencarb. The Henry's constant (H), which is the ratio of vapor pressure to aqueous solubility, is a measure of the potential of a chemical to volatilize from water into air (an H value below about 10^{-7} atm m^3/mole indicates non-volatility). The soil adsorption coefficient is a measure of the tightness of the chemical's binding to soil. The somewhat higher H value of molinate suggests that it would volatilize more readily than thiobencarb, while the latter's greater adsorption coefficient indicates somewhat tighter binding into soil and sediments. In a flooded rice field, the partitioning of molinate is predicted to be primarily from water into air, while that of thiobencarb from water into soil.

The environmental breakdown of the herbicides is based on their chemical reactivity. As both molinate and thiobencarb are esters, hydrolysis represents one possibility, and microbial degradation in soil is another. In addition, lack of ultraviolet absorption by solutions of either chemical in distilled water above the cutoff of sunlight energy (about 290 nm) suggests that photodegradation would not be significant. The practical use of physical and chemical properties is complicated by the range of differing values already existing in the literature.

However, field studies present a somewhat different picture. While the trend of the predicted partitioning is correct, the rates are not those expected. Soderquist et al. (*14*) measured a field half-life in water of as little as 2 days (Figure 2) and estimated that 85% of the total molinate dissipation could be due to volatilization. Ross and Sava (*15*) found that 9% of the applied molinate volatilized on the day of application, and an estimated 34% volatilized during the 4-day water holding period and estimated a water half-life of about 4 days. Crosby (*16*) reported a field half-life in water for thiobencarb of 5 to 7 days after water concentrations peaked (Figure 2), and based on laboratory data, considered that volatilization also would be a significant route of thiobencarb loss. However, Ishikawa (*17*) found that the addition of soil to an aqueous solution of thiobencarb reduced thiobencarb volatilization, presumably due to adsorption. Ross and Sava (*15*) found that less than 1% of the thiobencarb applied to a rice field volatilized on the day of application, and that this amount represented over 60% of the total amount volatilized during the 4-day period following application. Such information shows that the volatilization of thiobencarb, although measurable, is not a major component of the compound's dissipation in field water, while volatilization can dominate molinate's dissipation.

Another difference between field and laboratory data lies in the degradation routes. Although aqueous photolysis appears to be unimportant for degradation using laboratory data (Table II), sunlight proved to be a major factor for herbicide breakdown in field water. Soderquist et al. (*14*) found that natural "photosensitizers," such as tryptophan present in field water, caused the photodegradation of molinate to molinate sulfoxide and sulfone with a half life of 10 days, with eventual degradation to hexamethylenimine and ethanesulfonic acid. Draper and Crosby (*18*) demonstrated that the actual cause of degradation was highly reactive hydroxyl radicals generated in field water by sunlight. In addition, they observed the indirect photo-oxidation of thiobencarb to its sulfoxide, p-chlorobenzaldehyde, and ultimately to p-cholorobenzoic acid with a half-life of 6 days (Figure 3); this indirect photolysis has been reviewed by Mabury and Crosby (*19*).

Knowledge of these field dissipation half-lives proved essential to the development of a regulatory control strategy for rice herbicides in surface waters. The field half-life of molinate ranged from 2 days to as long as 5 days (*14,15,20,21*), and that of thiobencarb from 2.5 to 9 days (*15,16,22*), strongly depending upon temperature, sunlight intensity, wind speed, and depth of water (*20, 23,16*). These half-lives were important for the determination of an appropriate field strategy, but as water temperatures could vary from 20° to 35°C over the course of a single day, and sunlight intensity from none (night) to very intense (noon), considerable variation in persistence had to be expected and tolerated.

Regulation

Action Levels and Water Quality Goals. In response to detections of molinate and thiobencarb in the Sacramento River, the California Department of Health Services developed primary action levels for molinate and thiobencarb (Table III). These action levels were the initial step in developing maximum contaminant levels (MCLs) to protect human health and were based on the no observable effect levels for the most sensitive species. The action levels, developed for a 10 kg child who consumes one liter of water a day, were 20 and 10 μg/L for molinate and thiobencarb, respectively (*24*). In addition,

Table I. Use of major rice herbicides in California (kg, active ingredient) in 1982, 1988, and 1995.

Herbicide	1982	1988	1995
Bensulfuron-methyl	0	2	20,510
Bentazon	56,125	118,006	0
MCPA, dimethyl amine salt	184,078	105,610	52,551
Molinate	682,887	687,592	648,566
Propanil	38,030	36,139	18,239
Thiobencarb	306,224	195,516	261,167

Source: California Department of Pesticide Regulation, Pesticide Use Reports, Sacramento, CA

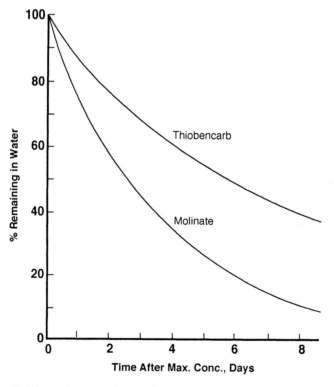

Figure 2. Dissipation rates for molinate and thiobencarb in rice field water.

Table II. Physical and chemical properties of molinate and thiobencarb.

Property	Molinate	Thiobencarb
Solubility at 20°C (mg/L)	800	30
Vapor Pressure at 25°C (Pa)	7.5×10^{-1}	2.9×10^{-3}
Henry's Law Constant (atm m^3 /mole)	1.5×10^{-6}	2.5×10^{-7}
Soil Adsorption (Koc)	186	1380
Aqueous Photolysis t ½ (days)	stable	190
Indirect Photolysis Field t ½ (days)	10 reference *14*	6 reference *18*
Hydrolysis, pH 5-9 t ½ (days)	>30	>30
Water Persistence Field t ½ (days)	2-5 references *14,15,20,21*	2.5-9 references *15,16,22*

All values are from reference 27, except where indicated.
All measurements made in the laboratory, except where indicated.

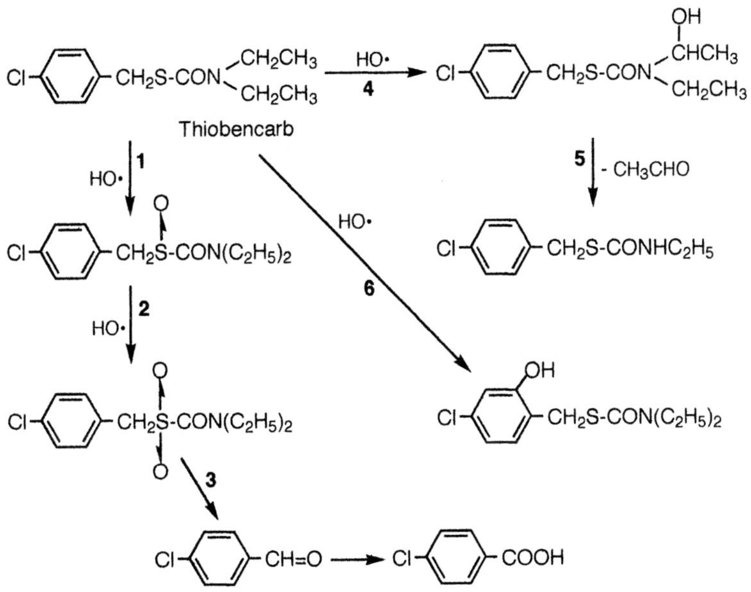

Figure 3. Reactions of hydroxyl radicals with thiobencarb.

a secondary action level of 1 µg/L for thiobencarb was developed to protect the taste of tap water consumed in the city of Sacramento. This action level was determined by correlating the number of taste complaints with concentrations of thiobencarb in the Sacramento River. The final MCLs were developed in 1988 and were 20 and 70 µg/L, respectively. The action level for thiobencarb was lower than the final MCL because missing data required the Department of Health Services to use a large safety factor until the data were acquired.

The California Department of Fish and Game recommended aquatic guidelines of 90 and 24 µg/L (instantaneous maxima) for molinate and thiobencarb, respectively, in 1983 (Table III, 24). These water quality goals were based on the no observable effect concentrations for the most sensitive aquatic species tested at that time (24,25,26). These values were each lowered by one half in 1987 because they were believed to have an additive effect when present together.

Limiting Pesticide Input to the Environment. At the time molinate and thiobencarb became an issue in surface water, water holding times were 4 and 6 days, respectively. Water holding times were a traditional practice to promote product efficacy. The holding time for thiobencarb was longer than for molinate because thiobencarb takes longer to dissolve from the granular formulation. Because molinate is relatively short-lived in rice field water, water management strategies which increase the residence time on the site of application were used to control off-site movement. Thus, growers using molinate were instructed to retain water on the field for a prescribed period of time, or contain it within a recirculating system prior to discharge into the watershed. The prescribed holding period was increased from the original 4-day hold to 8 days in 1984 and 14 days in 1988 (Table IV).

In contrast, the strategy to mitigate thiobencarb was different than for molinate because it is more persistent in rice field water and the secondary action level was low (1 µg/L). Chevron Chemical Company (the manufacturer of thiobencarb at the time) established a sales limitation, limiting sales to 1.81 million Kg of formulated thiobencarb in the Sacramento Valley, enough to treat about 40,470 ha, in 1984 (Table IV). This amounted to a 40% reduction in sales from 1982. It was hoped that a sales limit, leading to decreased use, would result in a reduction in concentrations detected in the watershed. In 1987, the sales limit was increased to 2 million Kg, but thiobencarb usage was restricted to areas with minimal amounts of water discharge to the Sacramento River. These areas, upstream of the city of Sacramento, were using water conservation practices, such as recirculating systems or ponding, which minimized water discharge to the Sacramento River. By using such practices, one water agency was able to reduce molinate and thiobencarb discharges by 95% from 1985 to 1986.

Watershed Monitoring: Were the Goals Met?

Watershed monitoring was conducted to determine the effectiveness of use restrictions originally set in place in 1984. Water samples were collected annually during May and June, the peak period for rice herbicide applications, at a number of sites in the watershed.

Table III. Regulatory goals.

	Fish and Wildlife	Year	Human Health	Year
Molinate	90 μg/L	1983	20 μg/L	1983
Revised	45 μg/L	1987	No Change	
Thiobencarb	24 μg/L	1983	10 μg/L	1983
Revised	12 μg/L	1987	70 μg/L	1988
Secondary Action Level to Protect Taste of Drinking Water			1.0 μg/L	1983

Table IV. History of restrictions on the use of molinate and thiobencarb.

Year	Molinate	Thiobencarb	
	Holding Period (days)	Holding Period (days)	Sales Limit (ha)
1983	4	6	None
1984	8	6	40,470
1985	8	6	36,423
1986	8	6	8,094
1987	12	30[a]	44,517[a]
1988	14		
1989	14		
1990	19		
1991	24		
1992	28[a]		

a. Use restriction currently in effect in 1998.

Water discharge information was also collected at certain locations. Data presented are for two sites: (1) the Colusa Basin Drain near the town of Colusa and (2) the Sacramento River at the city of Sacramento's water treatment facility (Figure 1). The data are displayed as concentrations compared with action levels and water quality goals (Figures 4 and 5).

Colusa Basin Drain. Concentrations of both herbicides are consistently higher in the agricultural drains than in the main stem of the Sacramento River (8). Prior to establishment of use restrictions, the maximum molinate concentration in the Colusa Basin Drain was 697 μg/L in 1982 (Figure 4, 10). In this year, an estimated 28,000 fish were killed in the Sacramento River watershed due to excessive molinate exposure. With extension of the 4-day holding period to 14 days, peak concentrations at this site decreased to 89 μg/L in 1988, and no fish kills attributable to molinate were reported after 1983. Thiobencarb concentrations in the Colusa Basin Drain also peaked in 1982, at 170 μg/L, and then fell to 1.0 μg/L in 1988 (Figure 4).

Water quality goals for molinate for the protection of fish and wildlife were originally set at 90 μg/L in 1983 (Table IV). This goal was met more consistently in the Colusa Basin Drain when the water holding period was increased to 14 days (Figure 4). In 1987 the goal was lowered to 45 μg/L, since molinate and thiobencarb are believed to act additively and occur together in surface water during the rice season. To meet this goal, the holding period was eventually increased to 28 days in 1992, however peak concentrations exceeded this goal in two of the past 6 years. After initial dramatic reductions in concentrations, longer holding periods had little additional influence on peak concentrations in the Colusa Basin Drain.

Water quality goals for thiobencarb were originally established at 24 μg/L in 1983 and reduced to 12 μg/L in 1987 (Table IV). Since the goals were established, they have been met in all but 2 years (Figure 4). The sales limitation proved effective for controlling thiobencarb residues in surface water, while the addition of a 30-day holding period in 1987 did not appear to promote further reductions at this site (Figure 4).

Sacramento River. In Sacramento River water, a maximum molinate concentration of 13 μg/L was observed in 1982 and declined to 4.8 μg/L in 1988, while thiobencarb concentrations fell from 6.0 to 0.21 μg/L between 1982 and 1988 (Figure 5). The MCLs to protect public health and the water quality goals to protect fish and wildlife were met in all years in the Sacramento River at this site. The secondary action level for thiobencarb to protect the taste of tap water was exceeded during the initial years of the regulatory program (Figure 5). Therefore, drinking water was treated with potassium permangate to remove thiobencarb sulfoxide. However, since implementation of a 30-day holding period in 1987 for thiobencarb, the secondary action level has not been exceeded, nor has the city received any taste complaints.

Conclusions

While the physical and chemical properties of pesticides, as measured in the laboratory, have an important and necessary place in the prediction of pesticide movement and

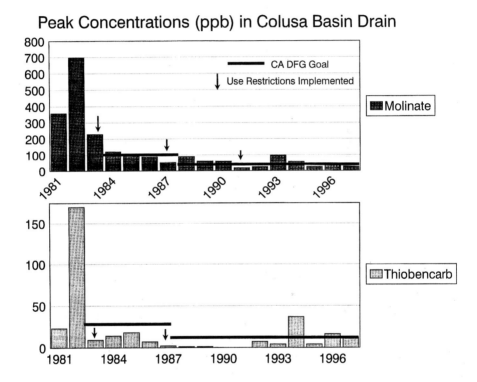

Figure 4. Peak molinate and thiobencarb concentrations in the Colusa Basin Drain. Arrows indicate when new use restrictions were implemented.

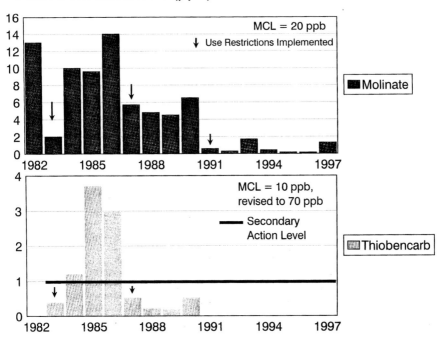

Figure 5. Peak molinate and thiobencarb concentrations in the Sacramento River. Arrows indicate when new use restrictions were implemented.

environmental fate, they alone are not sufficient for the establishment of regulatory measures. Environmental factors such as water temperature, ultraviolet intensity, and wind speed affect pesticide persistence, so measurements from actual field applications assume major importance in the development of control strategies.

Even with this information, one must expect variability that will depend on the timing and location of field experiments; a fairly narrow range of persistence values is the best that can be expected at this stage of our knowledge. This means that pesticide half-lives should be obtained from as many field trials or commercial applications as possible, although data from small field plots are generally all that are available.

Once the pesticide comes into general use, the final step is to monitor residue levels in local bodies of water for comparison with health standards or other criteria, followed by further restrictions on environmental input to the extent necessary to achieve compliance. This ideal may not be possible to attain in a single growing season, but repeated "titration"of environmental concentrations against regulatory standards should result in residue levels that are fair to both the growers and the public.

Literature Cited

1. State Water Resources Control Board. *Sacramento River Toxic Chemical Risk Assessment Project, Final Report*; State of California: Sacramento, CA, **1990**. Report No. 90-11WQ.
2. California Department of Pesticide Regulation. *Pesticide Use Report*; State of California: Sacramento, CA, **1983**.
3. United States Department of Agriculture. *California Field Crop Review*; California Agricultural Statistics Service: Sacramento, CA, **1990**; Vol 11(7).
4. California Department of Water Resources. *Low Applied Water on Rice*; Sacramento, CA, 1982.
5. Hill, J.E. *Proc. California Weed Conference*, Sacramento, CA. **1984**. 36th California Weed Conference.
6. Scardaci, S.C.; Hill, J.E.; Crosby, D.G.; Grigarick, A.A.; Webster, R.K.; and Washino, R.K. *Evaluation of Rice Water Management Practices on Molinate Dissipation and Discharge, Rice Pests and Rice Production*; Agronomy Progress Report; University of California Cooperative Extension, Davis, CA., **1987**, Report No. 200.
7. California Department of Pesticide Regulation. *Pesticide Use Report*; State of California: Sacramento, CA, **1983**.
8. Cornacchia, J.W.; Cohen, D.B.; Bowes, G.W.; Schnagl, R.J.; and Montoya, B.L. *Rice Herbicides Molinate (Ordram) and Thiobencarb (Bolero)*; State Water Resources Control Board: Sacramento, CA, **1984**; Report No. 84-4sp.
9. Finlayson B.J.; Nelson, J.L.; and Lew, T.L. Colusa Basin Drani and Reclamation Slough Monitoring Studies; California Department of Fish and Game, Sacramento, CA, 1982. Report No. 82-3.
10. Finlayson, B.J.; Lew, T.L. Rice Herbicide Concentrations in Sacramento River and Associated Agricultural Drains; California Department of Fish and Game, Sacramento, CA, 1983. Report No. 83-5

11. Peryam, D.R.; Swartz, V.W. *Food Technol.* **1950**, *4*, 390.
12. Roessler, E.B.; Pangborn, R.M.; Stone, J.L.; Sidel, H. *J. Food Sci.* **1978**, *43*, 940.
13. Mabury, S.A.; Cox, J.S.; Crosby, D.G., *Rev. Environ. Contam. Toxicol.*, **1996**, *147*, 71-117.
14. Soderquist, C.J.; Bowers, J.B.; Crosby, D.G., *J. Agric. Food Chem.*, **1977**, *25*, 940-945
15. Ross, L.J.; Sava, R.J. *J. Environ. Qual.*, **1986**, *15*, 220-225
16. Crosby, D.G. *IUPAC Pesticide Chemistry; Human Welfare and the Environment*; Editors P.C. Kearny and J. Miayamoto; Pergamon Press, New York, NY, 1983, pp. 339-346.
17. Ishikawa, K.; Nakamura, Y.; Kuwatsuka, S. *J. Pesticide Sci.*, **1977**, *2*, 127-134.
18. Draper, W.M.; Crosby, D.G. *J. Agric. Food Chem.*, **1981**, *29*, 699-702
19. Mabury, S.A.; Crosby, D.G. In *Aquatic and Surface Photochemistry*, Editors G.R. Helz, R.G. Zepp, and D.G. Crosby; CRC Press, Boca Raton, FL, 1994. pp.149-161.
20. Tanji, K.K; Biggar, J.W.; Mehran, M.; Cheung, M.W.; Henderson, D.W. *California Agric.* **1974**, *May*, 10.
21. Deuel, L.E.; Turner, F.T.; Brown, K.W.; Price, J.D., *J. Environ. Qual.*, **1978**, *7*, 373-377.
22. Yusa, Y.; Ishikawa, K. *Proc. Asian Pac. Weed Sci. Soc. Conf.* **1977**, *2*, 596
23. Lyman, W.J.; Reehl, W.F.; Rosenblatt, D.H., *Handbook of Chemical Property Estimation Methods*; ACS Books, Washington, DC, 1990.
24. Lee, J.M.; Ross, L.J.; Wang, R.G. In *Effective and Safe Waste Management*; Editors, R.L. Jolley and R.G.M. Wang; Integrating Environmental Toxicology and Monitoring in the Development and Maintenance of a Water Quality Program: California's Rice Herbicide Scenario; Lewis Publishers: Boca Raton, FL. 1993. pp 211-224.
25. Finlayson, B.J.; Fagella, G.A., *Trans. Am. Fish. Soc.*, **1986**, *115*, 882-890.
26. Fagella, G.A.; Finlayson, B.J., Hazard Assessment of Rice Herbicides Molinate and Thiobencarb to Larval and Juvenile Striped Bass. California Department of Fish and Game, Sacramento, CA, 1983. Report No. 87-2
27. Tomlin, C.D.S., *The Pesticide Manual*, 11th ed.; British Crop Protection Council, Farnham, UK, 1997.
28. Imai,Y., Kuwatsuka, S. *J. Pesticide Sci.*, **1982**, *7*, 487-497.

The Chesapeake Bay Watershed

Chapter 5

Potential for Herbicide Contamination of Groundwater on Sandy Soils of the Delmarva Peninsula

W. F. Ritter[1], A. E. M. Chirnside[1], R. W. Scarborough[1], and T. S. Steenhuis[2]

[1]Bioresources Engineering Department,
University of Delaware, Newark, DE 19717-1303
[2]Agricultural and Biological Engineering Department,
Cornell University, Ithaca, NY 14853

>Herbicide leaching was investigated in a field study, modeling study and rainfall simulation experiment on Coastal Plain sandy soils. In the field experiment, alachlor was detected more frequently than atrazine or metolachlor in the groundwater. The GLEAMS model simulations indicated that there is a great potential for herbicide leaching into the watertable aquifer. For the 100-year storm, the total annual leaching amounts for atrazine ranged from 0.4 to 5.9% of the applied herbicide for rainfall occurring 3, 7, 15, 30 and 90 days after herbicide application. Atrazine, alachlor, simazine, cyanazine and metolachlor all moved below the root zone after 75 mm of rainfall was applied in the rainfall simulation experiment. Alachlor was detected more frequently than the other herbicides in the lysimeters and ground water.

Poultry is a $1.58 billion business on the Delmarva Peninsula with 609 million broilers produced in 1997 (1). Since there is a tremendous demand for feed grains in the poultry industry, corn and soybeans are the major crops grown on the Delmarva Peninsula. The three most common herbicides used on corn nationally and on the Delmarva Peninsula are atrazine, alachlor and metolachlor. Many of the Coastal Plain soils on the Delmarva Peninsula are loamy sand or sandy loam. The Peninsula has an average annual rainfall of 110 cm. These soil and rainfall conditions make the watertable aquifer very vulnerable to groundwater contamination. The water-table aquifer generally fluctuates from 0 to 5 m below the surface on many of the Coastal Plain soils.

Since 1984, the Bioresources Engineering Department has been involved in pesticide transport research. Ritter et al. (2) found that atrazine, simazine, cyanazine and metolachlor were leached to the groundwater shortly after they were applied in 1987, but not in 1988. A total of 30.5 mm of rainfall occurred within 9 days of herbicide application in 1987, when the herbicides were detected in the groundwater. In 1988, no rainfall occurred within the first week after application. In another three-year study, dicamba leaching to groundwater was measured under no-tillage (NT) and conventional tillage (CT) (3). In the first year of the study, dicamba was detected in all monitoring wells at concentrations ranging from 2.0 to 37.0 ug/L 12 days after it was applied. A total of 54 mm of rainfall occurred in the 12 days. The other two years dicamba was detected infrequently in the groundwater because of the rainfall distribution.

Since 1989, the Bioresources Engineering Department has conducted a number of experiments on groundwater impacts by herbicides. This paper will summarize some of the results from a three-year field study, a modeling study and a rainfall simulation study.

Methods and Materials

Field Experiment. The research was conducted at the University of Delaware Research and Education Center near Georgetown on an Evesboro loamy sand soil (mesic coated, typic quartzipsamment) with an infiltration rate of 15 cm/h. The principal aquifer system beneath the research center consists of the sands of the Columbia formation. The aquifer varies in thickness from 27 to 61 m. Depth to the water table generally ranges from 1.5 to 3.1 m. The average transmissivity is about 1000 m^2/day. The average storage coefficient is around 0.05 (4).

The leaching of atrazine, alachlor, and simazine were investigated from 1989 to 1991. The treatments included NT with and without a rye cover crop and with a rye cover crop and CT with and without a rye cover crop. Conventional tillage consisted of chisel plowing and disking. Corn was planted each year on the plots and irrigated by solid set irrigation on a 9m x 9m spacing to maintain soil moisture at 50% or more of available moisture.

The rye winter cover crop was planted on the plots with a cover crop treatment on October 2, October 25 and September 24 in 1988, 1989 and 1990, respectively with a no-till drill. The rye winter cover crop was killed on April 13, April 13 and April 15 in 1989, 1990, 1991, respectively, with either paraquat or glyphosate (Roundup). The CT plots were tilled before the rye was planted. Atrazine, alachlor and simazine were applied broadcast each year at a rate of 2.24 kg/ha. Thirty- percent urea-ammonia-nitrate liquid nitrogen fertilizer was applied at a rate of 200 kg/ha.

The plots were approximately 0.25 ha in size and had very little slope, so no runoff occurred. Each plot had three monitoring wells at each depth of 3.0 and 4.5 m. Monitoring wells were constructed from 3.25 cm polyvinyl chloride (PVC) pipe with a screen length of 1.5 or 0.75 m and a slot size of 0.25 mm. All monitoring wells were installed by the auger drilling method. The top 1.5 m of the annulus around the casing of all wells was grouted with bentonite. The groundwater was sampled monthly from April 1989 to December 1991, except during the months of January and February with peristaltic pump. The wells were pumped dry and allowed to recharge before a sample was taken. The samples were transported on ice to the Bioresources Engineering Water Quality Laboratory in Newark and frozen at $-18°C$ until they were analyzed.

Modeling Study. The GLEAMS model was used to simulate leaching of atrazine and metolachlor in an Evesboro loamy sand. Twenty-four hour rainfall events of 2, 10 and 100 year return periods were used to model the herbicide leaching for rainfall occurring 3, 7, 15, 30 and 90 days after herbicide application.

In order to predict the outcome of different scenarios on herbicide leaching, past climatic data may not be useful. When long-term studies of many years are to be run, historical data would be acceptable, because any yearly abnormalities would be averaged-out. However, if only short-term simulations are to be run, actual data may not be appropriate due to the importance of rainfall in the GLEAMS model calculations. Any single year probably would not be a typical average year, and bias would be introduced in choosing the year to use. Therefore, in this study a first-order Markov chain and gamma distribution were used to generate climatic data for the GLEAMS model. The first-order Markov chain was used to determine the days of precipitation. A gamma distribution was used to estimate precipitation amounts on the days selected by the Markov chain. Both the Markov chain and gamma distribution are proven and accepted methods for analyzing climatic data (5).

The climatic data for calibration of the model were obtained from the National Oceanographic and Atmospheric Administration monitoring station located at the University of Delaware Research and Education Center at Georgetown, DE. These data included the daily temperature extremes and the daily precipitation. The monthly radiation, dew point, and wind movement values for the Georgetown area were incorporated into the program. Since the program is not highly sensitive to these parameters, average values are acceptable. For simulation runs, the modeled precipitation was used along the average daily temperatures as computed from historical data. The base permeability and sturated conductivity of the soil were taken from the USDA-SCS Soil Survey of Sussex County (6). Using the program for nutrient leaching, the authors had previously adjusted the soil parameters for GLEAMS to simulate most closely the conditions of southern Delaware. The parameters included soil porosity and saturated conductivity values which are the primary factors governing water movement in the soil. The values of 0.38 cm/cm and 2.36 cm/hr resulted in the highest correlation with actual values. The field capacity and wilting point were also available from the soil survey. The soil moisture content was assumed to be at field capacity at the beginning of each run (January 1).

The agronomic factors of the calibration trials included field preparation by means of chisel plowing followed by two diskings immediately prior to planting. A full season corn hybrid was planted with the herbicides being applied on the following day. The corn was irrigated, with the irrigation amount entered in the rainfall data as additional rain. The simulated crop data consisted of corn planted on April 20. The simulation field was prepared identically to the calibrated fields. Irrigation was applied by the program whenever the soil moisture fell below 50% of field capacity. Enough irrigation was applied to restore the soil moisture to field capacity. The crop was fertilized as recommended to simulate actual plant and root growth and evaportranspiration conditions.

For the calibration and simulated runs, no metabolites were considered. All application program runs used rates of 2.24 and 1.68 mg/ha active ingredients for atrazine and metolachlor, respectively,

A sensitivity analysis of the plant uptake variable indicated only a 1.5% change in the amount of both herbicides in the soil after 51 days. Therefore, plant uptake was set at one to show the best case scenario. The inputs for foliar application and foliar wash off were ignored because the herbicides were applied to bare soil. The method of application being considered was a preemergence broadcast surface spraying over the entire field one day after planting. The two variables that greatly affect the movement of a herbicide in the soil are its half-life and partitioning coefficient, or the ratio of the herbicide concentration on organic matter to its concentration in water.

Published half-life values vary from 18 to 120 days for atrazine and from 15 to 32 days for metolachlor (7). These values are greatly controlled by soil moisture, temperature, and microbial activity in the soil. The GLEAMS soil half-life calculation are based on the exponential decay formula:

$$C_t = C_o e^{-kt}$$

C_t – herbicide concentration at time t – (mg/kg)

t – time (days)

k – degradation rate constant (days $^{-1}$)

Therefore, these values could easily be determined separately from the GLEAMS model. Half-life values were determined from a previous research project of the authors where soil samples were taken from 0 to 150 cm in an Evesboro soil (2). A curve was fitted to the data using a first-order degradation rate. This produced a half-life of 31 days for atrazine and 28 days for metolachlor. These values were used for the computer simulations.

Rainfall Simulation Experiment. The research site was located at the University of Delaware Research Center in Georgetown, Sussex county, DE. Soils at the research site were classified as Evesboro, Fallsington and Rumford loamy sands, derived from Pleistocene fluvial deposits (6).

In the fall of 1990, the 2 ha field was divided equally into eight plots, which wee subjected in pairs to four tillage treatments. Ridges were established on two of the plots and a winter cover crop of rye and white clover was planted on two other plots in October of 1990.

Treatments used on the plots included ridge tillage (RT), CT, CT with a winter cover crop of rye and white clover and NT with poultry manure applied at a rate of 6 t/ha. The cover crop was planted each year in October and killed with paraquat the latter part of April. Previous research has shown that if sufficient rainfall occurs shortly after some herbicides are applied to sandy coastal plain soils they will move rapidly to the groundwater by macropore flow. The winter cover crop and poultry manure experiments were included because it was thought that increasing the soil organic matter may decrease herbicide leaching (2).

Corn was planted in May from 1991 to 1993. The herbicide alachlor, simazine, atrazine, metolachlor and cyanazine were applied preemergence after corn planting at rates of 2.24, 2.24, 2.24, 1.7 and 1.7 kg/ha, respectively. A solid set sprinkler irrigation on a 9 m x 9 m spacing was installed in the plots several days later. In 1991, five days after the herbicides were applied a total of 75 mm of water was applied to simulate a one hour, 10-year return period rainfall. In 1992, the same irrigation treatment was used five days after the herbicides were applied. Dry conditions occurred before the irrigation was applied in 1991, while in 1992 the soil moisture was at field capacity before irrigation. The 1993 treatment simulated a 14-day rainfall for a 10-year return period. The irrigation water was applied during three irrigations, with the first irrigation occurring five days after the herbicides were applied.

Each plot had three monitoring wells that were installed for a previous project. The wells were constructed from 3.25 cm diameter PVC pipe and installed to a depth of 4.5 m with the bottom 0.75 m of the wells screened with 0.75 mm slot size screen. A wick lysimeter was installed in each plot at 1 m depth in April of 1991. The wick lysimeter consisted of a 50 cm long wick attached to a lysimeter with a width of 10 cm and the length equal to 60 cm. There were three wicks on each lysimeter that drained to three sample bottles. The design and installation of the wick lysimeters are described by Boll et al. (8). From 1991 to 1993 samples were collected from the monitoring wells and lysimeters one day after the simulated rainfall.

Groundwater for the rainfall simulation studies were collected with a battery operated peristaltic pump. The monitoring wells were pumped dry and allowed to recharge before samples were collected. The samples were transported on ice to the Bioresources Engineering Department Water Quality Laboratory in Newark and frozen at $-18°C$ until they were analyzed.

Analytical Methods. The herbicides were analyzed by a method that was modified from a procedure developed by the pesticide Residue Laboratory of Cornell University at Geneva, NY (9). The method was developed to analyze large volumes of soil and water samples at a reasonable cost. The water samples were removed from the freezer and allowed to thaw at room temperature and then filtered through a 1.0 micron pore size glass fiber filter. Fifty ml. of the filtered samples were loaded onto a Sep-Pak cartridges at a rate of 3.0 to 4.0 ml/min. The Sep Paks were activated with 5 ml of high-pressure liquid chromotography (HPLC) grade methanol followed by 5 ml of double-distilled water. After the Sep-Paks was loaded and dried 5 minutes with an air aspirator, the herbicides were eluted from the Sep-Paks into a 15 ml graduated test tube with 12 ml of HPLC grade benzene. Samples were concentrated to dryness in a warm water bath under a hood with a gentle flow of high purity N_2 gas. Chlorothalonel spiked HPLC benzene (1.0 ml) was added to each dry test tube.

A 1 ml of sample was injected into a capillary column gas chromatograph with a ^{63}Ni electron capture detector. AN SPB-15 ml column with a 0.53 mm ID was used. The detector temperature was $300°C$. The inlet temperature was programmed to start at $120°C$ for 1 minute and rise at $8°C/minute$ to $240°C$ and hold for 10 minutes. Helium was used as the column gas at a flow rate of 1 m./min. Nitrogen was used as the carrier gas. The minimum detection limit for each herbicide was 0.1 ug/L.

Results and Discussion

Field Experiment. The summer of 1989 was very wet with the total rainfall from May 4 when the corn was planted until the end of August being 726 mm. In 1990, rainfall was 470 mm during the growing season, and in 1991 rainfall was 392 mm. Total irrigation water applied was 133,342 and 269 mm in 1989, 1990 and 1991, respectively. Results for the alachlor, atrazine and simonize are summarized in Tables I to III for the period of May to September. For the October, November, December,

Table I. Alachlor Detections in Groundwater at 3.0 and 4.5 Depths for May to September Samplings

Parameter	Depth (m)	No. Samples	No. Det.	Avg. Conc. (ug/L)	Conc. Range (ug/L)
1989[a]					
NT-CC	3.0	15	0	-	-
NT-CC	4.5	18	0	-	-
NT-NC	3.0	16	0	-	-
NT-NC	4.5	18	4	2.95	1.04-7.51
CT-CC	3.0	18	1	0.88	0.88
CT-CC	4.5	18	1	0.75	0.75
CT-NC	3.0	18	2	4.28	4.25-4.31
CT-NC	4.5	18	2	2.38	1.05-3.71
1990					
NT-CC	3.0	17	1	0.36	0.36
NT-CC	4.5	16	3	0.32	0.31-0.33
NT-NC	3.0	18	3	0.53	0.45-0.65
NT-NC	4.5	18	4	0.40	0.33-0.66
CT-CC	3.0	17	5	0.56	0.31-1.37
CT-CC	4.5	18	3	0.35	0.17-0.55
CT-NC	3.0	17	0	-	-
CT-NC	4.5	18	1	0.24	0.24
CT-NC	4.5	18	5	0.29	0.14-0.33
1991					
NT-CC	3.0	15	5	0.44	0.16-0.64
NT-CC	4.5	18	8	0.69	0.17-1.43
NT-NC	3.0	14	4	1.02	0.31-2.58
NT-NC	4.5	18	6	0.30	0.16-0.33
CT-CC	3.0	15	5	0.36	0.16-0.56
CT-CC	4.5	18	7	0.31	0.18-0.44
CT-NC	3.0	15	2	1.30	0.22-2.38
CT-NC	4.5	18	5	0.29	0.14-0.33

[a]NT- no-tillage, CT – conventional –tillage, CC- cover crop, NC – no cover crop

Table II. Atrazine Detections In Groundwater at 3.0 and 4.5 m Depths for May to September Samplings

Parameter	Depth (m)	No. Samples	No. Det	Avg. Conc (ug/L)	Conc Range (ug/L)
1989[a]					
NT-CC	3.0	15	0	-	-
NT-CC	4.5	18	0	-	-
NT-NC	3.0	16	0	-	-
NT-NC	4.5	18	2	2.75	2.01-3.49
CT-CC	3.0	18	2	2.07	0.17-3.97
CT-CC	4.5	18	0	-	-
CT-NC	3.0	18	0	-	-
CT-NC	4.5	18	0	-	-
1990					
NT-CC	3.0	17	1	4.35	-
NT-CC	4.5	16	0	-	-
NT-NC	3.0	18	2	0.34	0.17-0.51
NT-NC	4.5	18	0	-	-
CT-CC	3.0	17	1	3.46	3.46
CT-CC	4.5	18	1	0.38	0.38
CT-NC	3.0	17	1	2.03	2.03
CT-NC	4.5	18	1	1.34	1.34
1991					
NT-CC	3.0	15	11	2.90	0.38-7.30
NT-CC	4.5	18	15	2.93	0.18-9.18
NT-NC	3.0	14	9	4.18	1.45-6.93
NT-NC	4.5	18	11	2.88	0.64-8.37
CT-CC	3.0	15	10	2.57	0.96-7.74
CT-DD	4.5	18	13	2.05	0.17-5.98
CT-NC	3.0	15	10	1.25	0.15-5.85
CT-NC	4.5	18	13	1.04	0.19-2.62

[a]NT-no-tillage, CT-conventional tillage, CC-cover crop, NC-no cover crop

Table III. Simazine Detections in Groundwater at 3.0 and 4.5 m Depths for May to September Samplings

Parameter	Depth (m)	No. Samples	No. Det	Avg. Conc. (ug/L)	Conc Range (ug/L)
1989[a]					
NT-CC	3.0	15	0	-	-
NT-CC	4.5	18	0	-	-
NT-NC	3.0	16	1	0.75	0.75
NT-NC	4.5	18	1	1.82	1.82
CT-CC	3.0	18	0	-	-
CT-CC	4.5	18	1	2.11	2.11
CT-NC	3.0	18	2	1.58	0.79-2.37
CT-NC	4.5	18	1	0.90	0.90
1990					
NT-CC	3.0	17	0	-	-
NT-CC	4.5	16	0	-	-
NT-NC	3.0	18	0	-	-
NT-NC	4.5	18	1	6.52	6.52
CT-CC	3.0	17	1	0.38	0.38
CT-CC	4.5	18	0	-	-
CT-NC	3.0	17	0	-	-
CT-NC	4.5	18	0	-	-
1991					
NT-CC	3.0	15	6	3.66	0.11-5.85
NT-CC	4.5	18	10	3.61	0.37-10.8
NT-NC	3.0	14	8	2.49	0.53-6.60
NT-NC	4.5	18	11	2.24	0.28-5.37
CT-CC	3.0	15	9	3.06	0.32-15.8
CT-CC	4.5	18	10	1.78	0.47-3.63
CT-NC	3.0	15	8	2.57	0.30-5.50
CT-NC	4.5	18	11	2.50	0.37-8.44

[a] NT-no tillage, CT-conventional tillage, CC-cover crop, NC-no cover crop

March and April samples, the herbicides were only detected in the groundwater several times.

Alachlor was detected in the groundwater all three years. In 1989, alachlor was detected in both CT treatments but was only detected in the NT treatment without a cover crop. Four samples had alachlor concentrations above the E PA drinking water standard of 2.0 ug/L (10). In May of 1989, the water table was approximately 2.1 m below the surface.

Alachlor concentrations were lower in 1990 and 1991 than in 1989, but it was detected more frequently than in 1989. In 1990, alachlor was detected in a total of 11 NT samples and 9 CT samples. In 1991, alachlor was detected in 23 NT samples from May to September and in 19 CT samples. None of the concentrations in 1990 were above the EPA drinking water standard and only 2 samples were the above the standard in 1991. In 1989 and 1990 alachlor was only detected in the groundwater in one sample from October to December and March and April samplings. In 1991, alachlor was detected in approximately 20 percent of the samples from October to December with concentrations ranging from 0.10 to 2.08 ug/L.

The results regarding whether a winter cover crop affected the leaching of alachlor were mixed. In 1989, alachlor was detected more frequently in the no-cover treatment in both CT and NT treatments. In 1990, alachlor was detected in 8 samples with no cover crop planted and in 12 samples where a cover crop was planted. In 1991, on the no cover crop plots, 17 samples contained detectable concentrations of alachlor and the cover crop plots had 25 samples where alachlor was detected. The alachlor concentrations were in the same ranges for both the cover crop and no cover crop treatments.

Atrazine was detected much less frequently than alachlor from May to September over the three-year period. Only in 1991 was atrazine detected with any frequency in the groundwater. In 1991, at the 3 m depth, atrazine was detected in 21 out of 29 samples in NT and in 20 out of 30 samples in CT. Concentrations ranged from 0.38 to 9.18 ug/L for the NT and 0.15 to 7.74 ug/L for the CT. A total of 9 samples had atrazine concentrations above the EPA drinking water standard of 3.0 ug/L (10). At the 4.5 m depth, atrazine was detected in 26 out of 36 samples in NT with concentrations ranging from 0.18 to 9.18 ug/L and in 26 out of 36 samples in CT with concentrations ranging from 0.17 to 5.98 ug/L. A total of 7 samples had atrazine concentrations above the EPA drinking water standard of 3.0 ug/L. From October until the following April over the three-year period, atrazine was detected in the groundwater more frequently than alachlor. In three samples concentrations were above the EPA drinking water standard of 3.0 ug/L. From October 1989 until April 1990, atrazine was detected in 9 samples with concentrations ranging from 0.12 to 5.96 ug/L. Based upon the half-lifes of atrazine and alachlor, you expect atrazine to be detected more frequently than alachlor later in the season. Ritter et al. (2) found that the frequency of detection in groundwater of atrazine, simazine, cyanazine and metolachlor was directly related to the half-life of the herbicide.

Simazine was detected in the groundwater from May to September at about the same frequency as atrazine. In 1989 and 1990 atrazine was only detected eight times out of a total of 268 samples. In 1991 simazine was detected in 6 out of 15 samples for NT and cover crop and in 8 out of 14 samples for no cover crop at the 3.0 m depth. For CT, 9 out of 15 samples for the cover crop treatment had simazine and 8 out of 15 samples for the no cover crop had simazine detected at the 3.0 m depth. Concentrations ranged from 0.11 to 10.8 ug/L for NT and 0.32 to 15.8 ug/L for CT in the 3.0 and 4.5 m wells. A total of 10 samples had simazine concentrations above the EPA drinking water standard of 4.0 ug/L (10). From October until April over the three-year period, simazine was detected in the groundwater less frequently than atrazine.

Previous research has shown that the amount and intensity of the rainfall shortly after the herbicides are applied is important in determining leaching losses on the sandy soils (2). In 1989, the month of May had 86 mm of rainfall but was distributed throughout the month. Based upon the rainfall patterns in May and the early part of June in 1991, it is difficult to determine if the herbicides detected in the groundwater came from macropore flow from the 1991 application or if the herbicides were leached in the deeper vadose zone 1.0 to 2.0 m in 1990 and moved slowly to the groundwater (2.5 m). The half-life of alachlor is much greater in the subsoil than topsoil (11).

Modeling Study. The GLEAMS model was calibrated to fit the soil profile data for atrazine and metolachlor collected in 1987 (2). The GLEAMS output includes a herbicide mass output for twelve soil layers after any rainfall induces herbicide movement in the soil. The outputs were correlated with the corresponding soil sampling dates, and the soil layers were modified to correspond to the identical depths. A series of runs of the program were completed with differing values for the partitioning coefficient (KOC). An analysis of variance was carried out to select the value that best represented the actual pesticide profile in the soil. KOC values of 25.0 and 50.0 with correlation coefficients of 0.92 and 0.87 were selected from atrazine and metolachlor.

A summary of the simulated results for atrazine for base precipitation, 2-, 10- and 100-yr return periods are presented in Table IV. For all simulations an application loss of 5% is assumed.

Analysis of the base precipitation output revealed that a total of 122.8 cm of rainfall and irrigation occurred throughout the year with 41.8 cm of percolation below the root zone of 152 cm and, theoretically, into the water table. The percolation caused 3.53 g/ha or 0.2% of the atrazine to leach and 0.19 g/ha or 0.01% of the metolachlor leach into the water table. The first leaching episode did not occur until 40 days after application when a 3.8 cm rainfall event occurred. The amounts leached from that event were 0.44 and 0.11 g/ha of atrazine and metolachlor, respectively. This was the greatest single loss of atrazine for the year.

Three days after application, the base model indicated 87.7% of the atrazine and 86.3% of metolachlor remained in the soil. Simulations with all storm events increased degradation and other losses for atrazine to slightly more than 1%. Average degradation losses for all return periods was approximately 8 and 9% for the atrazine and metolachlor, respectively. Return period storms of 2, 10 and 100 years produced atrazine losses of 0.7, 2.0 and 3.9% of the applied rate, respectively. Metolachlor leaching rates ranged from 0.3 to 2.0% for the three storm scenarios.

Seven days after the application of the herbicide, the atrazine had degraded approximately 19% and metolachlor 21%. For the 2-, 10- and 100-year storms 7 days after 0.6%, 1.7% and 3.4% of the atrazine was leached for the 2, 10 and 100-year

Table IV. Atrazine Fate Based on the Interval From Application to Storm Event for Different Return Period Storms

		Base	2-Year	10-Year	100-Year
3 Day	% application loss	5.0	5.0	5.0	5.0
	% degraded + other loss	7.3	8.6	8.5	8.4
	% in soil	87.6	85.7	84.5	82.7
	% leached	0.0	0.7	2.0	3.9
7 Day	% application loss	5.0	5.0	5.0	5.0
	% degraded + other loss	17.9	19.7	19.6	19.5
	% in soil	77.1	74.8	73.7	72.2
	% leached	0.0	0.6	1.7	3.4
15 Day	% application loss	5.0	5.0	5.0	5.0
	% degraded+ other loss	35.2	35.0	34.9	34.9
	% in soil	59.8	59.4	58.4	57.1
	% leached	0.0	0.6	1.7	3.1
30 Day	% application loss	5.0	5.0	5.0	5.0
	% degraded + other loss	58.5	58.3	58.3	58.3
	% in soil	36.5	36.1	35.3	34.4
	% leached	0.0	0.6	1.4	2.3
90 Day	% application loss	5.0	5.0	5.0	5.0
	% degraded + other loss	89.3	89.1	89.1	89.1
	% in soil	5.7	5.8	5.6	5.3
	% leached	0.0	0.2	0.4	0.6
3 Day	Annual % leached		2.3	4.2	5.9
7 Day	Annual % leached		2.3	3.7	5.2
15 Day	Annual % leached		2.2	3.3	4.4
30 Day	Annual % leached		1.9	2.5	3.3
90 Day	Annual % leached		0.4	0.8	1.0
Base Value	Annual % leached	0.2			

storms, respectively. Similar trends in the data were found with 15 and 30 day internal storm. One exception occurred for the 2-year return period storm because base rainfall amounts had a greater influence on the 2-year storm than the 10- and 100-year storms.

The amount of herbicide left in the soil had decreased to around 5% of the application rate for all return period storms when the interval was 90 days. The potential leaching from a major storm event had become less than 0.6% of the applied atrazine and less than 0.3% of the applied metolachlor.

Total annual leaching losses of applied atrazine ranged from 0.2 (base value) to 5.9% (100-year storm three days after application). The base value for annual leaching losses of metolachlor was <0.1% and storm leaching potentials ranged from 0.1 to 3.4%.

There is also a common crop production practice that can compound the leaching problems. If there is not adequate weed control, often another application of herbicide is made. When these herbicides are applied pre-emergent, they function through root uptake or hindering germination of weed seeds (12). Therefore, the herbicide must be present in the uppermost 5 cm layer of the soil, where the weed seeds are located, at high enough levels to be effective. Soil profile simulation data showed that for a base rainfall simulation all the herbicide remaining in the top two layers through the first 15 days. A 2-year return period storm 3 days after application is capable of moving 65% of the available atrazine out of the effective soil layers (0-9 cm), while a 100-year storm will move over 76% of the atrazine away from the soil surface. These percentages remained consistent with the 7-and 15-day simulations. The values for metolachlor were slightly better with an average of 55% and 67% remaining in the top two layers after 2- and 100-year storms, respectively. The potential loss of the herbicides effectiveness are actually worse since the values are for the top 2 layers of output from GLEAMS (0 to 9 cm). However, the active zone of the pesticide is the top 3 to 5 cm (12). These loss values indicate potential lack of weed control and the probable need for another application of other herbicides to control the weeds, and the potential for more leaching of herbicides into the groundwater.

Rainfall Simulation Experiment. In 1991, samples were collected from the monitoring wells and wick lysimeters the day after 75 mm of simulated rainfall were applied when soil was at 50 to 60% of available soil moisture or 6 days after the herbicides were applied to the plots. No atrazine, cyanazine, simazine or metolachlor were detected in the groundwater. Alachlor was detected in 15 of the 24 wells at concentrations ranging from 0.46 to 1.69 ug/L. There was very little difference in alachlor concentrations among the different treatments.

All the herbicides were detected in the lysimeters. Alachlor and metolachlor were detected in all 18 samples collected. Alachlor concentrations ranged from 1.06 to 26.9 ug/L and metolachlor concentrations ranged from 0.60 to 4.72 ug/L. Cyanazine was detected in 7 samples with concentrations ranging from 0.17 to 12.1 ug/L, while atrazine was detected in 2 samples and simazine in a single sample.

In 1992, when soil moisture was at field capacity all herbicides were detected in the groundwater 6 days after they were applied and after 75 mm of simulated rainfall. Alachlor, cyanazine and simazine were detected most frequently in groundwater. Average alachlor concentrations ranged from 0.38 ug/L for CT with poultry manure to 1.91 ug/L for CT with a cover crop. Only two of the samples had concentrations above the EPA drinking water standard of 2.0 ug/L (10). Average cyanazine concentrations ranged from 0.22 ug/L for the RT to 1.14 ug/L for the CT and cover crop. Individual cyanazine concentrations ranged from 0.15 to 3.80 ug/L with 2 samples above the recommended maximum contaminant level (MCL) of 1.0 ug/L. Average simazine concentrations ranged from 0.52 ug/L for RT to 1.99 ug/L for the CT and a cover crop treatment. Only one sample was above the EPA drinking water standard of 4.0 ug/L (10). Atrazine was detected in 12 out of 24 samples with concentrations ranging from 0.10 to 6.31 ug/L. Four of the samples had concentrations above the EPA drinking water standard of 3.0 ug/L (10). Metolachlor was also detected in 12 out of 24 samples with concentrations ranging from 0.28 to 3.29 ug/L.

In 1992, a total of 16 samples were collected in the lysimeter bottles. Simazine was detected in all 16 samples with concentrations ranging from 0.36 to 24.7 ug/L. Metolachlor was detected in only 2 of the samples and cyanazine was detected in 9 of the samples. Cyanazine concentrations ranged from 0.21 to 5.74 ug/L. Atrazine and alachlor were detected more frequently than cyanazine and metolachlor. Alachlor was detected in 10 samples with concentrations ranging from 0.21 to 3.88 ug/L and atrazine was detected in 12 samples with concentrations ranging from 0.36 to 4.14 ug/L. There are several possible explanations as to why cyanazine and metolachlor were detected more frequently in the groundwater than the lysimeters. Some of the herbicide may have degraded sitting in the lysimeter bottles exposed to air before the samples were collected and frozen. The soil was disturbed to install the lysimeters, so the flow paths would not be the same as the entire plots. In general there is more variability in samples collected from lysimeters in the vadose zone than from groundwater monitoring wells.

In 1993 all the herbicides were detected in nearly all the monitoring wells after an input of the 229 mm of rainfall and irrigation water. Simazine and atrazine had the highest concentrations. Average concentrations for simazine ranged from 1.28 ug/L for the CT to 5.96 ug/L for the CT with poultry manure treatment. Average atrazine concentrations ranged from 1.56 ug/L for CT to 5.12 ug/L for the CT poultry manure treatment. Six samples had simazine concentrations above the EPA drinking water standard of 4.0 ug/L and 5 samples had atrazine concentrations above the EPA drinking water standard of 3.0 ug/L (10). The highest measured atrazine concentration was 19.1 ug/L and the highest simazine concentration was 12.9 ug/L. Metolachlor had the lowest concentrations in the groundwater. Average concentrations ranged from 0.16 ug/L for RT to 0.82 ug/L for the CT with poultry manure experiment. Alachlor concentrations ranged from 0.14 to 3.54 ug/L with the highest average concentration of 2.06 ug/L for the CT treatment. Average cyanazine concentrations ranged from 1.47 ug/L for RT to 3.86 ug/L for the CT with poultry manure treatment with 19 samples having concentrations above the MCL of 1.0 ug/L.

A total of 17 samples were collected from the lysimeters. There was a great variability in the herbicide concentrations for the different lysimeter bottles. Simazine and atrazine concentrations were the highest in the lysimeters. Simazine concentrations varied from 2.89 to 78.3 ug/L and atrazine concentrations ranged from 1.92 to 82.9 ug/L. Metolachlor was detected in 14 out of the 17 samples with concentrations ranging from 0.12 to 3.95 ug/L. Alachlor was detected in all the samples with concentrations ranging from 0.44 to 6.39 ug/L. The variability of cyanazine concentrations was similar to the variability for simazine and atrazine. Cyanazine concentrations ranged from 0.33 to 79.7 ug/L.

Over the three different rainfall simulation events there was no clear difference among the different treatments. In 1993, four of the herbicides had the highest concentrations in the CT with poultry manure treatment. In 1991, there was very little difference in the alachlor concentrations for the different treatments. In 1992, the CT with poultry manure treatment had some of the lowest herbicide concentrations.

These rainfall simulation studies show that if enough rainfall occurs shortly after herbicides are applied on sandy soils, they will move rapidly below the root zone and may be leached to the groundwater. Most of the movement is probably by macropore flow. Dye and tracer studies on the plots showed that up to depths of 0.6m, water and solutes moved according to the connective dispersive equation with some preferential movement due to root channels and perhaps some small perturbations of the wetting front. Below 0.6 m two types of preferential flow were active, preferential flow due to large root channels and to coarse sandy layers (13).

Conclusions

From the field studies, alachlor was detected more frequently in the groundwater than atrazine or simazine in a loamy sand Coastal Plain soil. The frequency of detection of the herbicide was directly related to the half-life of the herbicide. There is the potential for alachlor, atrazine and simazine to be found in the shallow groundwater at concentrations above the EPA drinking water standards.

The GLEAMS model simulations show the great potential for herbicides leaching into the water table of southern Delaware due to the combination of sandy soils, high water table, and intense crop production. There is no way to predict the possible occurrence of a major storm more than 7 days in the future, but attention should be paid to the weather forecast for the week following an intended application of herbicides. The effects of potentially high amounts of herbicide leaching should be combined with the very dynamic nature of the water table before any conclusions about contamination can be assumed. The surrounding areas, lateral movement of the aquifer, pumping, and other water budget considerations will greatly affect herbicide concentrations and movement in the water table. However, there is a high potential for problems on the micro-scale of a farmstead that draws water from the water table aquifer and is surrounded by production fields. Another question concerns multi-year scenarios. Since herbicide degradation in water is limited to chemical and biological processes, which can be extremely slow processes, the potential of every increasing levels of contamination in the aquifer is possible.

Herbicides moved below the root zone from an intense rainfall shortly after they were applied in the rainfall simulation studies. Alachlor was detected more frequently in the lysimeters and groundwater than metolachlor; atrazine, simazine and cyanazine but the maximum concentrations were generally lower. There appears to be no apparent relationship between herbicide movement and tillage or other cultural practices tested.

Literature Cited

1. Delmarva Poultry Industry. **1997**. Voice of the Delmarva's poultry industry. Georgetown, DE. Fact sheet.
2. Ritter, W.F., Scarborough, R.W., and Chirnside, A.E.M. **1994**. Contaminations of groundwater by triazines, metolachlor and alachlor. J. of Contaminat. Hydrology, 15:73-92.
3. Ritter, W.F., Chirnside, A.E.M. and Scarborough, R.W. **1996**. Leaching of dicamba in a coastal plain soil. J. of Env. Science & Health, Part A, A31:505-517.
4. Sundstrom, W.W., and Pickett, T.E. **1969**. The availability of ground water in eastern Sussex County. Water Resources Center, Univ. of Delaware, Newark, DE 123 pp.
5. Scarborough, R. W. **1995**. Modeling pesticide movement from major storm events. MS Thesis, University of Delaware, Newark, DE.
6. Ireland, W. and E.D. Matthews. **1974**. Soil survey of Sussex County, Delaware. United States Department of Agriculture, Soil Conservation Service.
7. Ware, G.W. **1992**. Review of environmental contamination and toxicology. Springer-Verlag, New York, NY.
8. Boll, J., Selker, J.S., Nijssen, B. M., Steenhuis, T.S., Van Winkle, J., and Jolles, E. **1991**. Water Quality sampling under preferential flow conditions. In: Lysimeters for Evapotranspiration and Environmental Measures, R.G. Allen, T.A. Howell, W. O. Pruitt, I.A. Walter and M.E. Jensen, Editors, Proceedings of American Society of Civil Engineers International Symposium on Lyimetry. Honolulu, HI. July 23-25, 1991. Pp. 290-298.
9. Spittler, T.D. **1987**. Analysis of ground water, soil and sediment from till and no-till corn production. Paper presented at American Chemical Society Rocky Mountain Conference, Denver, CO.
10. Maryland Department of Environment. **1994**. Facts about drinking water and your health. Fact Sheet, Maryland Dept. of Environment, Baltimore, MD. 4 pp.
11. Pothuluri, J.V., Moorman, T.B., Obenhuber, D.C. and Wauchope, R.D., **1990**. Aerobic and anaerobic degradation of alachlor in samples from a surface-to-groundwater profile. J. Environ. Qual., 19:525-530.
12. Hughes, H. A. **1982**. Crop Chemicals. Deere & Company, Moline, IL.
13. Ritter, W.F. and Steenhuis, T.S. **1994**. Preventing pesticide contamination of aquifers by best management practices. U.S. Geological Survey, Reston, VA. Award No. 14-08-0001-G1907. 45 pp.

Chapter 6

A Small Agricultural Watershed Study on Maryland's Outer Atlantic Coastal Plain

W. E. Johnson[1], L. W. Hall, Jr.[2], R. D. Anderson[2], and C. P. Rice[3]

[1]National Centers for Coastal Ocean Science, National Oceanic and Atmospheric Administration, Silver Spring, MD 20910–3281
[2]Wye Research and Education Center, Agricultural Experiment Station, University of Maryland, Queenstown, MD 21658
[3]Environmental Chemistry Laboratory, Agricultural Research Service, U.S. Department of Agriculture, Beltsville, MD 20705

> Seventeen pesticides or their transformation products, dissolved nitrogen and organic carbon were monitored in a stream leading to the Chesapeake Bay's Choptank River estuary during 1996. Pesticide concentrations closely followed the stream hydrograph during the spring and summer. The highest water concentrations measured during the study exceeded freshwater aquatic life guidelines for several compounds including atrazine, chlorothalonil, chlorpyrifos, cyanazine, 2,4-D, diazinon, malathion, metolachlor, simazine, and trifluralin. Bulk atmospheric deposition samples indicate that the atmosphere continues to be an important source of pesticides to the watershed. Sediment levels of current-use pesticides indicated that this medium could be a sink for pesticides in streams and a source of contamination to river estuaries downstream.

Nonpoint-source runoff of agricultural pesticides into the rivers and streams of Chesapeake Bay has long been suspected of adverse impacts to its living resources (1) but in only a few specific cases have direct connections been made (kepone, butyltins, and chemical spills). Pesticide monitoring in the Bay region began in earnest in the late 1970s and over twenty years of monitoring has produced little evidence that nonpoint-source contamination of the Bay and its tributaries from agricultural use of pesticides has contributed to the ill health of any of the Bay's living resources (2, 3). Despite the lack of evidence there remain skeptics who feel that the scientific community is not asking the right questions or obtaining the needed information (4). For example, most studies evaluate the aquatic toxicity of a single compound to a single organism. What is rarely considered are the potential impacts of a mixture of pesticides and other contaminants to a community of organisms and the resulting non-lethal and insidious effects that could occur over a period of years. A goal of this investigation is to identify and quantify the pesticides that should be targeted in

© 2000 American Chemical Society

subsequent studies which evaluate the potential effects of environmental exposure to pesticide mixtures in small (50 sq. km) agricultural watersheds.

Agricultural runoff can be studied at the regional scale (large rivers or river systems), landscape scale (small rivers or streams), and field scale or plot studies. Each of these can provide needed information to the complete understanding of agrochemical loss from the site of application and the ensuing fate and transport. Large-scale watershed studies in the Chesapeake Bay drainage basin have assessed both nutrient and pesticide loadings from the three major tributaries on the Bay's Western Shore (5, 6). Several small watershed studies on the Northern and Western Shores focused largely on nutrient runoff with only limited attention given to pesticides (7, 8). Similar investigations on the Eastern Shore have emphasized nutrient runoff (9, 10) or pesticides in groundwater (11, 12). Glotfelty and colleagues (13) conducted the first small watershed study on the Eastern Shore to assess pesticide runoff and water quality.

The hydrogeomorphology of the watersheds in the Bay's Northern and Western basins differ markedly from those of the Eastern Shore (14). The Northern and Western Shore watersheds are larger extending from the inner Coastal Plain two hundred miles through the Piedmont and into the Appalachian Province. These rivers contribute greater than 90 percent of the freshwater discharged to the Bay (15) and carry tons of pesticides and nutrients annually to the estuary (16, 17). In contrast, all of the Eastern Shore watersheds lie entirely within the Coastal Plain, their combined freshwater discharged to the Bay is only 8.2 percent of the total (15), and tidal influence affects a greater proportion of the drainage. In general, streams of the Coastal Plain do not form the integrated drainage networks characteristic of the Western Shore tributaries, but instead are small basins that drain directly to tidewater.

While the Eastern Shore runoff is a small part of the Chesapeake Bay total, over 50 percent of some pesticides used in Maryland are applied in the four counties comprising Maryland's upper Eastern Shore (Figure 1). Even if these pesticides do not affect the Bay proper, there is cause for concern due to the critical importance of Eastern Shore streams and rivers as spawning, and nursery areas for many fisheries of commercial and ecological significance.

This study investigates the effects of agricultural runoff on the water quality. Assessment of agricultural runoff in small streams, above the head of tide, may provide a clearer understanding of the effectiveness of agricultural management practices on water quality in the Chesapeake Bay region.

Approach

Site Geography and Land Use. German Branch flows into to the Choptank River, a major river estuary of Chesapeake Bay's Northern Eastern Shore, that lies entirely within the outer Atlantic Coastal Plain (Figure 2). The German Branch watershed also lies entirely within Queen Anne's County, Maryland, and drains 54 percent of its land area (Table 1). Total land area in the county, 237,990 acres, includes 161,000 acres used for agriculture. In 1996, 29.0 % percent of the agricultural land was planted in corn, 43.4% in soybeans, and 23.4% in wheat. The watershed drains 54 percent of the

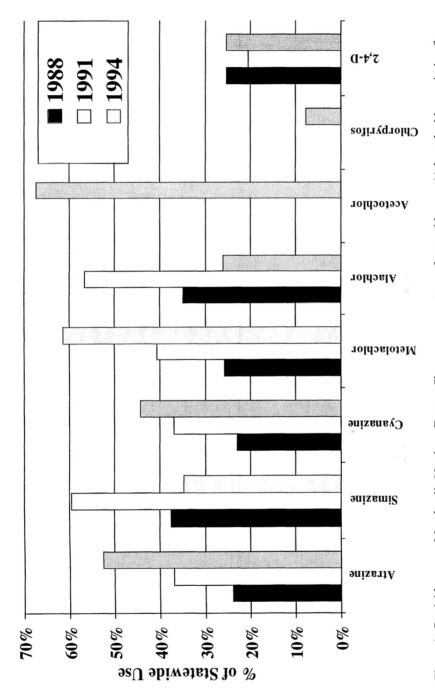

Figure 1. Pesticide use on Maryland's Northern Eastern Shore as a percentage of statewide use. (Adapted with permission from reference 21. Copyright 1995.)

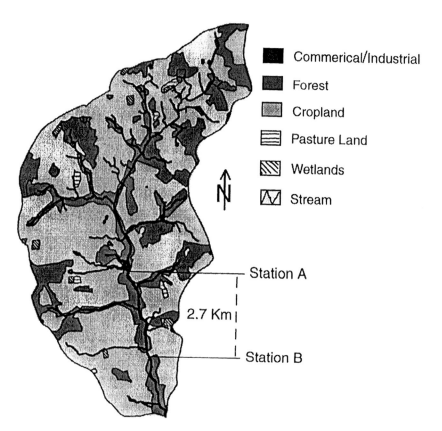

Figure 2. Land use in the German Branch watershed. (Adapted with permission from reference 21. Copyright 1995.)

County's land area and land use within the watershed is proportional to use countywide.

Pesticide use at the County scale is estimated by the State (18, 19, 20). We assumed pesticide use at the watershed scale (Table II) was proportional to pesticide use at the county scale. This assumption seemed reasonable given the historical record of land use and crop acreage reported at the County scale (22) and crop acreage reported at the watershed scale (23). Soils within the watershed are moderately to well drained (24). The mean annual precipitation is 43 inches, with monthly averages from 3 to 4 inches from April to September (Maryland State Climatologist, personal communication, University of Maryland, College Park, MD).

Table I. German Branch Watershed Characteristics.

Description	German Branch
Nominal watershed center	39° 01.00N / 75° 56.50' W
Tributary to :	Choptank River Estuary
Station A: Road crossing (upstream)	Deavers Branch Road
Station B: Road crossing (downstream)	MD Rt. 304
Average annual stream discharge[a]	35 (cm/yr., 1cm=100m^3/ha)
Land Use (ha)[a]:	
Total watershed size	12,948
Row crops:	8,676
soybean	3,765
corn	2,516
wheat	2,030
Forest	1,415
Fallow	115
Residential and Roads	121
Grassland	52
Pond	16

[a] Adapted from ref. 9.

Stream Hydrology and Sample Collection. Stream depth was recorded from a staff gage at least every two days and used to calculate stream flow from the 1995 stream flow-rating curve (Jordan, T.E. Smithsonian Environmental Research Center, Edgewater, MD, personal communication 1997). Surface water samples were collected weekly in glass bottles with Teflon-lined caps at two stations, A (upstream) and B (downstream) (Figure 2) from April to August of 1996. The samples were collected at the edge of the stream. We assumed that the stream was small enough (~5m width and <0.5 m depth at baseflow) to be well mixed throughout the cross-section.

Water samples were transported to the laboratory on ice, stored at -20 °C and extracted within seven days. Prior to refrigeration, two 10 ml aliquots were transferred to 20 ml glass vials with aluminum lined caps and frozen for subsequent nitrogen and dissolved organic carbon (DOC) analyses, and determination of the herbicide 2,4-D.

A single "critical" runoff event was monitored intensively at station B, every four hours, beginning May 21, 1996 and continuing for 5 days. A "critical" runoff event was defined as, 1 inch of rainfall within 24 hours following the period of peak pesticide application in the spring. A portable automatic water sampler (Isco model 3700, Isco, Inc., Lincoln, NE) was used during the intensive sampling period to collect water samples every 4 hours.

Table II. Pesticides analyzed in this study, their estimated usage (lbs. of a.i.) in the watershed, freshwater aquatic life criteria and recommended guidelines (ppb), and the highest concentration measure in this study (ppb).

Compound	Type[a]	Estimated Pesticide Usage[b]	Freshwater Criteria & Guidelines	Freshwater Guidelines CCREM[g]	Highest water conc. this study
Acetochlor	H	635			5.2
Alachlor	H	1,441			7.4
Atrazine	H	11,419	20.0[c]	2.0	59.8
Deethylatrazine	trans	NA			2.1
Deisopropylatrazine	trans	NA			1.2
Chlorothalonil	F	356		0.18	0.24
Chlorpyrifos	I	302	0.041[d]		0.45
Cyanazine	H	636		2.0	2.9
2,4-D	H	937		4.0	40.2
Diazinon	I	4.6	0.04[e]		0.23
Endosulfan (α, β)	I	7.5	0.056[d]	0.02	<0.01
Malathion	I	1,100	0.1[d]		0.50
Metolachlor	H	17,971	53.0[f]	8.0	30.2
Pendimethalin	H	728			0.54
Simazine	H	359		10.0	41.1
Trifluralin	H	91		0.1	0.81

[a] H = herbicide; F = fungicide; I = Insecticide; trans = transformation product.
[b] estimate of pesticide use (lbs of active ingredient) in the watershed based on County use estimates (20); NA = not applied.
[c] recommended chronic value: (29).
[d] US EPA aquatic life criteria, chronic value (28).
[e] recommended chronic value: (30)
[f] recommended acute value (31).
[g] recommended Canadian freshwater aquatic life guidelines, (27).

Sediment samples were collected in glass jars with Teflon-lined caps chilled and transported to the laboratory. Fine grain sediments were intentionally sought and the top 2-cm collected with a stainless steel spatula. The pore water was separated from sediments by filtration through porcelain Buchner-type funnels, then the sediments were returned to jars and frozen until extraction. The pore water was refrigerated, extracted within 7 days and analyzed for pesticides by the same procedure used for surface water samples.

Atmospheric deposition samples were collected from the German Branch watershed between the upper and lower sampling stations. The sampler consisted of a 20 cm diameter stainless steel funnel inserted into 4 liter brown glass bottle. It was deployed within a 24-hour period of expected rainfall and retrieved within 24 hours after the rain ended. The sample was refrigerated until extracted within seven days of collection.

Sample Analysis

Semi-Volatile Pesticide Extraction from Water. Water samples of 275 to 325 ml volume were filtered through glass fiber filters (0.7 µm nominal, Whatman GF/F, Whatman, Inc. with a Whatman GMF 1 µm glass fiber prefilter) then extracted through a octadecyl solid-phase extraction cartridge (tC18 Waters Sep Pak 1 gram, Waters Associates, Milford, MA). The octadecyl cartridges were air dried for 10 minutes then eluted with 3 ml each of ethyl acetate, ethyl acetate:dichloromethane (1:1), and dichloromethane (all solvents were pesticide grade Burdick and Jackson, Muskegon, MI). Water was removed from the organic extract by adding 1 g of 12-60 mesh anhydrous sodium sulfate (J.T. Baker, Phillipsburg, NJ). The extract was quantitatively transferred to either a storage vial for refrigeration at -20 °C or immediately concentrated to a volume of 0.5 ml under a gentle stream of nitrogen in a graduated glass centrifuge tube (Organomation Associates, Inc., South Berlin, MA). The resulting extract was spiked with internal standards (d_{10} diazinon and triphenylphospate) and analyzed by gas chromatography / mass spectrometry.

Each batch of samples extracted included at least one blank and matrix spike. A surrogate spike of ^{13}C-atrazine and ^{13}C-metolachlor was made to all samples prior to extraction. Duplicate samples were extracted except during the intensive sampling period. Break-through tests revealed no measurable level of pesticides in second-stage solid-phase extraction columns. Recoveries of the 17 pesticides from surface water ranged from 84.4 % to 117.4 %. Their concentrations are reported without correction.

Several water samples were spiked with a pesticide mixture and then filtered. The filtrate was then extracted to determine the extent of pesticide sorption to suspended particulates. No pesticides were detected in any of the filter extracts, therefore, no further filter analysis were performed.

Semi-Volatile Pesticide Extraction from Sediment. A modification of the method SW 846-8080 (25) was used for the analysis of pesticides in sediment. Percent moisture remaining in each filtered sediment samples was determined by drying triplicate aliquots of the sample at 105°C.

Gas Chromatography/Mass Spectrometry for 17 Compounds (Table II). Samples were analyzed by selective ion monitoring in the electron impact mode using a Hewlett Packard model 5989A mass spectrometer with a 5890 Series II gas chromatograph. The source and quad temperatures were 200 °C and 100 °C, respectively. A Durabond-5MS 30m x 0.2-mm (i.d.) 0.33film thickness fused silica column with helium carrier gas was used. The injection port was held at 225 °C with constant flow (161 KPA, initial value). Injections of 2 µl were made in the splitless mode with a Hewlett Packard 7673A automatic injector. The oven parameters were: initially 130°C for 1 min. then increased at 5 °C/min. to 225 °C, then increased 10 °C/min. to 260 °C, then increased 20 °C/min to 280 °C and held 5 minutes. Confirmation of the compound was based on the presence of the molecular ion and one or two confirming ions with a retention time match of +/- 0.2 percent relative to d_{10}-diazinon.

Enzyme Linked Immunosorbent Assay (ELISA) for 2,4-D. An ELISA technique (RaPID Assays®, Ohmicron Environmental Diagnostics, Newtown, PA) was used to monitor 2,4-D (2,4-dichlorophenoxyacetic acid) and related chlorophenoxy herbicides in water. Sample volumes of 250 µl were measured with a RPA-1 RaPID Analyzer (Ohmicron Environmental Diagnostics, Newtown, PA). Twenty percent of the samples were reanalyzed by LC/MS electrospray for confirmation.

Dissolved Inorganic Nitrogen Determination. Water samples for nitrogen determination were filtered though 0.7 µm glass fiber filters (Whatman Puradisc 25mm syringe filters) into 20 ml glass vials and preserved with concentrated sulfuric acid then refrigerated until analyzed. Analysis was done colorimetrically by automated cadmium reduction with flow injection analysis (Lachat Instruments) for nitrate plus nitrite, and ammonia was analyzed with the same system without cadmium reduction.

Dissolved Organic Carbon (DOC) Analysis. Water for DOC determination was filtered through 13mm, 0.2µm Teflon membrane syringe filters (Gelman Acrodisc CR PTFE, Gelman Sciences, Inc., Ann Arbor, MI) with a 25mm, 0.7µm glass fiber pre-filter (Whatman) and analyzed by the Combustion-Infrared Method (26). Samples were acidified with 2N HCl to convert the inorganic carbonates to CO_2, which was removed by purging with purified air before injecting into the heated reaction chamber.

Results

Rainfall and Hydrology. The total rainfall during the study (1Mar to 31Aug, 1996) was 28.25 inches or about 66 % of the twelve-month mean. Rainfall during the peak pesticide application period was higher then normal, 2.7, 7.5, and 4.0 inches for April, May, and June, respectively. The wet spring required many farmers in the region to delay planting by two weeks or more. This resulted in a bimodal planting season that probably resulted in lower than peak pesticide concentrations but lengthened the exposure period. Two "critical" runoff events occurred May 3rd and May 21st, delivering 1.77 inches and 1.02 inches of rain, respectively. We chose not to monitor the first event because much of the spring pesticide application had not been completed according to the County Extension Agent. The stream hydrograph increased as a function of both the amount and intensity of rainfall and also the preexisting soil moisture level during the study period (Figure 3).

Pesticides in Surface Water. Water quality criteria and guidelines are shown in Table II for comparison to water concentrations measured in this study. The Canadian Council of Resource and Environment Ministers (27) has recommended freshwater

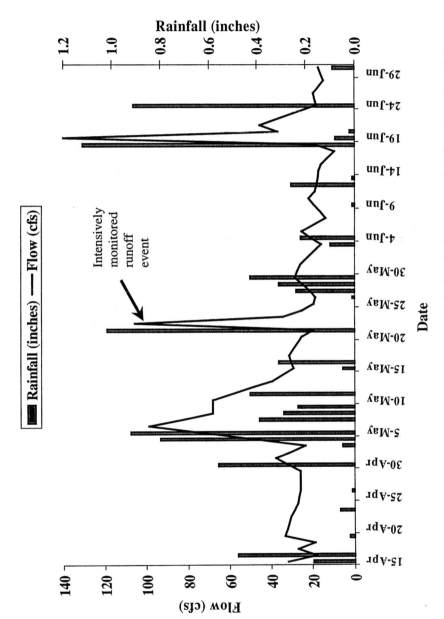

Figure 3. German Branch discharge (cfs) and rainfall (inches) in the watershed during the study period.

guidelines for the protection of aquatic life, which are listed in Table II. Similarly, the US EPA (28) has established aquatic life criteria (ALCs) but only a few for pesticides in current use. In addition, risk assessments for atrazine, diazinon, and metolachlor have recommended aquatic life guidelines of 20 µg/l, 0.04 µg/l, and 53 µg/l, respectively (29, 30, 31). German Branch is not used as a municipal drinking water source and therefore US EPA enforceable drinking water standards (28), maximum contaminant levels, (MCLs) and health advisory levels (HALs) were not used for comparison purposes.

All compounds were detected in surface water with the exception of alpha and beta endosulfan. Compounds detected at station B were not always detected at station A and concentrations were always higher at station B. The highest measured concentration of all compounds exceeded the recommended Canadian and US EPA freshwater guidelines and criteria, respectively (Table II).

The storm hydrograph for the intensive monitoring period rises rapidly, peaks at 18 hours, then is followed by a fairly rapid declined to base flow levels over a period of 4 days. The storm event concentrations of triazine, acetanilide herbicides, and 2,4-D are shown as chemographs in Figure 4 and Figure 5 relative to the stream hydrograph.

High pesticide concentrations in German Branch occurred in May, following rainfall and runoff events (Table II). The peak concentrations for most compounds were measured during the second "critical" runoff event (May 21st). High herbicide concentrations were also measured on May 8th during the first "critical' runoff event two days after the stream discharge peaked. Herbicide concentrations measured in the second runoff event peaked at nearly the same time as the stream hydrograph. Therefore, herbicide concentrations measured during the first runoff event (May 8th), two days post peak discharge, may underestimate peak concentrations of some pesticides by a factor of 2 or more. Peak acetochlor and metolachlor concentrations, 5.16 µg/l and 30.2 µg/l, respectively were higher in the first runoff event than they were in the second, intensively monitored (May 21st). Atrazine and simazine had concentrations of 43.8 µg/l and 28.5 µg/l, respectively which were almost as high as those recorded in the runoff event two weeks later.

Thirty-six water samples collected in May were analyzed for the chlorinated phenoxy herbicide 2,4-D including some during the intensively monitored event. With ELISA, 22 percent of the samples had detectable levels. Twenty percent of the samples were confirmed by LC/MS electrospray (Krynitsky, A.J. US Environmental Protection Agency, Beltsville, MD, personal communication 1996). The peak concentration of 40 µg/l occurred at peak stream discharge of the second "critical" runoff event.

Monthly time weighted mean concentrations (TWMC) were used to estimate pesticide exposure to aquatic life living in tributaries to Lake Erie (32). The TWMCs were calculated from the equation: TWMC=$\Sigma_i c_i t_i / \Sigma_i t_I$, where: c_i is the chemical concentration of the ith sample and t_i is the duration of the time that the ith sample is used to characterize the stream concentration and is equal to one-half the time interval between the sample immediately preceding and following the ith sample. In this study, TWMCs for the triazine and acetanilide herbicides (Figure 6) were highest in May. The May TWMCs of atrazine, simazine, and metolachlor were 18.5 µg/l, 9.7 µg/l

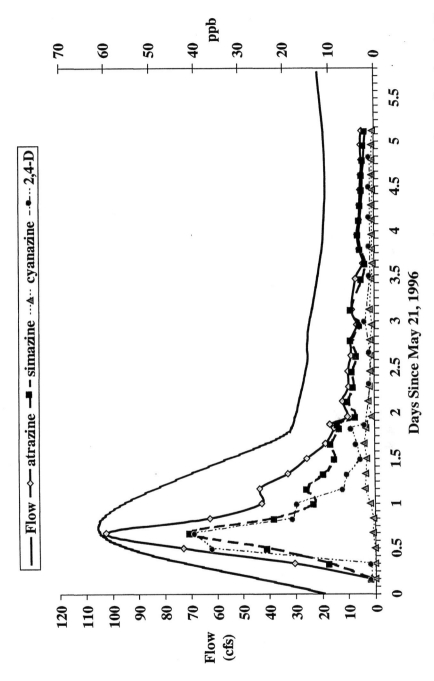

Figure 4. Triazine and 2,4-D herbicide concentrations (ppb) relative to the storm hydrograph (cfs) at German Branch station B.

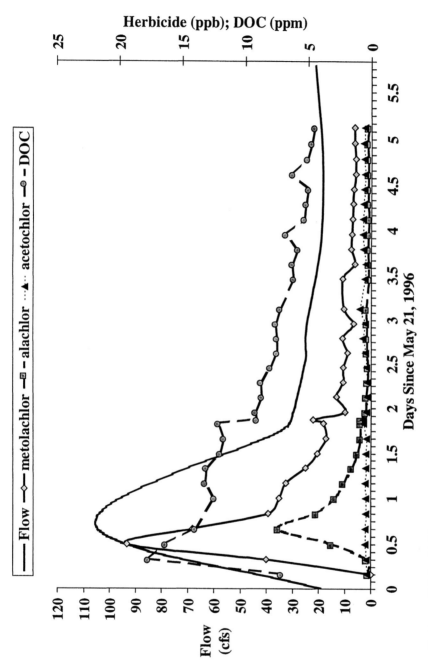

Figure 5. Acetanilide herbicides concentrations (ppb) and DOC concentrations (ppm) relative to the storm hydrograph (cfs) at German Branch station B.

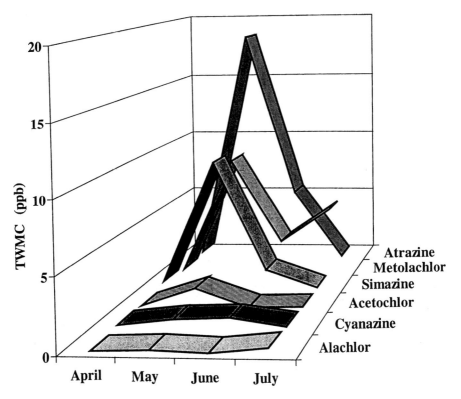

Figure 6. Monthly time weighted mean concentrations for the triazine and acetanilide herbicides at German Branch station B.

and 8.9 µg/l, respectively and nearly equaled or exceeded the recommended Canadian freshwater guideline (27) (Table II). The TWMC for metolachlor was bimodal, peaking first in May and then again in July. The second July maximum is presumed to result from a second application of metolachlor in June and early July when soybean planting after the wheat harvest occurs.

Similarly, monthly flow weighted mean concentrations (FWMC) were calculated from the equation: FWMC=Σ_i $c_i q_i t_i$ / Σ_i $q_i t_i$, where q_i is the instantaneous flow rate at the time the sample was collected. FWMCs for the triazine and acetanilide herbicides (not shown) also peaked during May and only the metolachlor FWMC (11.8 µg/l), exceeded the recommended Canadian freshwater guideline (Table II).

An ecological risk assessment of atrazine (29) reported that phytoplankton and macrophytes were reduced at atrazine concentrations greater that 20 µg/l. The peak atrazine concentration, nearly 60 µg/l, dropped rapidly to less than 10 µg/l within two days of the runoff event. Therefore long-term impacts to aquatic communities due to atrazine alone were unlikely. It is important to note however, that elevated levels of multiple pesticides are common but their combined effects on aquatic life are poorly understood.

Pesticide loading at station B was calculated for the triazines and acetanilide herbicides based on the dissolved phase concentration and the instantaneous stream flow for the period (April through July). Loadings for the triazines, atrazine, simazine, and cyanazine were 85.0 kg, 46.9 kg, and 4.3 kg, respectively. Loadings for the acetanilides, metolachlor, acetochlor, and alachlor were 50.1 kg, 6.6 kg, and 2.4 kg, respectively. Except for simazine, less than 2 percent of each herbicide was lost from the fields based on estimated application rates (Table II). The high loss for simazine (29%) probably reflects inaccuracy in the application estimate.

Deethylatrazine / Atrazine Ratio (DAR). The deethylatrazine to atrazine ratio (DAR) was first applied as an indicator of pesticide movement into surface streams from groundwater (33, 34). Thurman and Fallon, (35) also showed that the DAR may be a useful indicator of the first pulse of herbicide runoff from corn fields in the midwest. We measured a DAR value of about 2 in early April, which dropped to 0.02 by early May then gradually increased over the next two months to almost 1.8. Simultaneous with the drop in the DAR was a disproportional rise in both the deethylatrazine and atrazine concentrations. Groundwater discharge to the stream in the early spring contains deethylatrazine and other atrazine transformation products from the previous growing season. This along with fresh atrazine runoff from current application contributes to the low DAR measured in early May. The DAR values gradually increase to nearly 2 again as the 1996 atrazine application transforms.

Dissolved Nitrogen. The dissolved nitrogen concentrations were operationally defined as the ammonia nitrogen (NH_4-N) or nitrate plus nitrite nitrogen (NO_{23}-N), which passed through a 0.7 µm glass fiber filter. Weekly nitrogen concentrations ranged from 0.1 to 1.0 mg-N/l for NH_4-N and 1.6 to 5.5 mg-N/ml for NO_{23}-N. The average weekly concentration at station B was 0.2 and 4.1 mg-N/l, for NH_4-N and NO_{23}-N, respectively. The maximum concentrations of NH_4-N and NO_{23}-N (1.7 and

7.9 mg-N/l, respectively) were measured during the intensively monitored runoff event that began May 21st. None of the samples exceeded the US EPA water quality standard (MCL) for nitrate plus nitrite of 10 mg-N/l (28).

During the intensive event at station B, both NH_4-N and NO_{23}-N concentrations increased slowly and reached a maximum of 2.8 ppm and 2.0 ppm at 36 and 40 hours, respectively. Ammonia may be an indicator of surface runoff, as it has lower retention in soils than does nitrate. Total dissolved nitrogen levels (NH_4-N and NO_{23}-N) at station A showed no similarity to the trends observed at station B. NH_4-N and remained low throughout the intensive event, while NO_{23}-N levels, although larger than NH_4-N, increased only slightly to a maximum of 1.75 ppm.

Surface Water - Dissolved Organic Carbon (DOC). The average of weekly DOC levels was 8.0 mg-carbon/l and ranged from 4.42 to 14.8 mg-carbon /l. The maximum DOC concentration, 17.7 mg-carbon /l, occurred during the intensively monitored storm event. The DOC level reached a maximum prior to the peak of the storm hydrograph then decreased slowly over a four-day period. This slow decline suggests continued DOC input from gradual groundwater discharge to the stream after the initial surface runoff (Figure 5).

Sediment and Pore water. Sixteen pesticides were detected in German Branch sediment and pore water and reported as the mean of duplicate determinations at stations A and B (Figure 7). Pesticide concentrations in sediment exceeded pore water concentrations except in the case of trifluralin and acetochlor where sediment levels were undetected. Sediment concentrations were lower than surface water concentrations with the exception of alachlor, which had the highest sediment concentration, 29.3 ng/g (dry weight). Eskins et al. (36) analyzed for alachlor, atrazine, chlorpyrifos, metolachlor, and simazine (and others not analyzed in this study) in tidal tributaries of Chesapeake Bay. Their maximum concentrations were lower by a factor of 2 to 4 with the concentrations measured in this study. The exceptions were cyanazine, which were lower in this study by a factor of 6 and simazine and atrazine, which were not detected.

Pesticides in Rainfall. Atrazine and metolachlor were measured in bulk atmospheric deposition samples collected from the German Branch watershed between stations A and B. The temporal distribution of atrazine and metolachlor concentrations are shown in Figure 8. Concentrations ranged from 0.041 µg/l to 0.357 µg/l for atrazine and 0.075 µg/l and 0.632 µg/l for metolachlor. Mean concentrations were 0.135 µg/l and 0.233 µg/l, for atrazine and metolachlor, respectively. Glotfelty and others (37) measured bulk deposition in the Wye River watershed on the Eastern Shore a few miles southwest of German Branch. In midsummer they observed atrazine and metolachlor levels in the range of 0.03 µg/l to 0.48 µg/l and 0.046 µg/l to 0.29 µg/l, respectively. Atrazine and metolachlor atmospheric loadings to the German Branch watershed were calculated to be 6.3 kg and 11.6 kg, respectively.

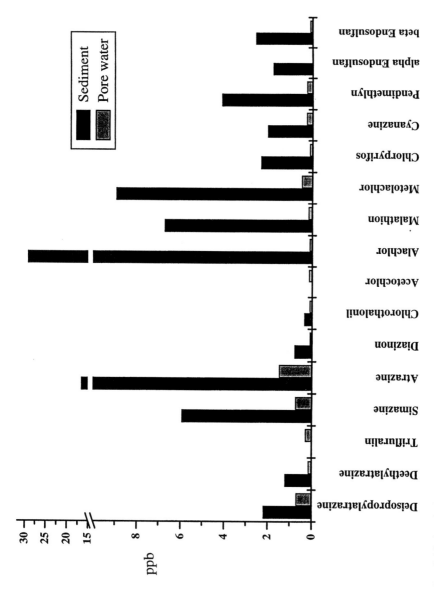

Figure 7. Pesticide levels in sediment (ppb dry-wt) and sediment pore water (ppb) collected at German Branch station B in June 1996.

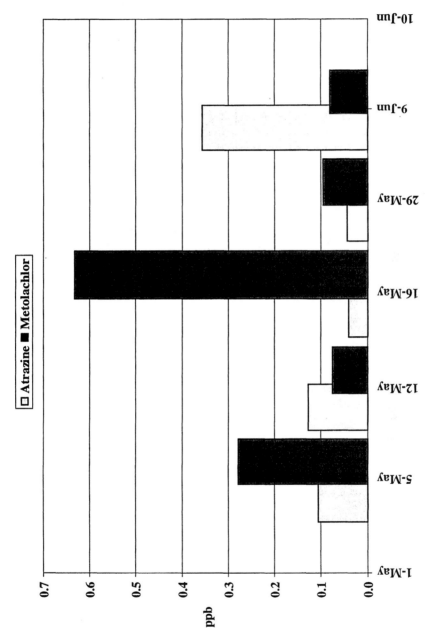

Figure 8. Atrazine and metolachlor concentrations (ppb) in bulk atmospheric deposition samples collected within the German Branch watershed.

Conclusions

This study identified and quantified seventeen pesticides in filtered water samples, sediment, sediment pore water, or rainwater from an agricultural watershed on Maryland's upper Eastern Shore. Pesticide, DOC, and nitrogen concentrations in German Branch pulsed in response to increased stream flow following rainfall events during the spring planting season. Peak concentrations in surface water occurred during runoff events. All compounds except endosulfan were detected simultaneously during runoff events occurring in May. The peak concentrations of 10 pesticides exceeded aquatic life criteria or recommended guidelines.

Five days of intensive monitoring at 4-hour intervals was sufficient to define the shape of the pesticide and nitrogen chemographs following a "critical" runoff event. Weekly concentrations of most pesticides were below water quality guidelines.
Further investigations should attempt to further define the concentrations and duration of multiple pesticide exposures by intensively monitoring as many runoff events during the planting and early growing season as possible, or as an alternative, increase the frequency of sampling from weekly to daily.

Concentrations and mass loading of atrazine and metolachlor from bulk atmospheric deposition indicated that the atmosphere is an important source of pesticides to this watershed. The portion of atmospherically deposited pesticides to the watershed that reaches the surface water of German Branch by direct deposition or runoff remains undetermined.

Future investigation should carefully consider the role of sediments as a sink and a source of current-use pesticides and related compounds to the water column and their potential for transport down stream. The low flushing characteristic of Eastern Shore river estuaries and our lack of information about the fate and transport of current-use pesticides once they enter stream and rivers demands further investigation to assure that current management practices are protective of the stream environment.

References

1. Klauda, R.J.; Bender, M.E. In *Contaminant Problems and Management of Living Chesapeake Bay Resources;* Majumdar, S.K.; Hall, L.W. Jr.; Austin, H.M., Eds.; Pennsylvania Academy of Science: Phillipsburg, NJ,1987, Chapter 15, pp321-372.
2. Johnson, W.E.; Plimmer, J.R.; Kroll, R.B.; Pait, A.S. In *Perspectives on Chesapeake Bay, 1994: Advances in Estuarine Sciences*; Nelson, S.; Elliott, P., Eds.; CRC Publication No 147; Chesapeake Research Consortium, Inc.: Edgewater, MD,1994, Chapter 4, pp105-146.
3. Johnson, W.E.; Kroll, R.B.; Plimmer, J.R.; Pait, A.S. Database of the Occurrence and Distribution of Pesticides in Chesapeake Bay. Online. U.S. Department of Agriculture, National Agricultural Library, 1996, Available: http://www.agnic.org/cbp. February 1999.
4. Day, D.E. In *Pesticide Transformation Products: Fate and Significance in the Environment;* Somasundaram, L.; Coats, J.R., Eds., ACS Symposium Series

459, American Chemical Society: Washington, DC, 1991, Chapter 16, pp 217-241.
5. Maryland Department of the Environment and Metropolitan Washington Council of Governments. *Chesapeake Bay Fall Line Toxics Monitoring Program: 1992-1993 Loading Report*. Final report to the US Environmental Protection Agency, Chesapeake Bay Program: Annapolis, MD, 1994.
6. Hainly, R.A.; Kahn, J.M. *J. Environ. Qual.* **1996**, 12, 330-340.
7. Wu, T.L.; Correll, D.L.; Remenapp, H.E.H. *J. Environ. Qual.* **1983**,12, 330-340.
8. Hainly, R.A.; Loper, C.A. *Water-quality assessment of the lower Susquehanna River Basin, Pennsylvania and Maryland: Sources, characteristics, analysis, and limitations of nutrient and suspended-sediment data, 1975-90*. Water-Resources Investigations Report 97-4209, U.S. Geological Survey: Lemoyne, PA, 1997.
9. Jordan, T.E.; Correll, D.L.; Weller, D.E. *J. Environ. Qual.* **1997**, 26, 836-848.
10. Bachman, L.J.; Phillips, P.J. *Water Res. Bul.* **1996**, 32, 779-791.
11. Koterba, M.T.; Banks, W.S.L.; Shedlock, R.J. *J. Environ. Qual.* **1993**, 22, 500-518.
12. Gallagher, D.L.; Deitrich, A.M.; Reay, W.G.; Hayes, M.C.; Simmons, G.M., Jr. *Ground Water Monit. Remediat.* **1996**, 16, 118-129.
13. Glotfelty, D.E.; Taylor, A.W.; Isensee, A.R.; Jersey, J.; Glenn, S.J. *J. Environ. Qual.* **1984**, 13, 115-121.
14. Phillips, P.J.; Bachman, L.J. *Water Res. Bul.* **1996**, 32, 767-778.
15. Schubel, J.R.; Pritchard, D.W. In *Contaminant Problems and Management of Living Chesapeake Bay Resources*. Majumdar, S.K.; Hall, Jr. L.W.; Austin, H.M. Eds.; The Pennsylvania Academy of Science: Phillipsburg, NJ, 1987, pp1-32.
16. US Environmental Protection Agency. *Chesapeake Bay Fall Line Toxics Monitoring Program - Final Report*, Chesapeake Bay Program, Annapolis, MD, 1994.
17. US Environmental Protection Agency. *Chesapeake Bay Basinwide Toxics reduction strategy reevaluation report*. CBP/TRS 117/94, US EPA: Chesapeake Bay Program, Annapolis, MD, 1994b.
18. Maryland Department of Agriculture. *Maryland Pesticide Use Statistics for 1988*, Annapolis, MD, 1990.
19. Maryland Department of Agriculture. *Maryland Pesticide Use Statistics for 1991*, Annapolis, MD, 1993.
20. Maryland Department of Agriculture. *Maryland Pesticide Use Statistics for 1994*, Annapolis, MD, 1996.
21. Maryland Office of Planning. Land use/land cover database for Queen Anne's County. Baltimore, MD, 1995.
22. Maryland Department of Agriculture. *Maryland Agricultural Statistics, Summary for 1996*; Annapolis, MD. 1997.
23. Jordan, T.E.; Correll, D.L.; Weller, D.E. *J. Amer. Water Resources Assoc.* **1997**, 33, 631-645.

24. Matthews, E.D.; Reybold, III, W.U. *Soil survey of Queen Anne's County, Maryland*. US Department of Agriculture, US Government Printing Office: Washington, DC, 1966.
25. US Environmental Protection Agency. *Test Methods for Evaluating Solid Waste, Physical/Chemical Methods, SW-846, 3rd Edition, Final Update 1*. US EPA: Washington, DC, 1990.
26. US Environmental Protection Agency. *Standard Methods for the examination of water and wastewater under Total Organic Carbon US EPA Methods for Chemical Analysis of Water*. EPA-600/4-7-020, US EPA: Cincinnati, OH, 1983.
27. Canadian Council of Resource and Environment Ministers. *Canadian Water Quality Guidelines*. Task Force on Water Quality Guidelines, Environment Canada: Ottawa, Ontario, Canada. 1993.
28. Federal Register. *Part IV: Environmental Protection Agency, National recommended water quality criteria; Notice; Republication*. 1998, Vol.63, No.237. pp68354-68364.
29. Solomon, K.R.; Baker, D.B.; Richards, R.; Dixon, K.R.; Klaine, S.J.; La-Point; T.W.; Kendall, R.J.; Weisskopf, C.P.; Giddings, J.M.; Giesy, J.P.; Hall, L.W., Jr.; Williams, W.M. *Environ. Toxicol. Chem.* **1996**,15, 31-76.
30. Menconi, M.; Cox, C. *Hazard assessment of the insecticide diazinon to aquatic organisms in the Sacramento/San Joaquin river system*. California Department of Fish and Game: Sacramento, CA, 1994.
31. Solomon, K.R.; Chappel, M.J. *An ecological risk assessment of metolachlor residues found in various surface water locations. Draft Report by the Center for Toxicology*, University of Guelph: Ontario, Canada. 1996
32. Baker, D.B.; Richards, R.P. In *Long range Transport of Pesticides*, Kurtz, D.A. Ed.; Lewis Publishers: Chelsea, MI, 1990, pp241-270.
33. Thurman, E.M.; Goolsby, D.A.; Meyer, M.T.; Kolpin, D.W. *Environ. Sci. Technol.*. **1991**, 25, 1794-1796.
34. Thurman, E.M.; Goolsby, D.A.; Meyer, M.T.; Mills, M.S.; Pomes, M.L.; Kolpin, D.W. *Environ. Sci. Technol.* **1992**, 26, 2440-2447.
35. Thurman, E.M.; Fallon, J.D. *Intern. J. Environ. Anal. Chem.* **1997**, 65, 203-214.
36. Eskins, R.; Roland, K.; Allegre, D.; Magnien, R. *Contaminants in Chesapeake Bay Sediments: 1984-1991, Final Report;* US EPA: Chesapeake Bay Program, Annapolis, MD, 1994.
37. Glotfelty, D.E.; Williams, G.H.; Freeman, H.P.; Leech, M.M. In *Long-range transport of pesticides*; Kurtz, D.A., Ed.; Lewis Publishers: Chelsea, MI, 1990, pp 199-222.

Chapter 7

Watershed Fluxes of Pesticides to Chesapeake Bay

Gregory D. Foster and Katrice A. Lippa[1]

Department of Chemistry, MSN 3E2, George Mason University, Fairfax, VA 22030

Input mass budgets of pesticides and other organic contaminants in Chesapeake Bay are being used to identify the most important source areas for chemical contamination. The concentrations and fluxes of selected current-use pesticides, such as simazine, prometon, atrazine, and metolachlor, were determined above the fall lines of nine of the major tributaries of Chesapeake Bay during base flow hydrologic regimes in spring (May) and autumn (November) of 1994. The watershed fluxes of the pesticides showed a high degree of spatial variability across the Chesapeake Bay region. The greatest fluxes occurred for most of the pesticides above the river fall lines of the Choptank River, an eastern shore tributary, and the Susquehanna River, the largest tributary, in the spring and above the river fall line of the Patuxent River in autumn.

Chesapeake Bay is a collection of delicate ecosystems, many of which have been greatly perturbed through the mismanagement of resources within the Bay as well as by continued pollution of anthropogenic substances from the surrounding watershed. A critical step in understanding the effects of toxicants on the Bay's ecology is knowing the types and quantities of substances being delivered to the estuary. The fluxes of current-use pesticides above the river fall lines of the three largest tributaries of Chesapeake Bay have been quantified in various years since 1990 through the Chesapeake Bay Fall Line Toxics Monitoring Program (FLTMP). The fall line is defined as the physiographic boundary between the Piedmont and Coastal Plain Provinces in the eastern United States or above the head of tide in coastal streams. The primary goal of the FLTMP has been to estimate the annual riverine fluxes of contaminants to tidal Chesapeake Bay, including pesticides, from the non-

[1] Current address: Department of Geography and Environmental Engineering, The Johns Hopkins University, Baltimore, MD 21218

tidal riverine source areas above the fall lines *(1-5)*. Riverine fluxes estimated through the FLTMP are used to determine mass budgets of contaminant inputs to the Bay from a variety of identified sources, including, among others, riverine discharges, atmospheric deposition, and urban runoff *(6)*.

The focus of the FLTMP since its inception has been on the Susquehanna, Potomac, and James Rivers. Recent findings from the FLTMP have shown that the magnitudes of annual fluxes of several current-use pesticides, such as atrazine, metolachlor, and cyanazine, estimated for the three tributaries were variable but generally correlated with annual river discharges at the fall lines *(3,5)*. To more comprehensively compare the spatial variability in fall line loadings throughout the Chesapeake Bay watershed, a number of additional bay tributaries were sampled synoptically in the spring and fall of 1994. The new tributaries included in the FLTMP, in addition to the three tributaries above, were the Patuxent, Rappahanock, Pamunkey, Mattaponi, Choptank, and Nanticoke Rivers located in Maryland and Virginia. The above nine tributaries provide approximately 80% of all the freshwater flow to Chesapeake Bay *(7)*. The objective of the present report is to compare the fluxes of selected pesticides estimated for the nine major tributaries across the Chesapeake Bay watershed in 1994 through the synoptic sampling study.

Materials and Methods

River Fall Line Sampling. River water was collected from the nine tributaries twice during 1994 employing a synoptic sampling design. The first sampling occurred during the spring flush (May 5- 12) and the second in autumn (November 18-22) when flows were lowest. Each tributary (Figure 1) was sampled along the fall line reach or just above the head of tide depending on the nature of the stream or river using a Fultz submersible pump (Fultz Pumps, Inc., Lewistown, PA). Surface water was collected from each river by immersing the pump at least one-third of a meter below the surface at the approximate center of flow of the river and pumping water into a 35-L stainless steel beverage container. The container was filled with river water, capped tightly, and transported to the laboratory. In the laboratory, the samples were immediately placed in a cold room (1 °C) and stored for no more than 48 hrs. prior to the analysis of nine pesticides (Table I).

Sample Processing and Analysis. The surface water samples were filtered using a stacked arrangement of 15-cm disks of Whatman GF/D overlying GF/F glass fiber filters (Whatman Inc., Clifton, NJ), housed in a 142-mm Millipore (Millipore Corporation, Bedford, MA) stainless steel filtration apparatus, at 1 L/min using a positive displacement pump (Model QB, Fluid Metering Inc., Oyster Bay, NY). The filtered water was collected in separate precleaned 37.5 L containers and held for surrogate spiking and extraction. The filters were wrapped in precleaned aluminum foil envelopes and stored at -25 °C for subsequent chemical analysis.

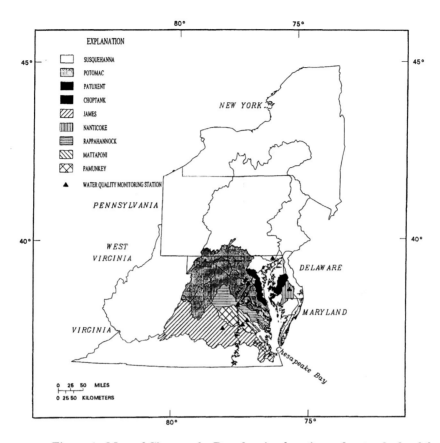

Figure 1. Map of Chesapeake Bay showing locations of watersheds of the nine tributaries monitored in the synoptic study in 1994. (This map was provided courtesy of the USGS in Baltimore, MD.)

Filtered water was extracted with dichloromethane (DCM) using a Goulden large-sample extractor (GLSE) to isolate the pesticides. The GLSE extraction procedures reported by Foster et al. (8) were used with the following modifications: (i) a customized distillation device was added to recover DCM from the waste stream of the GLSE for reuse in the extraction and (ii) the sample flow rate in the GLSE was decreased to 110 mL/min to accommodate the solvent recovery system. Sample volumes processed through the GLSE ranged from 25 to 35 L. Following extraction, DCM from the GLSE was passed through anhydrous sodium sulfate and solvent exchanged with n-octane during solvent-volume reduction to 0.5 mL by using rotary flash evaporation and nitrogen gas blowdown.

The GLSE extracts were subsequently analyzed for the pesticides using GC/MS. The filters were intermittently analyzed for the pesticides but were below method detection limits in all cases.

Table I. List of pesticides monitored in water samples collected from the tributary fall lines.

Pesticide	Abbreviation	Chemical Class	CAS Registry Number
Simazine	sima	triazine	122-34-9
Prometon	prom	triazine	1610-18-0
Atrazine	atra	triazine	1912-24-9
Diazinon	diaz	organophosphorus	333-41-5
Alachlor	alac	chloroacetamide	15972-60-8
Malathion	mala	organophosphorus	121-75-5
Metolchlor	meto	chloroacetamide	51218-45-2
Cyanazine	cyan	triazine	21725-46-2
Hexazinone	hexa	triazine	51235-04-2

The pesticides were analyzed in the GLSE extracts using a 5890A Hewlett-Packard (HP) GC (Hewlett-Packard, Wilmington, DE) coupled to a Finnigan MAT Incos 50 mass spectrometer (Finnigan MAT, San Jose, CA). Sample injection (2 µL) was performed using an HP 7673A autoinjector in the splitless mode, with the split and purge vent flow rates adjusted to 30 and 3 mL/min, respectively. The GC/MS was fitted with a 30 m X 0.25 mm (id) DB-5 fused-silica capillary column (0.25 µm film; J&W Associates, Folsom, CA) and operated using the following temperature program: 100° (5 min); 100°-120° at 7°/min; 120° (0.1 min); 120°-180° at 1.25°/min; 180° (0.1 min); 180-290° at 20°/min; and 290 (10 min) for a total run time of 74.2 min. All GC/MS acquisitions were performed using multiple ion detection (MID) in the electron impact ionization mode (70 eV) with the electron multiplier voltage ranging from 1100-1300 V. Three characteristic ion masses were selected for each analyte in MID quantitation. Quantitation was accomplished using internal injection standards consisting of naphthalene-d8, phenanthrene-d10, and chrysene-d12, which were added to all the sample vials immediately before GC/MS injection.

Quality Assurance. Quality assurance (QA) samples included laboratory and field blanks and the addition of terbutylazine and fluoranthene-d10 as surrogate standards to each river fall line sample prior to GLSE extraction. The monitored contaminants were surrogate normalized to the overall mean surrogate recoveries for all the fall line samples when the surrogate recovery (%rec) in the sample was outside of the established range of performance (determined from numerous extractions of the surrogates as the mean %rec ± 0.15 X mean %rec for both surrogate standards). Method quantitation limits were determined as three times the signal found in the GC/MS chromatograms of laboratory blank samples, and ranged from 0.4 – 2.1 ng/L for the nine pesticides.

Flux Estimates. Instantaneous fluxes of the pesticides were determined for each tributary as the measured concentration times the instantaneous river discharge at the point of sampling along the fall line. Fluxes were normalized by dividing instantaneous fluxes by the land areas within each tributary basin classified as agricultural. For pesticides which were below the method quantitation limits in the river fall line samples, fluxes were estimated using the method quantitation limit concentrations.

Results and Discussion

Hydrologic conditions during the spring and fall synoptic sampling were at base flow. No storms occurred during the river water collections in the spring although the fall line discharges were above their historical averages since sampling was near the end of a rainy spring season. River discharges measured during the 1994 spring and autumn synoptic collections are listed in Table II. The spring collection followed the heaviest pesticide field application period (between March to July in the Chesapeake Bay basin (9)) during the spring flush, providing a good potential to capture the effects of pesticide runoff in the rivers. Hydrologic conditions for the autumn synoptic collections were near base flow, which is typically the season with lowest river flows, but mild storms occurred during river water collections from the Virginia rivers (i.e., the Rappahannock and James Rivers). The influence of the storms did not markedly affect river flow conditions in these tributaries. The river discharges during the autumn synoptic study were lower than those in spring for all nine rivers.

Concentrations of the pesticides measured in the river fall line samples are summarized in Table III for the nine tributaries monitored in the spring and autumn synoptic studies. The concentrations of the pesticides at the tributary fall lines showed a wide range of spatial variability. There have been many geologic, hydrologic, and geochemical variables postulated which affect the seasonal concentrations of pesticides in river runoff. Among those variables proposed to influence the fluvial transport of moderately polar pesticides such as those presented in this study include pesticide soil/water sorption constants *(10,11)*, agricultural growing season *(12,13)*, rainfall frequency and quantity *(14,15)*, soil texture and moisture *(13,15)*, watershed bedrock lithology *(12)*, landuse *(12,13)*, and cropping patterns *(12,13)*. The limited synoptic sampling design in 1994 did not accommodate a statistical evaluation of these parameters.

The Choptank and Nanticoke Rivers showed the highest concentrations of simazine and metolachlor in spring, and furthermore the Choptank River showed the highest concentrations of all monitored pesticides except hexazinone in non-tidal river water (Table III). The watersheds of these two eastern shore tributaries lie entirely in the Coastal Plain and the relatively high concentrations observed at their fall lines, especially in the Choptank River, arose, in part, from their low stream order (leading to less dilution), permeable soils *(16)*, and close proximity to the agricultural fields. All of the monitored western shore tributaries flow through the Atlantic Piedmont to reach Chesapeake Bay and are higher order rivers than those on the eastern shore. Of the western shore rivers the Susquehanna River showed the greatest concentrations of pesticides in the spring. There is extensive farming in the Piedmont region of the Susquehanna River basin, and most of the pesticide runoff occurs in this region of the basin *(13)*.

Table II. Spring and autumn river discharges and basin areas above the fall lines of the nine monitored tributaries of Chesapeake Bay for 1994.

Tributary	Spring River Discharge, m^3/s	Autumn River Discharge, m^3/s	Agricultural Land Area,[a] km^2
Pamunkey	25	6.9	980
Mattaponi	19	5.6	436
James	174	48	3726
Rappahannock	43	10	1835
Potomac	626	69	10870
Choptank	4.7	1.1	161
Nanticoke	5.3	0.96	195[b]
Patuxent	8.8	4.0	414
Susquehanna	1730	144	21890

[a]Land area within the watersheds designated as agricultural; [b]area of the entire Nanticoke basin above the head of tide.

In the autumn synoptic study, the highest or next highest concentrations for many of the pesticides were found in the Patuxent River (Table III). The Patuxent River basin has 35% of its watershed area classified as urban, which is considerably greater than any of the other eight tributary basins. Urban application of pesticides may provide considerable inputs into nearby rivers, and non-agricultural use of pesticides occurs from spring through winter *(17)*, correlating with the high pesticide concentrations seen in the Patuxent River. Simazine and prometon concentrations in the river fall lines samples were greater in five of the nine tributaries and atrazine in four of the nine tributaries in fall relative to spring sampling, indicating the existence of important non-agricultural sources for these pesticides in autumn runoff.

Pesticide fluxes estimated for each river basin are listed in Table IV. Area-normalized fluxes provided a more clear comparison of individual basin dynamics in watershed runoff because instantaneous loads (i.e., µg/s) were found to follow the order of descending stream discharges above the fall lines. The magnitude of instantaneous load estimates are governed predominantly by river discharge.

Estimates of fluxes above the river fall lines provided a comparison among the Bay's major tributaries of the integrative effects of all source inputs on the scale of the individual watershed. A great deal of spatial variability was found in fluxes among the nine tributaries (including the Susquehanna River) for both spring and autumn

Table III. Summary of pesticide concentrations measured in the river fall line samples collected through the spring and fall synoptic study.

Tributary	Pesticide								
	sima	prom	atra	diaz	alac	mala	meto	cyan	hexa
Spring (26 April – 6 May)									
	Concentration in ng/L								
Pamunkey	20.3	7.0	20.6	1.3	2.3	2.9	15.5	4.8	12.7
Mattaponi	1.6	4.3	6.6	2.7	0.9	2.9	15.4	4.0	0.8
James	13.8	3.0	5.8	1.3	0.9	2.9	3.0	109	3.8
Rappahannock	13.3	6.2	14.1	1.3	0.9	2.9	15.3	0.9	0.6
Potomac	29.1	6.9	28.5	1.4	8.6	3.1	46.0	38.9	1.2
Choptank	171	37.2	630	31.2	177	23.8	452	239	2.9
Nanticoke	157	16.2	46.2	24.2	7.7	3.2	69.5	29.0	2.1
Patuxent	1.6	1.7	1.4	1.3	0.9	2.9	0.8	0.9	2.9
Susquehanna	89.3	8.0	99.9	4.9	23.1	2.9	101	49.3	<QL[a]
Autumn (11 November – 18 November)									
Pamunkey	30.1	7.0	12.2	1.0	<QL	7.8	4.0	<QL	17.2
Mattaponi	53.0	5.6	9.7	<QL	<QL	<QL	3.0	<QL	9.5
James	12.2	6.6	3.1	0.7	<QL	<QL	1.3	118	6.9
Rappahannock	49.7	8.5	24.0	0.7	<QL	1.1	21.2	<QL	<QL
Potomac	53.1	20.0	46.3	0.7	1.6	<QL	19.6	88.6	2.0
Choptank	28.5	7.8	12.0	<QL	4.7	<QL	24.0	<QL	2.9
Nanticoke	19.7	9.6	11.7	0.6	46.7	<QL	49.0	<QL	3.0
Patuxent	66.6	18.9	65.4	23.5	3.6	<QL	47.0	<QL	10.1
Susquehanna	nd	9.5	31.2	1.9	4.5	5.6	19.6	48.5	<QL

[a]Pesticide concentration was less than the method quantitation limit in the sample.

collections. The Susquehanna River basin, by far the largest in the Chesapeake watershed, did not consistently show the greatest area-normalized fluxes, indicating that river discharge and total agricultural land area were not directly related to measured fluxes.

One of the primary motives in determining basin specific fluxes has been to evaluate the feasibility of using semi-annual sampling data in mass budget calculations for riverine fluxes. An intensive sampling of the Susquehanna River fall line was also conducted in 1994, in conjunction with the synoptic study, to determine accurate annual fluxes from this tributary (3). Comparisons of the intensive (35 samples collected at the fall line in 1994) and synoptic (2 samples collected in 1994) estimates of annual fluxes are illustrated in Figure 2. Re-expression of the pesticide fluxes listed in Table IV on an annual time scale allowed for a comparison to those determined

Table IV. Estimated fluxes of the pesticides above the fall lines of major tributaries of Chesapeake Bay in 1994.

Tributary	Pesticide								
	sima	prom	atra	diaz	alac	mala	meto	cyan	hexa
Spring synoptic	flux in µg/km²/s								
Pamunkey	0.18	0.06	0.18	3e-3	0.02	0.01	0.14	0.04	0.11
Mattaponi	6e-3[a]	0.05	0.08	0.03	3e-3	0.01	0.19	0.05	0.01
James	0.15	0.03	0.06	4e-3	3e-3	0.01	0.03	1.2	0.04
Rappahannock	0.14	0.06	0.15	4e-3	3e-3	0.01	0.16	3e-3	0.01
Potomac	0.61	0.14	0.60	0.03	0.18	0.06	0.96	0.81	0.02
Choptank	2.8	0.60	10	0.50	2.8	0.38	7.3	3.9	0.05
Nanticoke	4.2	0.44	1.2	0.66	0.21	0.09	1.9	0.78	0.06
Patuxent	5e-3	6e-3	4e-3	4e-3	3e-3	0.01	2e-2	3e-3	0.01
Susquehanna	2.2	0.20	2.5	0.12	0.56	0.02	2.5	1.2	*0.02*[b]
Autumn synoptic									
Pamunkey	0.08	0.02	0.03	3e-3	*1e-3*	0.02	0.01	*1e-3*	0.04
Mattaponi	0.19	0.02	0.04	*1e-3*	*1e-3*	*3e-3*	0.01	*1e-3*	0.03
James	0.04	0.02	0.01	2e-3	*1e-3*	*2e-3*	4e-3	0.35	0.02
Rappahannock	0.12	0.02	0.06	2e-3	*1e-3*	3e-3	0.05	*1e-3*	*2e-3*
Potomac	0.12	0.05	0.11	2e-3	4e-3	*2e-3*	0.04	0.20	5e-3
Choptank	0.10	0.03	0.04	*1e-3*	0.02	*3e-3*	0.09	*1e-3*	0.01
Nanticoke	0.10	0.05	0.06	3e-3	0.23	*4e-3*	0.24	*1e-3*	0.02
Patuxent	0.30	0.08	0.29	0.11	0.02	*4e-3*	0.21	*1e-3*	0.04
Susquehanna	1e-3	0.02	0.06	4e-3	0.01	0.01	0.04	0.10	*2e-3*

[a]All values <0.01 are expressed as scientific notation (e.g., 6 X 10⁻³); [b]fluxes estimated by using method quantitation limits are shown in boldface, italic.

through the intensive study. Estimates obtained through the synoptic study were either less than or greater than estimates made through the intensive study with no apparent systematic bias. For all of the pesticides except prometon, cyanazine, and hexazinone the synoptic study provided flux estimates within a factor of two of the intensive study. Assuming the magnitudes of differences in sampling approaches displayed in Figure 2 hold for the fall lines of other watersheds of Chesapeake Bay, pesticide fluxes measured through the synoptic study will be useful in determining mass budgets, provided additional uncertainties are considered, for tributaries not previously sampled intensively as was done for the Susquehanna River in 1994.

The major tributaries of Chesapeake Bay showed basin-specific concentrations and fluxes at the river fall lines for nine current-use pesticides. An eastern shore tributary, the Choptank River, had the highest concentrations and basin fluxes of the

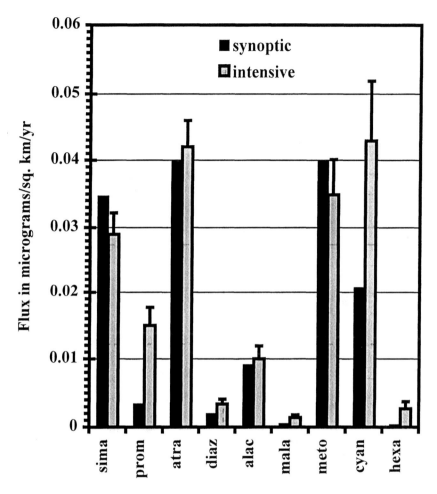

Figure 2. Comparison of estimated pesticide fluxes above the fall line of the Susquehanna River for the synoptic (2 samples/yr) and intensive (35 samples/yr.) studies in 1994. The error bars associated with the intensive study estimates represent ± 1 standard deviation.

pesticides above the river fall line in spring, while a more urbanized western shore tributary, the Patuxent River, had the highest overall concentrations and fluxes of the pesticides above the river fall lines in autumn.

Acknowledgments

We sincerely appreciate the efforts by Dr. Cherie Miller, Mr. Maston Mount, and Ms. Brenda Feit of the USGS in Baltimore, MD in the collection of the river fall line samples. Financial support for this study was provided by the Maryland Department of the Environment (contracts 61-C-MDE95 and 313-C-MDE95) and the U.S. Environmental Protection Agency.

References

1. *Chesapeake Bay Fall Line Toxics Monitoring Program: 1990-1991 Loadings. Final Report to the Chesapeake Bay Program Office*; CBP/TRS 98/93; Chesapeake Bay Program Office: Annapolis, MD, 1993.
2. *Chesapeake Bay Fall Line Toxics Monitoring Program 1992 Final Report*; CBP/TRS 121/94; Chesapeake Bay Program Office: Annapolis, MD, 1994.
3. *Chesapeake Bay Fall Line Toxics Monitoring Program 1994 Final Report;* CBP/TRS 144/96; Chesapeake Bay Program Office: Annapolis, MD, 1996.
4. Godfrey, J.T., Foster, G.D. and Lippa, K.A. . *Environ. Sci. Technol.* **1995**, *29*, 2059-2064.
5. Foster, G.D. and Lippa, K.A.. *J. Agric. Food Chem.* **1996**, *44*, 2447-2454.
6. Velinksy D. *A Chemical Contaminant Mass Balance Framework For Chesapeake Bay;* CBP/TRS 176/97 and EPA 903-R-97-016; Chesapeake Bay Program Office, Annapolis, MD, 1997.
7. Langland, M.J. A synthesis of nutrient and sediment data for the Chesapeake Bay drainage basin. USGS-Water Resources Investigations Report.
8. Foster, G.D., Foreman, W.T.. Gates, P.M., McKenzie, S.A., and Rinella, F.A. *Environ. Sci. Technol.* **1993,** *27*, 1911-1917.
9. Pait, A.S., DeSouza, A.E., and Farrow, D.R.G. *Agricultural Pesticide Use In Coastal Areas: A National Summary.* Report by the Strategic Environmental Assessments Division, Office of Ocean Resources Conservation and Assessment; National Oceanic and Atmospheric Administration: Rockville, MD, 1992..
10. Squillace, P.J. and Thurman, E.M. *Environ. Sci. Technol.*, **1992**, *26*, 538-545.
11. Pereira, W.E. and Rostad, C.E. *Environ. Sci. Technol.*, **1990**, *24*, 1400-1406.
12. Hainly, R.A. and Kahn, J.M. *Wat. Resour. Bull.*, **1996**, *32*, 965-984.
13. Breen, K.J., A.J. Gavin, and R.R. Schnabel. *Occurrence And Yields Of Triazineherbicides In The Susquehanna River And Tributaries During Base-Flow Conditions In The Lower Susquahanna River Basin, Pennsylvania And Maryland, June 1993.* Editor Hill, P. and Nelson, S.; Toward A Sustainable Watershed--The Chesapeake Experiment. Proceedings of the 1994 Chesapeake Research Conference; Chesapeake Research Consortium Publication No. 149: Norfolk, VA, 1994. p. 312-328.
14. Klaine S.J., Hinman,, M.L., Winkelmann, D.A., Sauser, K.R., Martin, J.R., and Moore, L.W. *Environ. Toxicol. Chem.,* **1988**, *7*, 609-614.

15. Schottler, S.P., Eisenreich, S.J. and Capel, P.D. *Environ. Sci. Technol.*, **1994**, *28*, 1079-1089.23.
16. Markewich, H.W., Pavich, M.J., and Buell, G.R. *Geomorphology*, **1990**, *3*, 417-447.
17. Barbash, J.E. and Resek, E.A. *Pesticides in Groundwater: Distribution, Trends, and Governing Factors*; Pesticides in the Hydrologic System; Ann Arbor Press: Chelsea, MI, 1996; Vol. 1, pp 115-125.

Chapter 8

Watershed Monitoring in Sustainable Agriculture Studies

T. D. Spittler[1], S. K. Brightman[1], M. C. Humiston[1], and D. R. Forney[2]

[1]Cornell Analytical Laboratories, New York State Agricultural Experiment Station, Cornell University, Geneva, NY 14456
[2]DuPont Agricultural Products, Chesapeake Farms, Chestertown, MD 21620

To obtain maximum efficiency and sensitivity in pesticide residue studies of various cropping systems on multiple watersheds at Chesapeake Farms, a Dupont environmental research center, a series of triazines and acetanalides were first analyzed by gas-chromatography with electron capture detection (GC-ECD) at sensitivities selected to give the maximum number of direct readings with minimal dilutions (0.5-1.5 ppb). Samples having non-detects for specific compounds in the multi-residue series were individually reanalyzed by enzyme immunoassay at a sensitivity of 0.1 ppb. Two synthetic pyrethroids were also analyzed using similar GC-ECD parameters, and a separate flow injection liposomal immunoassay (FILIA) was developed for imazethapyr.

Maintaining the ability to effectively analyze a number of pesticides in a series of samples having varying matrices and extremes in concentration is a desirable goal, but an organizational challenge. In this study we undertook a significant fraction of the analytical burden in a four-year sustainable agriculture program being conducted at Chesapeake farms, a Dupont-owned environmental research center on the eastern shore of the Chesapeake Bay near Chestertown, MD. Chemicals were being applied in accordance with four corn/soybean/forage production protocols to both replicated plots and production-scale fields. From the former, pan- and suction lysimeter water samples were obtained; runoff and well samples were derived from the latter which were sited on four measured and monitored watersheds. Full information on the field portion of the study is contained elsewhere in this volume (1).

Analytical Requirements and Strategy

The pesticides applied in each of the four protocols are in Table I, along with the general class of compound to which each belongs. Dupont undertook the analyses of all sulfonyl urea herbicides at their own facilities, requiring that all water samples be split upon collection for shipment to the respective laboratories. Of the five classes of compounds assigned to our laboratories, the triazines and acetanalides were recognized as appropriate for a multi-residue scheme coupling solid phase extraction and gas-chromatography with electron capture detection (SPE/GC-ECD) we had developed in 1985 and used in a variety of studies since then (2-4).

Table I. Pesticides Applied

A Plots	B Plots
Atrazine - tz	Atrazine - tz
Cyanazine - tz	Cyanazine -
Nicosufuron - su	Metribuzin - tz
Esfenvalerate - py	Metolachlor - ac
Tefluthrin - py	Alachlor - ac
Chlorethoxyfos - op	Chlorimuron Ethyl - su
	Nicosulfuron - su
	Trifensulfuron Methyl - su

C Plots	D Plots
Atrazine - tz	Nicosulfuron - su
Chlorimuron Ethyl - su	Trifensulfuron Methyl - su
Nicosulfuron - su	Tribenuron Methyl - su
Trifensulfuron Methyl - su	Imazethapyr – im
Tribenuron Methyl – su	
tz - triazine	ac - acetanalide
su - sulfonyl urea	op - organo phosphate
py - synthetic pyrethroid	im - imidazolinone

Attaining maximum sensitivity for all pesticides in a multi-residue scheme can be important, but maintaining apex performance in the face of variable supply lots, fluctuating sample cleanliness, background/matrix effects and analyte concentration extremes is the primary consideration. The ability to function consistently at a sensitivity that yields a maximum number of direct readings with minimal dilutions for linear range, coupled with a default step for samples requiring additional sensitivity, significantly enhances efficiency. To this end we have found that enzyme linked immunosorbent assays (ELISA), available from a variety of vendors, can complement a multi-residue scheme in a manner not usually encountered. Frequent reference is made to the use of ELISA techniques for rapid screening of numerous samples followed by more specific or more accurate determinations by GC-mass spectrometry or related techniques (5,6). We have employed this strategy ourselves, frequently (7). When the sample load can have large, unpredictable variations in concentration, operating a GC detector at maximum sensitivity (or some other low level of quantitation, LOQ) requires that many samples be diluted and rerun to bring the signal within the linear dynamic range of the detector. ELISA is also capable of good sensitivity for those compounds for which a system exists, but its response range is generally more narrow than for a chromatographic detector—ca. a factor of ten vs. two to three orders of magnitude. It was advantageous, therefore, to set our GC-ECD level of quantitation in the range where the maximum number of readable concentrations would occur, then redetermine those data points below the GC LOQ by ELISA. The ELISA LOQ is defined primarily by the system being utilized, unless pre-concentration is employed (8). Esfenvalerate, tefluthrin and chlorethoxyos were all sensitive to election capture, also, and thus allowed further utilization of our GC-ECD set-up. The published methods for imazethapyr required complicated and expensive HPLC-MS equipment or extensive derivatization. We therefore developed a flow injection liposomal immunoassay (FILIA) that is described in another chapter of this volume (9).

Experimental

The collection of water samples had started over a year before the establishment of the analytical portions of the project. Samples had been stored at 0-4°C in glass. While we prefer to immediately freeze water samples in plastic bottles and maintain them at $-10°C$, or colder, a requirement included in the analytical protocol, the chilled samples represented a valuable data series if they could be established as still viable. Accordingly, a storage stability study was initiated wherein standard solutions were both frozen and held at 1°C for analyses at six-month to one-year intervals, until the age of the oldest samples run had been duplicated. This would not only give a comparison of individual compound stabilities at the two temperatures, it would give correction factors for storage intervals if they were found to be significant.

Sample Preparation. Samples were thawed and up to 200 mL were filtered through Gelman A/E Glass filters (1.0M), with the filtrate being returned to the original cleaned container, or to a new Nalgene polypropylene bottle if sample still remained in the original container. A portion of the sample in that 25% requiring pyrethroid analyses was decanted, but left unfiltered, because preliminary recovery work showed an unacceptable loss of esfenvalerate and tefluthrin on the glass filters. Sample volumes were occasionally small (<50 mL) in lysimeter samples, not enough for SPE prior to GC-ECD analyses. In those instances, all analyses were performed by ELISA, only 1-2 mL being needed for each determination. This was also done when most of the sample would be required for imazethapyr determination. As was the case with the pyrethroids, only one field protocol used imazethapyr, so only 25% of the samples required analysis for that compound.

Solid Phase Extraction. Employing a 12-station vacuum manifold and Waters C_{18} Sep-Pak Classic 310 mg cartridges (Millipore Corp.), devices were preconditioned by first washing with 10 mL HPLC-grade benzene, followed by 10 mL of HPLC grade acetonitrile. They were then activated with 20 mL of HPLC-grade methanol and washed with 25 mL of HPLC-grade water. (Once the methanol was added the cartridges could not be allowed to run dry at any time until after samples were completely loaded.) Fifty-mL subsamples were loaded onto the SPE cartridges at a rate of 5-mL min^{-1}; when all samples on the manifold were finished, they were dried briefly under vacuum.

Loaded SPE cartridges were eluted with 2 mL of HPLC-grade acetonitrile, followed by 2 mL of HPLC-grade benzene, into 13-mL glass centrifuge tubes. The sample eluents were refrigerated overnight to allow separation of any residual aqueous material. After warming to room temperature, the lower aqueous phase of each eluent was removed by a disposable glass Pasteur pipette. The organic phase was blown down using a gentle stream of high purity grade nitrogen and a warm (tepid) water bath until just dry. Using a Class A glass volumetric pipette, each dried eluate was redissolved in 2.0 mL of HPLC-grade benzene containing a chlorpyrifos-methyl external standard at a concentration of 0.01 µg/mL. Samples were then ready for gas chromatographic analysis. When sample dilution was necessary, chlorpyrifos-methyl spiked benzene was used as the diluent for sample extracts to dilute target analytes into the range of the calibrators. When GC analyses were complete, sample extracts were transferred to 2 mL silanized glass autosampler vials with crimp seals (Supelco, Bellefonte, PA). Each extract and its corresponding Sep-Pak (air-dried overnight) were stored frozen in 20 mL borosilicate glass scintillation vials.

Solvent Extraction for GC-ECD. Decanted but unfiltered 50 mL aliquots of samples slated for synthetic pyrethroid analyses were extracted with 50 mL of hexane:acetone (3:1). The separated organic phase was dried under nitrogen and redissolved in 2.0 mL hexane with pronamide external standard.

Solvent Extraction for FILIA. Filtered samples (100 mL) were adjusted to pH 1.9 and extracted with 125 mL of CH_2Cl_2. After separation and drying of the organic phase under nitrogen, the sample was redissolved in ten mL of tris-buffered saline for FILIA analysis as described in Lee, et al. (9).

GC-ECD Determination of Triazines and Acetanides. SPE prepared samples from all four plot types (Table I) were analyzed as follows:

Gas Chromatograph:	Hewlett Packard Model HP 5890 Series II Plus
Detector:	ECD, 349°C
Column:	30 m HP-5 fused silica capillary column with 0.32 mm ID and 0.25 µm phase film thickness (Hewlett Packard)
Injection:	1.0 µL at 280°C, splitless with 1.0 min purge delay
Carrier gas:	Helium (ultra-high purity), flow rate of 34.6 cm sec^{-1} at 190°C
Carrier gas head pressure:	13 psi (tank pressure: 40 psi)
Makeup gas:	Nitrogen (ultra high purity), flow rate of 30.4 cm sec^{-1} (tank pressure: 70 psi)

Temperature program: Initial temperature set at 130°C with 1.0 min hold, 10°C min^{-1} to 190°C with 7.0 min hold, 35°C min^{-1} to 280°C with final 10 min hold
Equilibration time: 2.0 min
Total run time: 26.57 min

Because baseline noise early in the runs frequently caused interference with the atrazine peak, the LOQ (Level of Quantitation) was set at 1.5 ppb for atrazine, but maintained at 0.5 ppb for cyanazine, metrabuzin, alachlor and metolachlor.

GC-ECD Determination of Esfenvalerate and Tefluthrin. Solvent extracted samples from plot series A only were analyzed as below:

Gas Chromatograph: Hewlett-Packard Model HP5890 Series II Plus
Detector: ECD, 350°C
Column: 30 m HP-5 fusel silica capillary column with 0.32 mm ID and 0.25 mm phase filter thickness (Hewlett-Packard)
Injection: 1.0 mL at 280°C, splitless with 1.0 min purge delay
Carrier gas: He, 2 mL/min
Carrier gas head pressure: 12.1 psi
Makeup gas: Nitrogen, 45 mL/min
Temperature program: Initial temperature set at 130°C with 1.0 min hold, 10°C/min to 250°C with 5 min hold, 20°C/min to 280°C with 5 min hold
Equilibration time: 2.0 min
Total run time: 26.5 min

The LOQ for both synthetic pyrethroids was set at 0.4 ppb for extraction of a 50 mL water sample. A lower sensitivity would have required using more sample volume than was always available for a given sample. Samples in sets 9 and above containing esfenvalerate and chlorethoxyphos were lowered in priority and not run.

EIA of Triazines and Acetanalides. Samples in which specific analytes were determined to be below their GC-ECD levels of quantitation were subjected to reanalysis using the RaPID Assays (Rapid Pesticide Immuno Detection Assays) system from Ohmicron. The procedure for any of five compounds was the same. Disposable ca. 5-mL glass tubes were placed in a special rack to which 200 µL of either standard, control or sample was added (all determinations run in duplicate). An enzyme conjugate of the analyte and peroxidase was added to all tubes (250 µL): this would compete for binding sites with the analyte in the standards or samples. A suspension of magnetic particles with attached analyte antibodies (500 µL) was added to each tube and all were incubated for 15 min. The rack was then attached to its magnetic base and all particles were held in the bottom of the tubes when the rack was inverted to pour off reagents. One-mL of washing agent was added, after two min. the rack was inverted to remove wash. The rack was removed from the magnetic base and 500 µL of Chromogen-peroxide solution were added to all tubes. After a 20 min. incubation, 500 µL of acidic stopping solution was added and the optical density of the color in the tubes was read at 450 nm. Color formation was inversely proportional to the analyte in the sample blank or standard. Duplicates were averaged.

RESULTS AND DISCUSSION

Storage Stability at 0°C and −20°C. There was concern at the start-up of the analytical phase because the first three sets of samples had been taken the year prior and were being held in glass at 0°C. Plus, by the time they were to be analyzed, they would be 24-26 months old. Subsequent sets were frozen at −20°C, but were still as old as 17 mo. for triazines and acetanalides. Pyrethroid-containing sets ranged from 14-36 mo. old before their analyses commenced, and the imazethapyr samples were 1-21 mo. old when determined. Table II gives the time of arrival of each sample set, the storage conditions and the analysis date and interval.

Table II. Receipt to Analyses Intervals and Storage Temperatures

			Triazine & Actanalides		Pyrethroids		Imazathapyr	
Set*	Received	Temp	Anal'd	Int(mo)	Anal'd	Int(mo)	Anal'd	Int(mo)
1	5/94	0°C	7/96	26	5/97	36		
2	8/94	0°C	9/96	24	5/97	33		
3	8/94	0°C	9/96	24	5/97	32	1/97	28
4	3/95	-20°C	8/96	17	5/97	26	2/97	23
5	8/95	-20°C	9/96	13	5/97	21	3/97	19
7	12/95	-20°C	12/96	12	5/97	18	3/97	16
8	3/96	-20°C	3/97	12	5/97	14	4/97	13
9	6/96	-20°C	1/98	19			4/97	10
10	8/96	-20°C	3/98	19			5/97	9
11	1/97	-20°C	4/98	14			5/97	5
13	4/97	-20°C					5/97	1

*Sets 6 and 12 were bulk water for validation studies

Triazines and Acetanalides. Spiked water samples (20 ppb) were placed in storage at both 0°C and −20°C to determine upon analyses at later intervals if significant degradation had taken place under either storage protocol. Two months after the first set was stored, a question arose concerning solubility of some analytes in the water only stock standards, therefore, a second set was produced from stock standards in 0.2% MeOH to ensure complete dissolution. Table III gives the average recovery (five replicates) for each analyte at 14 and 30 months. Differences between water and 0.2% MeOH stock solution were negligible, so the values (and two month time interval) were averaged.

Table III. Storage Stability of Triazines and Acetanalides (%)

Storage Temp	Time Interval	Atrazine	Cyanazine	Metrabuzin	Alachlor	Metolachlor
-20°C	14 mo	124	70	70	70	81
0°C	14 mo	118	74	74	73	82
-20°C	30 mo	80	68	68	64	70
0°C	30 mo	64	62	64	66	60

On the bases of these results, the data from old or unfrozen samples sets were deemed acceptable: correction factors for minor differences in stability vs. time could be applied to modeling data if needed.

Synthetic Pyrethroids and Organophosphates. The request to include these three soil insecticides in the analytical scheme was made later in the program, and it was decided to run the most recent of the samples to establish if there were any gleanable data to which correction factors determined in a storage stability study could be applied. Analysis of sets 8 and 7 gave no indication of either esfenvalerate or tefluthrin above the LOC of 0.4 ppb. This was not surprising considering the propensity of many pyrethroid materials to rapidly bind to organic matter (10). Further consideration of these compounds was dropped: furthermore, in the absence of these data for comparison, the potential value of chlorethoxyfos data was greatly diminished—thus it was not pursued.

Imazethapyr. This herbicide was reported to be stable in water and under frozen storage so a stability study was not initiated (9).

Validation. Study protocol required that 15-20% of all determinations be validation data. In the initial methods development, analyte recoveries were run in sets of five replicates at at least four concentrations, plus, replicate blanks of both HPLC grade water and blank water from an untreated Chesapeake watershed were determined. During the study, every tenth sample was either a replicate, a blank, or a recovery at one of several repeating levels, Triazines and Acetanalides, 0.1-30 ppb (GC-ECD) 0.05-1.0 ppb (EIA): Pyrethroids, 0.4-2.0 ppb: Imazethapyr 0.01-10 ppb). Recoveries were required to be between 70 and 120% or determinations were repeated.

EIA Reruns and Verification of GC-ECD Data. Over 1000 samples were run for triazines and acetanalides by GC-ECD. In the first sets, all data points below the LOQ were rerun by EIA, even when the analyte in question was not part of the treatment scheme for the sample's plot of origin. This was undertaken to determine if there was cross-contamination between plots, or residual materials carried over from prior years' work on those fields. Later, EIA analyses to a lower level of sensitivity were selectively performed when the data were desirable or potentially significant. EIA analyses were also valuable for verifying GC-ECD results, e.g. when matrix interference with the first-emerging atrazine peak occasionally gave questionable results. Enzyme immunoassay was not susceptible to the matrix interference being picked up by GC-ECD.

There was also the consideration, however, of cross-reactivity between the five EIA systems. Ohmicron reports the levels of quantitation possible with each of their kits for the title analytes, and any related compounds that show significant cross-reactively as LDD, Least Detectable Dose. Table IV summarizes those cross-reactivities relevant to our study. Table V gives the calculated amount of cross-reactive compound that would have to be present to give an EIA false positive at our arbitrary Levels of Quatitation of 0.1 ppb. These values would also allow correction for cross-reactively if two related materials were present at levels causing mutual interference. No such routine corrections were required in our study.

Table IV. EIA Sensitivity and Cross-Reactivity

Analyte & LLD*(ppb)		Cross-reactant & LLD*(ppb)		% Cross-reactivity
Atrazine	0.048	Cyanazine	1.0	4.8
Metrabuzin	0.04	Cyanazine	5000	0
Alachlor	0.05	Metolachlor	6.0	0.8
Metolachlor	0.05	Alachlor	1.3	3.8
Cyanazine	0.035	Atrazine	200	<0.05

*Least Detectable Dose

Table V. EIA False Positive Threshold at Protocol LOQ's

Analyte LOQ		Equivalent Cross-reactant
Atrazine	0.1 ppb	2 ppb Cyanazine
Alachlor	0.1 ppb	12 ppb Metolachlor
Metolachlor	0.1 ppb	2 ppb Alachlor
Cyanazine	0.1 ppb	200 ppb Atrazine
Metrabuzin	0.1 ppb	none

Conclusions

Making the best use of analytical resources committed in support of ongoing field research requires the capability to assess the potential contribution from large blocks of samples, to exploit fully those that screening indicates will yield significant data when pushed to lower sensitivity, and to terminate sample series or lines of inquiry when it is discerned that only trivial information remains. LOQ's need only be adequate to address the needs of the study: needlessly pushing to lower levels does not necessarily improve data, and may result in its being reported in a form incompatible with that being contributed by collaborators. We selected our LOQ's for the several analytical facets to produce the maximum usable data consistent with the internal guidelines of the sponsor and the needs of the field researcher and modelers (11).

Literature Cited

1) Forney, D. R., Strahan, J., Rankin, C., Steffin, D., Peter, C. J., Spittler, T. D., and Baker, J. L. in *Agrochemical Movement: Perspective and Scale.* Steinheimer, T. R., Ross, L. J., and Spittler, T. D., Ed.; ACS Symposium Series, American Chemical Society, Washington, DC (in press).
2) Bourke, J. B. Leichtweis, H. C., Snook, D. L., and Spittler, T. D. Agrochemicals Division, 191st ACS National Meeting, New York, **1986** Abs No. 54.
3) Hall, L. W., Jr., Ziegenfuss, M. C., Anderson, R. D., Tierney, D. P., Spittler, T. D., and Lavin, L. *Chemosphere* **1995**, 31:3, 2919-44.
4) Hall, L. W., Jr., Ziegenfuss, M. C., Anderson, R. D., Spittler, T. D., and Leichtweis, H. C. *Estuaries.* **1994**, 17 181-6.
5) VanEmon, J. M. and Gerlach, C. L. *Environ. Sci. Technol.* **1995**, 29:7, 312-17.
6) Watts, D. W., Novak, J. M., and Pfeiffer, R. L. *Environ. Sci. Technol.* **1997**, 31:4, 1116-19.
7) Lavin, L., Young, B. S., and Spittler, T. D. *Immunochemical Detection of Residues in Foods*, Beier, R. C. and Stanker, L. H., Ed. ACS Symposium Series No. 621, American Chemical Society, Washington, DC. **1996**. 150-66.
8) Hottenstein, C. S., Rubin, F. M., Herzog, D. P., Fleeker, J. R., and Lawruk, T. S. *J. Agric. Food. Chem.* **1996**, 44 3576-81.
9) Lee, M. Durst, R. A., Spittler, T. D. and Forney, D. R. *Agrochemical Movement:Perspective and Scale.* Steinheimer, T. R., Ross, L. J. and Spittler, T. D. Ed. ACS Symposium Series, American Chemical Society, Washington, DC. (in press)
10) Spitter, T. D., Argauer, R. J., Lisk, D. J., Mumma, R. O., Winnett, G., and Ferro, D. N. *J. Assoc. Off. Anal. Chem.* **1984**. 67:4, 824-6.
11) Erhardt, J. M. and Wheeler, J. K. *Environmental Studies-Policy for Quantitation and Detection Limits of Residue Methods and Studies.* **1995**. Dupont Agricultural Products, Wilmington, DE.

Chapter 9

FILIA Determination of Imazethapyr Herbicide in Water

M. Lee[1,3], R. A. Durst[1], T. D. Spittler[1], and D. R. Forney[2]

[1]Cornell Analytical Laboratories, New York State Agricultural Experiment Station, Cornell University, Geneva, NY 14456
[2]DuPont Agricultural Products, Chesapeake Farms, Chestertown, MD 21620

In response to the need for a rapid, economical method for determination of the herbicide imazethapyr at low concentrations in water, the capillary FILIA (flow injection liposome immunoanalysis) system has been applied. A capillary tube (57 cm x 0.45 mm i.d.) with immobilized imazethapyr antibody was used as the immunoreactor column in the flow injection system. The assay is based on sequential competitive binding between imazethapyr and imazethapyr-tagged liposomes for a limited number of antibody binding sites. Subsequent rupture of the liposomes by injection of a detergent (n-octyl β-D-glucopyranoside) releases carboxyfluorescein which elutes and is measured fluorometrically. Water samples from wells, lysimeters and run-off were collected from test plots and monitored watersheds following imazethapyr application at the Chesapeake Farms environmental research center (Chestertown, MD). Imazethapyr residues in water samples were concentrated 10 times by partitioning into methylene chloride, which was then evaporated. The residue was dissolved in TBS (Tris-buffered saline) solution and injected onto the immunocolumn. The analysis provides a limit of detection of 0.01 µg/L and a working range of 0.02-10 µg/L imazethapyr.

Imazethapyr [5-ethyl-2-(4-isopropyl-4-methyl-5-oxo-2-imidazolin-2-yl)nicotinic acid] belongs to the imidazolinone class of herbicides and is the active ingredient of Pursuit (Figure 1) herbicide. This herbicide is being developed for use with corn, soybeans, peanuts, beans, peas, alfalfa, and other leguminous crops (*1*). It controls annual and perennial grasses and broad-leaved weeds in such crops by inhibiting acetohydroxy acid synthase, the feedback enzyme in the biosynthesis of the branched-chain essential amino acids (*2*). This enzyme is not present in animals. In general, this herbicidal selectivity between weed species and crops is attributable to the differential metabolic rates, or in some cases to the absorption rate at different growth stages, rather than differential sensitivity of the target site (*3*).

The wide use of agrochemicals has caused concern for soil and ground water contamination. The increasing concern of contaminants and their movement into the ecosystem has accelerated the development of methods for the measurement of contaminant residues in the environment which are rapid, inexpensive, sensitive, and specific. The

[3]Current address: Department of Chemistry, Tufts University, Medford, MA 02155

conventional monitoring methods are chromatographic techniques like high performance liquid chromatography (HPLC) and gas chromatography (GC). However, these traditional analytical methods are time-consuming and expensive because they need extensive extraction steps and sophisticated instrumentation. In order to overcome these problems, many studies have been done utilizing an immunoassay for monitoring agrochemicals (4, 5). Most assays have chosen the enzyme-linked immunosorbent assay (ELISA) format. In general, this method is carried out on microtiter plates and requires extensive pipetting, causing the increased possibility of experimental error. Also, a drawback of this approach is the long analysis time due to the required incubation step and the difficulty in automating. Automated ELISA is now commercially available, but the cost of the equipment is very high and limits its usage. As a new immunoassay format, a flow-injection system presents considerable promise for detecting environmental pollutants due to its automation control and low cost (6, 7).

Liposomes which entrap fluorescent molecules as the detectable label provide instantaneous, rather than time-dependent enhancement. Liposomes are formed when phospholipid molecules spontaneously self-assemble in aqueous solution to form a spherical structure enclosing an aqueous core. They can be made immunogenic by incorporating a phospholipid that has an antigen or antibody associated with its head group. This immunogenic specificity of liposomes has been exploited in flow-injection immunoanalysis (8, 9).

Recently, many studies have reported that capillary immunoassays have a number of advantages in comparison with conventional immunoassay format (10, 11). In our previous study, we showed the successful demonstration of capillary immunoassay in a flow-injection analysis system for the determination of imazethapyr in TBS (Tris-buffered saline) solution and spiked clean water samples (12). In this research, we report that capillary column FILIA (flow injection liposome immunoanalysis) can be readily used for the analysis of real environmental samples. Water samples from runoff, wells and lysimeters were collected from test plots and monitored watersheds following imazethapyr application as part of the Chesapeake Farms Project, a multi-faceted study of sustainable corn/soybean/forage production agriculture.

Materials and Methods

Materials. Monoclonal antibody to imazethapyr and dipalmitoylphosphatidylethanolamine conjugate (DPPE conjugate) were kindly provided by the American Cyanamid Co. (Princeton, NJ). 5-(and 6-)-carboxyfluorescein was purchased from Molecular Probes, Inc. (Eugene, OR). Dipalmitoylphosphatidylcholine (DPPC) and dipalmitoylphosphatidylglycerol (DPPG) were obtained from Avanti Polar Lipids (Alabaster, AL). Polycarbonate syringe filters were purchased from Poretics (Livermore, CA). Protein A, immobilized on porous silica glass, was purchased from Bioprocessing Inc. (Princeton, NJ), and dimethyl pimelimidate (DMP) was from Pierce (Rockford, IL). 3-Glycidoxypropyltrimethoxysilane (GPTMS) was purchased from Aldrich (Milwaukee, WI), undeactivated fused-silica capillary (0.45 mm i.d.) was from Alltech Associates (Deerfield, IL), and all other chemicals were from Sigma Chemical Co. (St. Louis, MO).

Apparatus. A schematic diagram of the FILIA system is shown in Figure 2. The injector (Rheodyne Model 7725, Rainin, Emeryville, CA) had a 2-mL sample loop. The flow system was constructed with two three-way 12 V solenoid pinch valves (Biochem Valve Corp., East Hanover, NJ) to control the flow of reagents. All connecting poly(ether ether ketone) (PEEK) tubing (0.02 in i.d.) was purchased from Upchurch Scientific Inc. (Oak Harbor, WA). A HPLC pump (Varian, Walnut Creek, CA) at the outlet of the system was used to maintain a flow rate of 1.0 mL/min in conjunction with 8 psi N_2 head pressure on each of the mobile-phase bottles. To give sufficient back pressure, an old HPLC column

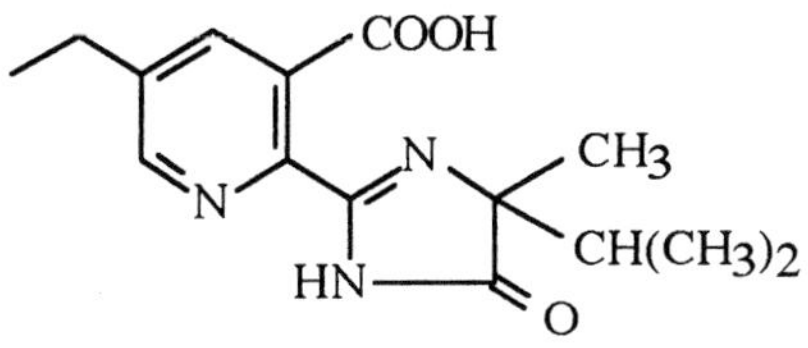

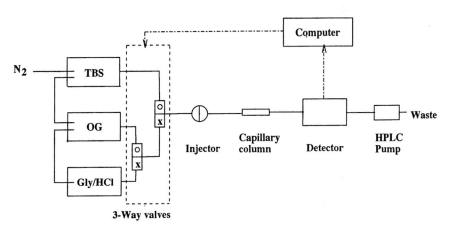

Figure 1. Structure of imazethapyr.

Figure 2. FILIA system. (TBS: Tris buffered saline with 0.01% sodium azide, pH 7.0; OG: 30 mM n-octyl-β-D-glucopyranoside in TBS; Gly/HCl: 0.1 M glycine/HCl, pH 3.0)

was linked to the pump. A MacIntegrator I data analysis package, used for valve operation and data collection and integration, was purchased from Rainin. The fluorescence detector was a Model 240 obtained from Perkin-Elmer (Norwalk, CT).

FILIA Method. The imazethapyr sample was introduced onto the immunoreactor column by manual injection. This was followed by injection of the liposomes. These two steps were done in the flow of the TBS carrier solution. After all unbound liposomes were eluted, a detergent solution (30 mM n-octyl-β-D-glucopyranoside in TBS) was passed through the column. The detergent caused rupture of the bound liposomes, and the fluorescence of the release dye was generated and measured. Finally, 0.1 M glycine/HCl buffer solution (pH 3.0) was passed through the column to regenerate the antibody binding sites by dissociating bound analyte, and the column is then reconditioned for the next analysis by returning the mobile phase to TBS.

Purification of the Monoclonal Antibody. The monoclonal antibody was purified from ascites by passing through a protein A column. The manufacturer's instructions were followed for this procedure. The ascites was loaded onto the column after dilution with 1 M glycine/0.15 M NaCl buffer (pH 8.6). The column was then washed with the glycine buffer, and the bound monoclonal antibody was eluted with 0.1 M citrate buffer (pH 3.0). Finally, the antibody concentration was determined by using a protein test kit from Bio-Rad (Hercules, CA).

Production of Liposomes. Liposomes were formed according to the reversed-phase evaporation method (13, 14) from a mixture of DPPC, cholesterol, DPPG, and DPPE conjugate in a molar ratio of 5:5:0.5:0.02. This mixture was dissolved in a solvent system containing chloroform, isopropyl ether and methanol (6:6:1, v/v). The dye solution of 200 mM carboxyfluorescein (1.4 mL) was added with swirling. This mixture was sonicated for 5 min under a low flow of nitrogen. The organic phase was removed under vacuum on a rotary evaporator at 45 °C. An additional aliquot of the dye solution (2.6 mL) was added, and the liposomes were then extruded through three polycarbonate filters of decreasing pore sizes of 3, 0.4 and 0.2 μm. To remove any unencapsulated dye the liposomes were gel filtered on a 1.5 x 25 cm Sephadex G-50-150 column and finally dialyzed overnight against TBS at 4 °C.

Preparation of the Capillary Column. The inside wall of the capillary was first coated with protein A to which the antibody could be covalently attached without masking antigen binding sites. The fact that the antigen binding region is not affected has made protein A useful as an antibody binding protein in immunoassays. The procedure for modification of the capillary is as follows. Introduce the solutions into the capillary with a syringe. Pretreat the fused-silica capillary with 1 M NaOH overnight followed by rinsing with 1 M HCl and doubly distilled water. Fill the capillary with 10% GPTMS solution in 0.1 M acetate buffer (pH 3.5) and heat at 90 °C for 2 h for silylation. Rinse the capillary with water and treat with 10 mM sulfuric acid at 90 °C for 10 min to convert residual epoxy groups to diols. After flushing with water, treat the capillary with 20 mM sodium metaperiodate containing 2 nM potassium carbonate at room temperature for 2 h. [In this step, aldehyde groups are formed by the periodate cleavage. Protein A is then conjugated through the aldehyde groups by a modification of the Larsson (15) and de Frutos et al. (10) methods in which the capillary is filled with 1 mg of protein A and 0.5 mg of sodium cyanoborohydride (to reduce the Schiff base) in 200 μL of 0.1 M phosphate buffer (pH 6) and incubated overnight at room temperature.] Treat the capillary with a solution of sodium borohydride in 0.1 M phosphate buffer (pH 8) for 1 h to reduce remaining aldehyde groups to their alcohols, and rinse with doubly distilled water and 1 M glycine buffer (pH 8.6). Fill the capillary with antibody in glycine buffer and incubate at room temperature for 1 h followed by 8 h at 4 °C. After washing with glycine buffer and 0.2 M triethanolamine (TEA) buffer (pH 8.2), fill the capillary with DMP solution (6.6 mg of DMP in 1 mL of TEA buffer with the pH readjusted to pH 8.2) and incubate overnight at room temperature. Rinse the capillary sequentially with TEA buffer, 1 M NaCl, 0.1 M glycine/HCl buffer (pH 2.5) and TBS, and store at 4 °C until used.

Water Sample Analysis. After filtering through 0.2 μm nylon membranes, water samples were adjusted to pH 1.9, and 100 mL aliquots were transferred to separatory funnels. Imazethapyr was partitioned into 125 mL of methylene chloride, which was then evaporated on a rotary evaporator. The residue was dissolved in 10 mL of TBS solution and analyzed in FILIA. Blank water samples were spiked with imazethapyr and extracted by the same method to obtain a standard curve. A standard curve was generated each day, and imazethapyr concentrations for water samples were interpolated from the standard curve.

Results and Discussion

FILIA Assay. A typical diagram for an analytical run is shown in Figure 3. 20-μL of liposomal suspension in TBS was injected to give the optimal response in the detector. This amount was calculated to contain ca. 10^{10} liposomes tagged with imazethapyr. Peak A represents the signal from unbound liposomes and traces of free dye passing through the fluorescence detector. Peak B is the analytical signal produced by the lysis of bound liposomes and consequent release of the fluorescent dye. Imazethapyr in the standard and samples is first introduced onto the column and it binds to the antibody sites in proportion to its concentration. After 2 min, when the analyte-tagged liposomes are injected, they bind to unoccupied antibody sites. Since the binding of liposomes depends on the number of remaining antibody sites available, the area of peak B is inversely proportional to the amount of imazethapyr present in the standards and samples.

Dose-Response Relationship. Figure 4 shows a dose-response curve for imazethapyr from the spiked blank water samples. The fluorescence data were recorded as arbitrary units using MacIntegrator software and transformed as in equation (1), where F is the fluorescence at a given concentration of imazethapyr, F_0 is the fluorescence at zero imazethapyr, and F_∞ is the fluorescence in the presence of excess imazethapyr.

$$\%B/B_0 = (F - F_\infty/F_0 - F_\infty) \times 100 \quad (1)$$

$\%B/B_0$ represents the percentage of liposomes bound to the immunoreactor column.

As shown in the figure, 0.01 μg/L of imazethapyr could be easily detected by this system. But some water samples were dirtier than the blank water and might give more matrix effect on the results. Therefore, imazethapyr concentrations from the water samples were interpolated from this standard curve using a working range of 0.02-10 μg/L where 0.02 μg/L corresponded to 80% B/B_0. When clean TBS buffer was used as the sample solutions, the capillary column could be reused for up to 150 analysis runs. However, with the water samples, some contaminants were co-extracted with imazethapyr and were detrimental to the stability of the capillary column. The column lost antibody activity faster than with clean TBS. As a result, the column was reused for up to 50 analysis runs.

Detection of Imazethapyr in Water Samples. Analyte water samples were collected from wells and run-off after imazethapyr application. However, the samples had a yellowish brown color due to humic substances from soil. These humic substances were not removed by a simple filtration step and gave unpredictable responses in the assay.

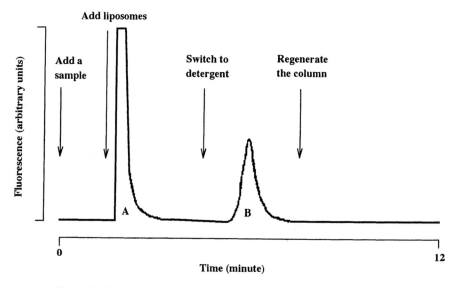

Figure 3. Typical FILIA run. One analysis can be performed in 12 min.

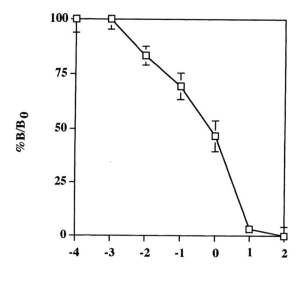

Figure 4. Dose-response curve for imazethapyr which was extracted from spiked blank water samples and concentrated 10 times. Each point represents the mean of three measurements; error bars represent ±1 SD.

Blank water samples were also collected from wells which were not affected by the imazethapyr application for use in construction of the standard curves. As a result, an extraction step for further cleanup was introduced after the filtration, and through this extraction step, imazethapyr residue in water samples was concentrated 10-fold for recovering low concentrations. Table 1 is a daily data summary (4/1/97) of FILIA determinations for a five-point standard curve plus eleven actual field samples, reported as averaged duplicates. The typical FILIA run presented for Sample #10863 (Figure 5) shows negligible interference, however, the outlying data point for #10833 could be either random error or a negative bias introduced by a heavily contaminated sample. Limited sample volumes restricted reanalysis or extensive investigation of questionable points. Caution should be exercised to verify any bias introduced if a reliable blank water source is unavailable for curve generation.

Imazethapyr is a low-use-rate herbicides for the protection of a wide variety of agricultural commodities and must be monitored at low ppb levels in soil and at sub-ppb levels in water to determine the compound's persistence in the environment. Current analytical methodologies for the determination of imazethapyr in water are by liquid chromatography with a limit of quantitation (LOQ) of 5 ppb after processing several hundred milliliters of water through a series of solid-phase extraction (SPE) cartridges and solvent-partitioning steps. Alternatively, the final extract can be methylated and analyzed by gas chromatography with nitrogen/phosphorous detection which yields LOQ of 100 ppb. Recent report (16) has indicated that liquid chromatography/electrospray ionization mass spectrometry (LC/ESMS) can monitor imazethapyr in water at the 1 ppb level with only a simple filtration prior to analysis. In the previous report, we also reported that FILIA could be used for the measurement of imazethapyr with a working range of 0.1-100 µg/L. Among these methods, FILIA has the lowest detection limit and is a simple, inexpensive method. As a result, this method was the most suitable for the off-site determination of imazethapyr in water samples requiring the least concentration of samples and was chosen to analyze the samples in this study. Water samples ranging from <0.02 ppb to >10 ppb have been successfully analyzed by this system. Two thirds of samples fell between the upper and lower limits of the method and provided excellent modeling data for use in devising future application strategies that maximize production return at minimal environmental cost.

Conclusions

This work demonstrates the feasibility of using capillary flow injection liposome immunoanalysis for the determination of imazethapyr in water samples. A limit of detection of 0.01 µg/L was achieved by this method with 10 time preconcentration of samples, and a broad working range of 0.02-10 µg/L was obtained. A single assay could be performed in 12 min.

Acknowledgments

The financial support of Dupont Agricultural Products and the Chesapeake Farms Project are gratefully acknowledged. The authors thank Dr. Rosie Wong (American Cyanamid Co.) for immunoassay reagents and valuable discussion.

Table 1. FILIA data summary for April 1, 1997 for standard curve plus field samples.

(a) Standard curve

Imazethapyr concentration spiked (ppb)	0	0.01	0.1	1	10
Averaged peak area	6264561	5999115	5637374	5288194	5143042

(b) Field samples

Sample No.	10832	10833	10862	10863	10864	10865	10871	10879	10881	10883	10885
Averaged peak area	5497613	7087225	4340275	5755392	5747672	5184925	5465235	5617125	5490984	5785886	5782472
Imazethapyr concentration (ppb)	0.25	n.d.	>10	0.05	0.05	5.15	0.31	0.11	0.26	0.04	0.04

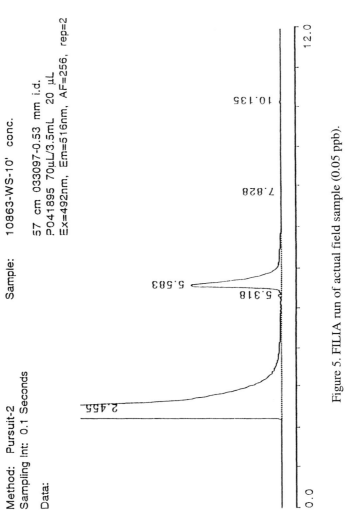

Figure 5. FILIA run of actual field sample (0.05 ppb).

Literature Cited

1. *The Imidazolinone Herbicides*; Shaner, D. L; O'Connor. S. L. Eds.; CRC Press: Boca Raton, FL, 1991.
2. Anderson, P. C.; Hibberd, K. A. *Weed Sci.* **1985,** *33*, 479.
3. Shaner, D. L.; Mallipudi, N. M. In *The Imidazolinone Herbicides*; Shaner, d. L., O'Connor, S. L., Eds.; CRC Press: Boca Raton, FL, 1991.
4. Schwalbe, M.; Dorn, E.; Beyermann, K. *J. Agric. Food Chem.* **1984,** *32,* 734.
5. Thurman, E. M.; Meyer, M.; Pomes, M.; Perry, C. A.; Schwab, A. P. *Anal. Chem.* **1990,** *62,* 2043.
6. Wittmann, C.; Schmid, R. D. *J. Agric. Food Chem.* **1994,** *42,* 1041.
7. Krämer, P.; Schmid, R. *Biosensors & Bioelectronics* **1991,** *6,* 239.
8. Locascio-Brown, L.; Plant, A. L.; Horváth, V.; Durst, R. A. *Anal. Chem.* **1990,** *62,* 2587.
9. Rule, G. S.; Palmer, D. A.; Reeves, S. G.; Durst, R. A. *Anal. Proc. Anal. Commun.* **1994,** *31,* 339.
10. de Frutos, M.; Paliwal, S. K.; Regnier, F. E. *Anal. Chem.* **1993,** *65,* 2159.
11. Jiang, T.; Halsall, H. B.; Heineman, W. R. *J. Agric. Food Chem.* **1995,** *43,* 1098.
12. Lee, M.; Durst, R. A. *J. Agric. Food Chem.* **1996,** *44,* 4032.
13. Szoka, F.; Olson, F.; Heath, T.; Vail, W.; Mayhew, E.; Papahadjopoulos, D. *Biochim. Biophys. Acta* **1980,** *601,* 559.
14. O'Connell, J. P.; Campbell, R. L.; Fleming, B. M.; Mercolino, T. J.; Johnson, M. D.; McLaurin, D. A. *Clin. Chem.* **1985,** *31,* 1424.
15. Larsson, P. *Methods Enzymol.* **1984,** *104,* 212.
16. Stout, S. J.; daCunha, A. R.; Picard, G. L.; Safarpour, M. M. *J. Agric. Food Chem.* **1996,** *44,* 2182.

THE MIDWESTERN PLAINS

Chapter 10

Spring Season Pattern of Nitrate-N and Herbicide Movement in Snowmelt Runoff from a Loess Soil in Southwestern Iowa

T. R. Steinheimer and K. D. Scoggin

National Soil Tilth Laboratory, Agricultural Research Service, U.S. Department of Agriculture, 2150 Pammel Drive, Ames, IA 50011

In the Loess Hills along the Missouri River valley of southwestern Iowa, field studies are underway to determine the impact of continuous corn production on both surface water and groundwater quality. The landscape is characterized by gently sloping ridges, steep side slopes, and well-defined alluvial valleys often with incised channels that usually terminate at an active gully head. Surface water quality is evaluated by analyzing field runoff for nitrate-N, atrazine and metolachlor. A comparison among four field-sized watersheds under several different tillage practices reveals different responses to an early spring runoff event. In each case, the long-term impact resulting from more than 20 years of nitrogen fertilization and more than 15 years of atrazine and metolachlor application is assessed. Beneath the snow cover of March 1993, a diurnal freeze-thaw cycle was observed to cause displacement of both nitrate-N and parent herbicides (from past and current applications) in surface runoff generated by melting snowcover. During the five days of repetitive events, the nitrate-N and herbicides exhibited different displacement patterns, perhaps as a result of their fundamentally different chemical properties.

In 1991, Congress funded the Presidential Initiative to Enhance Water Quality, which was implemented in five midwestern states. Two central objectives are to (1) measure the impact of farming systems on ground water and surface water chemistry and agroecosystem resources, and to (2) identify the factors and processes controlling the fate and transport of fertilizers and pesticides. In Iowa and four other midwestern states, field research is conducted under the Management System Evaluation Area (MSEA) Program, a federal interagency, state, academia, cooperative study of Best Management Practices (BMP) and water quality. In Iowa, BMP are defined in the context of combinations of tillage, crop rotation or sequencing, and both fertilizer and pesticide usage; whereas, in Nebraska and Minnesota, BMP's also include water application. The Iowa MSEA program includes three areas under study which represent diverse scales, landscapes, soil associations, and management practices. Fields under controlled agronomic practices are extensively instrumented for monitoring environmental conditions. Throughout the year, both soil and water samples are collected for laboratory determination of chemical residues as the legacy from both fertilizer and pesticide-derived formulations. Detailed studies of the extent of agrochemical contamination across varied geohydrologic settings in the upper midwest are best described only in recent reports (*1-6*). The Iowa MSEA has three research sites: Treynor, near Council Bluffs, Walnut Creek, near Ames, and Nashua, 40 miles north of Waterloo. Thirty-five percent of the corn and corn/soybean rotation acreage in Iowa is characterized by the agronomic managment practices at these sites. Late in 1991 field studies began focusing on water quality issues at Treynor as a consequence of the intensive row cropping practiced on this somewhat environmentally sensitive landscape, and recently reported on the legacy of long-term nitrogen fertilization on both surface and groundwater quality (*7,8*). In this chapter we compare patterns of surface runoff quality resulting from late-spring snowmelt. Patterns of movement for both fertilizer and herbicide chemicals are compared for four field-sized watersheds representing several different tillage practices. Observations confirm that the diurnal freeze-thaw cycle displaces both nitrate-N and parent herbicides in runoff generated from melting snowcover.

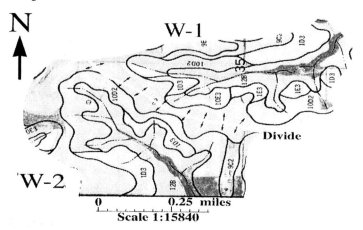

Figure 1. Aerial Photographs Showing Field Boundaries and Soil Associations for Watersheds 1 and 2 (W-1 and W-2).

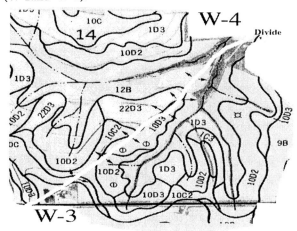

Figure 2. Aerial Photographs Showing Field Boundaries and Soil Associations for Watersheds 3 and 4 (W-3 and W-4).

Site Description Aerial photographs showing boundaries and soil associations are presented in Figures 1 and 2 for each of the watersheds at the Deep Loess Research Station (DLRS) near Treynor, Iowa. Distinct topographic features, such as slope and soil type, define the hydrogeologic boundary conditions which delimit these fields as small watersheds. The cultivated areas range between 30-60 ha. The topography of each watershed delimits its drainage pattern for both direction and intensity of surface water flow. Across these fields total relief is greater than 25 m. The Loess Hills of Iowa and Missouri are characterized by gently sloping ridges, steep side slopes, and well-defined alluvial valleys often with incised channels that usually terminate at an active gully head (9). Soil association is Monona-Ida-Napier with the Monona (Series 10D2, 9-14% slope with moderate erosion potential) predominating on the ridge tops and upper side slopes, the Ida (Series 1D3, 9-14% slope with severe erosion potential) predominating on the steeper side slopes, and Napier (Series 12B, 2-5% slope with only slight erosion potential) predominating on the toe slopes and in the valleys. The loess soils are composed of wind-deposited silt that is principally closely packed quartz grains admixed with lesser amounts of sand and clay. Thickness of this material within 3-16 km of the Missouri River valley is typically >20 m, with some deposits of 50-60 m recorded locally. It is porous, lightweight with low bulk density, and easily eroded. Although very cohesive when dry, loess materials are vulnerable to collapse when wetted. Loads of eroded sediment carried to streams in drainage from cultivated fields in the region are among the highest recorded in the U.S. (10). Table I provides a comparison of the four watersheds in terms of geographic characteristics, soil map units and recent agronomic practices. All are similar in slope and area drained. Watersheds 1 and 2 are adjacent fields located about 2.5 miles southwest of the adjacent watersheds 3 and 4.

Table I. Comparison of the Four Watersheds in Terms of Topographic Characteristics, Erosional Properties, and Agronomic Practices Imposed on the Landscape for 1992.

	Watershed 1	Watershed 2	Watershed 3	Watershed 4
Size	30.1 ha, 6% Grassed Waterways	33.5 ha, 5% Grassed Waterways	43.3 ha, 6% Grassed Waterways	60.7 ha, 7% Sod Backslopes
Tillage	Conventional Corn	Conventional Corn	Ridge-Till Corn	Terraced Ridge-Till Corn
Nitrogen Fertilizer	185 kg ha^{-1}	185 kg ha^{-1}	164 kg ha^{-1}	164 kg ha^{-1}
Herbicide, Date, Applied	Metolachlor 5/13/92, 2 kg/ha a.i. Atrazine 6/03/92 3.25 kg/ha a.i.	Metolachlor 5/13/92, 2 kg/ha a.i. Atrazine 6/03/92, 3.25 kg/ha a.i.(15 ha)	Metolachlor 5/13/92, 0.76 kg/ha a.i. Atrazine 5/29/92, 3.36 kg/ha a.i.	Metolachlor 5/13/92, 0.89 kg/ha a.i. Atrazine 5/29/92, 3.36 kg/ha a.i.
(Slope%) Area%	(9-14%),42% (2-5%),35% (5-9%),17% (14-18%),4%	(9-14%),44% (2-5%),30.3% (5-9%),25.3% (14-18%),0%	(9-14%),42.5% (2-5%),25.5% (5-9%),28.8% (0-2%),3.2%	Double Spaced (88 m) Back Sloped Parallel Terrace, W\ Surface Inlets and Subsurfaces Conduits
Sheet Rill	2832 kg/ha (avg.)	2280 kg/ha (avg.)	224 kg/ha (avg.)	123 kg/ha (avg.)

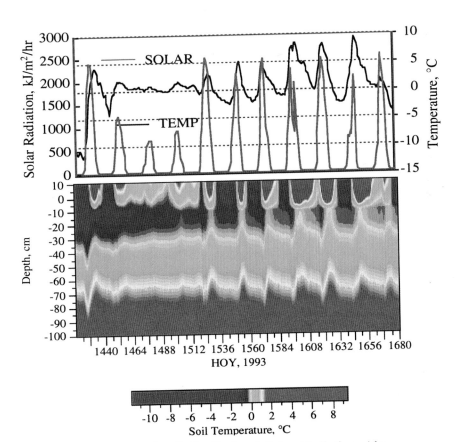

Figure 4. Representative Daily Cycle of Solar Radiation, Air Temperature at the Soil Surface, and Soil Temperature at the DLRS Between March 1-10, 1993.

Field and Laboratory Measurements

Temperature Measurement. Air and soil temperatures during this period were measured at 5-minute intervals and recorded on automatic dataloggers during the study. Ambient air temperature was measured at an elevation of 200 cm above the surface. Soil temperature was measured by thermocouple probes at 10, 20, 30, 50 and 100 cm below an undisturbed soil surface. Each watershed showed no significant deviations in meteorological conditions over the period of study.

Surface Water Sampling. Each watershed was instrumented for the collection of surface runoff. With the onset of a runoff event automatic sampling of surface water began and continued at 10-minute intervals for a maximum of 4 hr using a Model 3700FR refrigerated automatic sampler (ISCO Environmental Division, Lincoln, NE.). The autosampler intake was positioned in the drainage waterway along the thalweg of the valley approximately 3 m upgradient and above the incised channel of the headcut (see Figure 3). In an incised gully at the base of each watershed about 200 m below the seepage face (headcut), the outlet is instrumented with a broadcrested V-notch weir and water-stage recorder which continuously measures storm surface runoff from the field and seepage/spring flow from the gully.

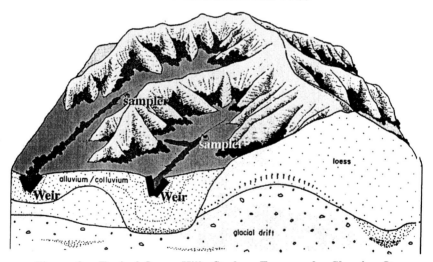

Figure 3. Typical Loess Hills Surface Topography Showing Its Relationship to the Sampling Location for Surface Runoff.

Chemical Determination. Surface water quality was evaluated by analyzing runoff for nitrate-N, atrazine, metolachlor and several of their major metabolites. All nitrate determinations employed the flow-injection colorimetry method (*11*). Depending upon the matrix background of the water sample, the minimum detectability was approximately 0.5 µg/mL at signal-to-noise ratio less than 3. An HPLC method was used for quantitative determination of deisopropylatrazine (DIA) [2-chloro-4-amino-6-ethylamino-*sym*-1,3,5-triazine], deethylatrazine (DEA) [2-chloro-4-amino-6-(1-methylethyl)-*sym*-1,3,5-triazine], atrazine (ATR) [2-chloro-4-ethylamino-6-(1-

methylethyl)amino-*sym*-1,3,5-triazine], cyanazine (CYN) [2-chloro-4-ethylamino-6-(2-methylpropionitrilo)amino-*sym*-1,3,5-triazine], metolachlor (MET) [2-chloro-*N*-(2-ethyl-6-methylphenyl)-*N*-(2-methoxy-1-methylethyl)acetamide], and morpholino-metolachlor (METMOR) [4-(2-ethyl-6-methylphenyl)-5-methylmorpholin-3-one]. These parent herbicides and metabolites were determined on all samples collected from watershed 3. The method employed solid-phase extraction on cyclohexyl cartridges, LC separation on a reversed-phase column, and UV detection with a photodiode-array detector (*12*). Based upon a 0.5 L sample volume, the minimum detection limit is 0.04 µg/L. All samples for HPLC determination were spiked with the surrogate terbuthylazine at the 1.0 µg/L just prior to analysis. Calculated recovery of this surrogate analyte provided a QA/QC control check on both the extraction method and the instrumental calibration. Surrogate recoveries consistently fell within ±15% of theoretical (*12*). Samples collected from watersheds 1, 2, and 4 were analyzed for atrazine and metolachlor only using a robotic protocol (*13,14*).

Results and Discussion

Environmental Conditions. By early spring 1993 a series of runoff events were produced by the snow melting from an accumulated base of 8-20 cm. The environmental condition at the soil surface is illustrated in the compilation of the three information sets shown in Figure 4 on the the adjacent page. The scale at top left shows the solar radiation and at the top right shows air temperature, both for the first ten days of March. Previous to this, the last week of February brought overnight low temperatures that reached -10° C. The lower graph indicates the air temperature just above the soil surface and the soil temperature by depth. Zero depth corresponds to the soil temperature at the land surface. The freeze/thaw zone is emphasized in this compartmentalized chart. Daily changes in soil temperature within the top 15 cm of soil delineate the changing freeze/thaw zone. Following 5 days of cloudy and overcast weather, on March 5 skies clear and a diurnal pattern of solar radiation produces sufficient heat to drive the thawing of the soil. The combined effects of crop residue cover, stored thermal energy, and warmer near freezing overnight temperatures result in sequentially less and less refreezing of the soil overnight. These effects are cumulative with respect to stored heat.

Surface Hydrology. Runoff resulting from snowmelt occurs whenever solar-radiant heat penetrates the snow and heats the soil surface beneath it. This occurs because the snow has a much higher infrared transmitivity and much lower thermal heating capacity than does the soil material. Consequently, a thawed zone is created at the soil surface which produces surface runoff. This simultaneous warming of the soil surface and melting of the snowpack creates a void space through which water can flow. However, it is unlikely that the soil is being thawed to any significant depth below the surface because excess heat is removed by the flowing runoff. Thus, under these conditions, infiltration is retarded by the frozen soil just beneath the surface.

Mobilization of Agrochemicals. During periods of rapidly melting snowpack and subsequent surface water movement, agrochemicals are displaced from each watershed. The mechanisms by which herbicides and nitrate are released from the shallow soil zone and subsequently removed in runoff are complex. Furthermore, they are very different for marginally soluble nonionic organic compounds, such as several classes of corn and soybean herbicides, than they are for infinitely soluble ionic materials, such as nitrate. The nitrate ion is unique compared to soil applied nutrients, because its an anion which forms few insoluble compounds with soil constituents.This anomalous behavior may be peculiar to nitrate ion and different from that observed for other fertilizer-type nutrients; e.g., ammonium or orthophosphate ions. Herbicide release from soil is compound specific. Nevertheless, it is generally assumed that water movement at the surface produces a thin soil-zone undergoing turbulent mixing with release of sorbed chemicals into the overland flow (*15*). When this occurs the two physical processes by which chemicals are released are dispersion and diffusion. The mixing is accelerated by the kinetic energy of the runoff, but is also dependent upon soil properties and microclimate conditions. As the wetting front infiltrates deeper into the soil volume, thus thawing more soil, more surface area is exposed to diffusion into the flow path. During runoff events, water travels down-gradient along paths defined largely by tillage practices and the landscape contours, or can pool within the tillage contour, such as with ridge-till, until breakthrough or overflow.

Field drainage at the watershed outlet carries a nitrate load which is determined by both the distance traveled over the field to reach the thalweg area and its time-of-travel through the watershed. These factors are ultimately dependant on depth of thawed soil, kinetic energy, and duration of contact as well as on antecedent soil moisture. For the soluble nitrate ion, transfer to runoff from below the surface can occur either from dispersion within the turbulent mixing zone or from diffusion from subsurface soil. In frozen soils this is limited by the depth of the thawed zone. Therefore, as soil erosion progresses through a runoff event, the main mixing zone of interaction can contain an increasing portion of the subsoil material. This has been verified for severe storm events (16). During runoff on the highly-erodible loess material, erosion is limited by the thawed depth which determines the amount of exposed soil available for diffusion from the mixing zone and into the flowpath.

Temporal Variation in Concentration of Agrochemicals

Conventional Tillage. Watersheds 1 and 2 are modified conventional tillage systems comparable in size and management. Modified conventional tillage uses deep-disking and is much less destructive to the physical attributes of soil than traditional tillage that uses the moldboard plow. It is evident from the 30-year record of hydrologic events that surface water movement, whether snowmelt or rainfall, is very similar for both fields. With respect to agrochemicals (see Table 1), nitrogen fertilizer rates are identical; while atrazine loading to the surface is markedly different. While watershed 2 received the same rate of atrazine as Watershed 1, it received it on slightly less than half of its acreage. Figure 5a and b shows the runoff hydrographs from watersheds 1 and 2 with herbicides concentration data for the period of March 2-7, 1993, observed as hour of year (HOY).

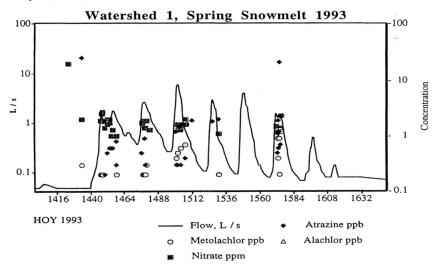

Figure 5a. Snowmelt Runoff Hydrographs from Watershed 1 with Fitted Nitrate and Herbicides Concentration Data for the Period March 2-7, 1993

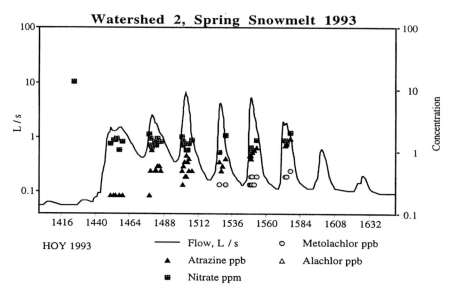

Figure 5b. Snowmelt Runoff Hydrographs from Watershed 2 with Fitted Nitrate and Herbicides Concentration Data for the Period March 2-7, 1993.

While the number of data points acquired for nitrate is somewhat greater for watershed 2 than for 1, no significant differences are observed for nitrate concentration in water moving offsite. This suggests that the two watersheds are responding very similarly to the combined effects of tillage, N fertilization, and runoff hydrology with respect to nitrate concentrations in snowmelt. For the herbicides the differences in watershed response are more pronounced. Ten months following the most recent application, atrazine and metolachlor are detected in snowmelt runoff from both watersheds. Herbicides were detected in each of 6 consecutive daily events, with the exception of the 5th day where the sampler failed for watershed 1. In contrast, atrazine was detected each day while metolachlor is detected during only 3 consecutive daily events, and those all later in the week. While the atrazine concentrations observed for watersheds 1 and 2 are most likely related to the May 1992 application, the source of the metolachlor is not necessarily the application made the previous May. 1992 marked the 16th year for application of atrazine on these fields, and the 12th year for application of metolachlor. Laboratory studies with soil microorganisms indigenous to loess materials suggests that biodegradation occurs more slowly for acetamide herbicides than it does for triazine herbicides. Although the rate constants for the component processes may be different, it appears that for both chemicals the adsorbed herbicide is mobilized and liberated by the combination of the thermodynamics of the system and by the kinetics of the adsorption/desorption process during the freeze/thaw cycle. Alternatively, adsorbed metolachlor may be released from the sorbed soil fraction by the energy associated with the physical process of sheering of soil particles associated with the expansion of water as it alternately freezes and thaws. Generally, the half-life in silt and silt loam soils for metolachlor is 4 weeks compared to 6-10 weeks for atrazine (17). The turbulent mixing which results in desorption/diffusion can influence the timing and amount of agrochemical released. Conventional tillage, in contrast to conservation tillage practices, produces more sheet rill erosion (18); thus, a more direct influence on concentration. For the herbicides, dilution resulting from very high flow tends to diminish diffusion/dispersion effects and not affect concentrations in runoff. This is not observed for the nitrate ion, apparently due to differences in sorptive affinity for nonpolar herbicides compared nitrate. For each daily runoff event, herbicide concentrations increase as flows increases. This is a function of the kinetics of mixing and stirring rather than to the diffusion depth, which is related to the depth of the thawed soil.

Ridge-Tillage. Watershed 3 is a ridge till system with 30-50% plant residue remaining on the surface following Fall harvest. In this system, small micro-terraces (96 cm width ridge to ridge, 25 cm depth) are laid out along landscape contours across the field. Figure 6 shows the runoff hydrograph from watershed 3 with fitted nitrate and herbicide concentration data for the same period. While desirable for reducing erosion, the ridge-till practice imposes unusual hydrologic conditions. As ridges fill with water, infiltration increases, and, at steeply sloping positions, inter-flow between ridges occurs. This is one factor responsible for the higher nitrate concentrations in snowmelt runoff from watershed 3 in comparison to watersheds 1 and 2. Another may be related to the influence of the soil microbial biomass, specifically, aerobic heterotrophs, which decompose and mineralize plant material organic matter, with the formation of soil nitrate. As the runoff peaks, nitrate concentration is diminished as a result of the reduced availability of diffusible or dispersible material entrained in the frozen soil leaving a limited mass remaining after the initial pulse. During time intervals between the peak events the soil thawed snowmelt is pooled between ridges across the watershed. This impounded water is enriched with nitrate diffused from the thawed soil. The snow acts as an insulator by slowly and steadily increasing the soil temperature with each succeeding diurnal freezing. This sequence of events does not occur on watersheds 1 or 2, where there is no pooling of water.

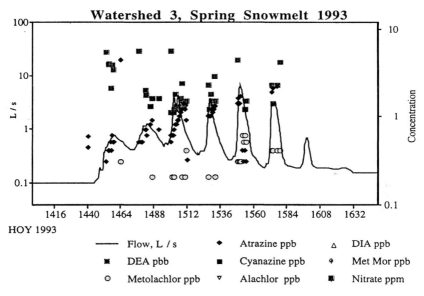

Figure 6. Snowmelt Runoff Hydrograph from Watershed 3 with Fitted Nitrate and Herbicides Concentration Data for the Period March 2-7, 1993.

Among the 4 watersheds monitored, watershed 3 was the most extensively sampled during the 5 days of study. Herbicides included atrazine, cyanazine, metolachlor, and alachlor. Atrazine-metabolites, desethylatrazine and desisopropylatrazine, along with the metolachlor degradate, morpholino-metolachlor, were also measured .Other studies on watershed 3 have shown that atrazine is metabolized readily to DEA and DIA. Both of these compounds leach through the soil profile to be detected in pore water below the root zone and in well water (*19*). Atrazine application on watershed 3 is slightly higher than watersheds 1 and 2; consequently, more atrazine, which was previously not leached, microbially degraded, or adsorbed to the soil matrix, is displaced in runoff. The long-term ridge till system, with its higher crop residue compared to conventional till, produces higher levels of organic matter over time, with the potential to retain sorbed atrazine. This indicates that the most recent previous application and, to a lesser degree, earlier applications are responsible under the freeze/thaw process for having greater potential for re-releasing atrazine in higher concentrations than the conventional till practices on watersheds 1 and 2. Crop residue on the surface may reduce soil temperatures sufficiently to retard microbial degradation of the herbicides, thus making more herbicide mass available for early spring runoff and leaching. The contrast between the nitrate and atrazine concentrations measured indicates that diffusion/dispersion processes and/or sorption kinetics operate more slowly for atrazine than for nitrate. Atrazine concentrations increase with increasing runoff flow rate corresponding to an increasing thawed volume of soil. This pattern repeats itself each day of the 6 day study. Lower concentrations for metolachlor detects on watershed 3 compared to watersheds 1 and 2 results from the lower application rate applied on watershed 3, approximately one-third of that applied on watersheds 1 and 2 during the same time period.

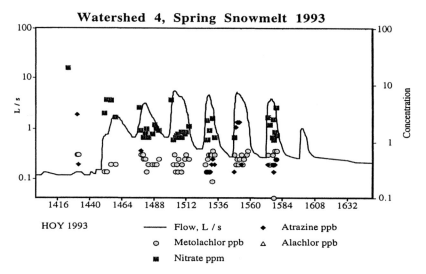

Figure 7. Snowmelt Runoff Hydrograph from Watershed 4 with Fitted Nitrate and Herbicides Concentration Data for the Period March 2-7, 1993.

Terraced Ridge-Tillage. Watershed 4 is a terraced landscape with ridge till management. Surface runoff in the terraces is drained by a tile at the lowest corner of each backslope drainage. Water is pooled whenever maximum flow from this tile is exceeded in the back slope. Consequently, the runoff hydrograph shown in Figure 7 has wider peaks and less recessive baseflow in comparison to all other watersheds. This also substantially changes the chemical movement profiles. Pooling of water also influences the concentration of both herbicides and nitrate moving off-site. The diurnal cycle produces an increase in concentration as the tile becomes a direct conduit. As pooling occurs, the available chemical mass is diluted at maximum flow. As the flow subsides, mobilized chemical results in a concomitant increase in concentration. The net result is an averaging of concentration over the daily runoff event which is more pronounced than that observed for the other watersheds. Furthermore, it is site specific in that it is characteristic of a particular terrace landscape, the number of terraces on a slope, and the drainage profile for each terrace.

Desorption Characteristics. Each watershed exhibits compound-dependent concentration profiles for which movement can be described by the advection-diffusion expression given in equation (1)(20).

$$\left[\frac{\partial C}{\partial t} + u\frac{\partial C}{\partial x} + v\frac{\partial C}{\partial y} + w\frac{\partial C}{\partial z}\right] = \frac{\partial}{\partial x}\left(E_x \frac{\partial C}{\partial x}\right) + \frac{\partial}{\partial y}\left(E_y \frac{\partial C}{\partial y}\right) + \frac{\partial}{\partial z}\left(E_z \frac{\partial C}{\partial z}\right) - KC \quad (1)$$

In equation (1) u, v, w = velocity components (m hr^{-1}) in three coordinate directions x, y, z (m) ; C = concentration (µg L^{-1}) in the turbulent flow; E_x, E_y, E_z = nonisotropic, nonhomogeneous turbulent diffusivities in the three directions. It is assumed that there is a first-order constant with coefficient K (µg L^{-1}) and where t = time. For a point in time on a plane surface the solution to this expression is given in equation (2).

$$C(x,t) = -Kt + \frac{\text{Exp } M}{2A\sqrt{\pi}\sqrt{tE_L}} - \frac{(tU - x + x_1)^2}{4tE_L} \quad (2)$$

M = the instantaneous mass of compound (µg)
E_L = longitudinal dispersion coefficient (m^2 hr^{-1}); E_L assumed small
U = average velocity of the stream (m hr^{-1})
A = the cross-sectional area (m^2)
$x + x_1$ = distance traveled from plane source

The velocity vector is parellel to the x axis; x is a point in a plane and water is initially clean (C=0 at t=0). By dropping the last two terms in equation (2) when E_L is small, the advection in this series of runoff events is linear first order. Thus, we observe different advection-desorption coefficients (K) related to the each compound. The desorption coefficient (K) characteristics were tested as linear, and explored in nonlinear Freundlich, and Langmuir form equations. Each equation was tested for best goodness of fit, and, in general, the linear model provided best correlation for watersheds 1, 2, and 3. This indicates the empirical solution models the observed concentration distributions. The Freundlich equation assumes that concentration associated with both the solid phase and the liquid phase are known instantaneously. This is rarely the case in field studies and was not the case here. However, the results given in Table II from field observations did lead to time-dependent desorptive-phase distribution relationships for (K); not to be confused with a Freundlich isotherm determination.

Table II. Time Dependent Desorption-Phase Distribution Coefficients for Atrazine and Metolachlor Release from Watersheds 1-4.

Watershed	Atrazine		Metolachlor	
	K(t)	Std Error	K(t)	Std Error
1	8.57E-3	2.35E-3	2.45E-3	8.71E-4
2	8.5E-3	9.14E-4	3.2E-3	9.17E-4
3	9.75E-3	1.41E-3	3.10E-3	8.79E-4
4	2.66E-3*		5.78E-5*	

* E_L becomes dominant factor at U

Concurrently the average depth of thawed soil is changed by daily thermal heating. Equation (3) relates the average depth of thaw over this period of daily heating to the hyperbolic function.

$$\text{Depth} = a*t / (b + t) \quad (3)$$

a = continuously thawing depth (cm)
b = average thawing/refreeze rate constant of soil (cm hr^{-1})
t = time from initial diurnal event start (hr)

During these six daily events refreezing occurred to a depth of 3.02 cm (**a**) and at a thaw rate constant of 0.558 cm hr^{-1} (**b**). For this case, the thermal hysteresis of the freeze-thaw cycle is averaged over the period until the soil was completely thawed. At that point contribution from a lower depth to the chemical load is negligible. This indicates the depth of soil and the magnitude of the advective substrate become the limiting factors for chemical release.

The terraced field at watershed 4 produced kinetic results similar to a batch equilibrium, where time dependence becomes significant in the third term of equation (2). The dispersion E_L and rate of runoff velocity U in equation (2) interact with **K**, which dominates at lower flow at the beginning and end of each hydrograph runoff event. The longitudinal dispersivity (E_L) is responsible for dilution at high flow velocity (U). The final events tend to remove concentrations nearly equivalent to prior events. This may relate directly to the general concentration distribution in the soil, where typically the highest concentration for atrazine and metolachlor is within the top 30 cm of the soil surface.

Watersheds 1, 2, and 3 have similar water profiles. Application rates for atrazine and metolachlor on watersheds 1 and 2 were the same. In 1992, watershed 3 received about one-fourth of the metolachlor applied to watersheds 1 and 2. For atrazine, the desorptive-phase distribution relationships for **K** varies between watersheds, but not statistically significantly. This is probably attributed to the gradient slope factor differences between the conventional tillage watersheds and the ridge tillage. The rate of metolachlor concentration magnitude **K** change is less to that of atrazine. In this case, differences are statistically significant for each watershed. The rate of desorption is generally slower for metolachlor and increases as soil volume is thawed. Relating this to the thaw depth would concurrently indicate the possibility for more labile metolachlor to be released from a greater depth than that for atrazine. These field runoff events indicate higher desorption rate **K** for atrazine than for metolachlor **K**. This is also validated by the affinity for absorption K_{OC} for atrazine (39-155) and metolachlor (120-307) *(21)*.

Summary

Spring snowmelt transporting agricultural contaminants in runoff contributes to degrading surface water quality. Management practices directly influence temporal concentrations moving to the edge of the field and off site. During five days of repetitive events, the nitrate-N and herbicide displacement patterns are different for the conventional vs the ridge-till fields. Daily changes in the frostline depth together with the residual legacy of chemical residues from previous year's applications of herbicides account for differences in the mass of each pesticide made available for transport. For the nitrate ion, residual fertilizer together with a contribution from the nitrification of organic nitrogen in soil or plant residue, account for the nitrogen profiles observed. Early spring runoff events may reduce levels of residual herbicide sufficiently to minimize or prevent crop damage during spring planting the following year.

Acknowledgments

We wish to thank Dr. John H. Prueger and his staff for installation of sensors with dataloggers for meteorological and soils data. We are indebted to Larry A. Kramer and Michael D. Sukup for sample collection. We thank Dr. Richard L. Pfeiffer, Ms. Amy Morrow, and the staff of the analytical services unit of the National Soil Tilth Laboratory for assistance with analyses.

Disclaimer

Mention of specific products, suppliers, or vendors is for identification purposes only and does not constitute an endorsement by the U.S. Dept. Of Agriculture to the exclusion of others.

Literature Cited

1. Baker, D. B. and Richards, R. P. In *Long Range Transport of Pesticides*; Kurtz, D. A., Ed.; Lewis Publishers: Chelsea, MI., 1990; pp. 241-270.
2. Thurman, E. M.; Goolsby, D. A.; Meyer, M. T. and Kolpin, D. W. *Environ. Sci. Technol.*, **1991**, *25*, 1794-1796.
3. Thurman, E. M.; Goolsby, D. A.; Meyer, M. T.; Mills, M. S.; Pomes, M. L. and Kolpin, D. W. *Environ. Sci. Technol.*, **1992**, *26*, 2440-2447.
4. Goolsby, D. G.; Boyer, L. L. and Mallard, G. E. In *Selected Papers on Agricultural Chemicals in Water Resources of the Midcontinental United States*. USGS Open-File Report 93-418, U. S. Geological Survey, Denver, CO., 1993, pp. 89.
5. Schottler, S. P.; Eisenreich, S. J. and Capel, P. D. *Environ. Sci. Technol.*, **1994**, *28*, 1079-1089.
6. Schottler, S. P. and Eisenreich, S. J., *Environ. Sci. Technol.*, **1994**, *28*, 2228-2232.
7. Steinheimer, T. R.; Kramer, L. A. and Scoggin, K. D. *Environ. Sci. Technol.*, **1998**, *32*, 1039-1047.
8. Steinheimer, T. R.; Kramer, L. A. and Scoggin, K. D. *Environ. Sci. Technol.*, **1998**, *32*, 1048-1052.
9. Prior, J. C. *Landforms of Iowa* ; University of Iowa Press, Iowa City, IA., 1991, pp. 48-57.
10. *Land Resource Regions and Major Land Resource Areas of the United States*, USDA Agriculture Handbook 296. U. S. Soil Conservation Service, Washington, D.C. 1981, pp. 75-79 and 133.
11. *QuikChem Method 10-107-04-1-E*, Lachat Instruments, Milwaukee, WI.; adapted from USEPA Method 353.2, in *Methods for the Chemical Analysis of Water and Wastes*, USEPA, Washington, D.C., 1992.
12. Steinheimer, T. R. *J. Agric. Food Chem.* **1993**, *41*, 588-595.
13. Koskinen, W. C.; Jarvis, L. J.; Dowdy, R. H.; Wyse, K. L.; and Buhler, D. D., *Soil Sci. Soc. Am. J.*, **1991**, *55*, 561-2.
14. Pfeiffer, R. L., In *Proceedings of the International Symposium on Laboratory Automation and Robotics*, Zymark Corporation, Hopkinton, MA., 1992, 531-542.
15. Ahuja, L. R., In *Advances in Soil Science*, B. A. Stewart, Ed., Volume 4, Springer-Verlag, New York, NY, 1986, 149-182.
16. Smith, S. J.; Sharpley, A. N. and Ahuja, L. R. *J. Environ. Qual.*; **1993**, *22*, 474-80.
17. *The Agrochemicals Handbook*, Third Edition, Update 5, The Royal Society of Chemistry, Cambridge, England, 1994.
18. Kramer, L. A. and Hjelmfelt, Jr., A. T., In Proc. Effects of Human-Induced Changes in Hydrologic Systems, Am. Water Res. Asso., June 24-26, Jackson

Hole, WY., 1994, pp. 1097-1106 ; Spomer, R. G. and Hjelmfelt, Jr., A. T., **1986**, Trans. ASAE, 29(1), 124-7.
19. Steinheimer, T. R. and Scoggin, K.D.; *J. Environ. Monit. ; in press.*
20. Hubber, Wayne C. In *Contaminant Transport in Surface Water: Handbook of Hydrology*, Maidment David R. Ed. McGraw-Hill Inc. 1993, 14.16-14.46.
21. CIBA-GEIGY CORPORATION DATA; USDA-ARS, Remote Sensing and Modeling Lab Pesticide Data Base, (1989)

Chapter 11

Runoff Losses of Suspended Sediment and Herbicides: Comparison of Results from 0.2- and 4-ha Plots

L. M. Southwick[1], D. W. Meek[2], R. L. Bengtson[3], J. L. Fouss[1], and G. H. Willis[1]

[1]Soil and Water Research Unit, Agricultural Research Service, U.S. Department of Agriculture, 4115 Gourrier Avenue, Baton Rouge, LA 70808
[2]National Soil Tilth Laboratory, Agricultural Research Service, U.S. Department of Agriculture, 2150 Pammel Drive, Ames, IA 50011
[3]Department of Biological and Agricultural Engineering, Louisiana State University Agricultural Center, Baton Rouge, LA 70803

>Over the last dozen years we have conducted field studies of runoff losses of herbicides and suspended sediment. In this paper we investigate the possible influence of plot size on our results by comparing data from plots of 0.21 and 4.4 ha. At the areal extent of our studies, this assessment fails to reveal a consistent plot size influence on runoff losses of atrazine and metolachlor, chemicals of moderate to high water solubility. The data may indicate more efficient extraction from soil into runoff of these herbicides in the larger plots. The runoff results for trifluralin and pendimethalin, of low water solubility, may indicate a plot size influence but also can be explained by a year effect related to application differences. Sediment yield in runoff has been sensitive to the plot size differences in our work: from the smaller plots we have observed three times larger sediment yields.

In investigations and descriptions of the contribution of runoff from agricultural fields to nonpoint source pollution problems, the effect of study area size on the applicability of results at the watershed and basin scale becomes an important concern. Spatial variability of rainfall and other weather factors, soil characteristics, topography, geology, vegetation, and drainage patterns lead to uncertainties when applying results from field and small plot studies to larger geographical areas (*1, 2*)
  In studies of pesticide concentration patterns in the Lake Erie Basin starting in 1981, Baker and coworkers (*3*) have made the following scale effect observations: peak observed concentrations increase as watershed size decreases; and the average length of time during which intermediate pesticide concentrations are continuously exceeded tends to increase with watershed size. Richards and Baker state (*3*, p. 21) that scale effects "can be expected to be present all the way down to the plot scale."

© 2000 American Chemical Society

An additional scale effect that has been reported is that sediment yield (sediment/unit area) decreases as watershed size increases (4, 5).

Since 1985 we have been conducting field studies of runoff losses of various herbicides and insecticides, nitrate, and suspended sediment. Our early work was performed on 2- to 4-ha plots (6); since 1994 we have carried out investigations on 0.2-ha study areas (7). The earlier larger plot work was done on Commerce clay loam (fine-silty, mixed, nonacid, thermic Aeric Fluvaquents), a Mississippi River alluvial soil graded to 0.1% slope; the later smaller plot work was performed 0.2-0.5 km away on Commerce silt loam (fine-silty, mixed, nonacid, thermic, Aeric Fluvaquents) graded to 0.2 % slope. Corn was the crop in both cases. In this paper we compare/contrast runoff results of herbicides (application rates are listed in Table I) and suspended sediment in order to test for trends that might be assignable to the differences in plot sizes in our field work. We look at results for atrazine and metolachlor in runoff from 4.4-ha plots in 1987 (6) and from 0.21-ha plots in 1995 (8) and 1996. In addition, we compare/contrast runoff of trifluralin from the 4.4-ha plots in 1992 (9) with results for pendimethalin from the 0.21-ha plots in 1996. We also discuss the suspended sediment losses in runoff observed in these studies (8, 10).

Table 1. Herbicide Application Rates

Herbicide	Year	Rate*
Atrazine	1987	1.63†
	1995	0.80†
	1996	1.49†
Metolachlor	1987	2.16†
	1995	1.00†
	1996	1.91†
Trifluralin	1992	1.12‡
Pendimethalin	1996	0.92†

* kg/ha; † unincorporated; ‡ incorporated

Table II. Herbicide Properties*

Herbicide	S_w	K_{oc}	V_p	$t_{1/2}$	Ref.
Atrazine	33	100	2.9×10^{-7}	35	15, 16
Metolachlor	530	200	3.1×10^{-5}	23	15, 16
Pendimethalin	0.3	5000	9.4×10^{-6}	55	8, 15
Trifluralin	0.3	8000	1.1×10^{-4}	31	9, 15

* S_w, water solubility, mg/L; K_{oc}, soil organic carbon sorption coefficient, mL/g; V_p, vapor pressure, mm Hg; $t_{1/2}$, soil half life, days.

Although this work has been conducted on plots with and without subsurface drains, we consider in this paper results only from plots without subsurface drains. An additional interest in our comparison is an effort to assess the size of an "elemental area" (*11*) in these field studies. We and others have reported work on smaller field plots (*12*, 0.0013 ha; *13*, *14*, 0.00070 ha). Table II lists relevant properties of the chemicals treated in this paper.

Concentrations of Atrazine and Metolachlor in Runoff

Disappearance of pesticides from soil normally shows an exponential decrease with time and often approximates to first order kinetics, even though the mechanism of disappearance is a complicated mixture of physical and chemical processes such as irreversible sorption to the soil, volatilization, leaching, runoff, and chemical/microbiological decompositon (*17, 18*). Similarly, concentrations of pesticides in runoff also diminish exponentially with time, reflecting the corresponding decrease in concentration in the runoff-active zone of the soil (19).

Atrazine. Our field studies with atrazine have routinely shown that the concentration of this herbicide in runoff quickly drops with increase in elapsed time after application (Figure 1). This disappearance of atrazine in runoff closely fits modified first order

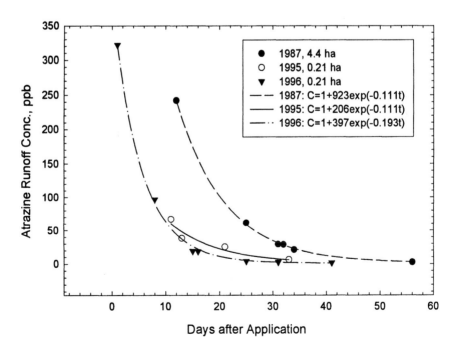

Figure 1. Concentration of atrazine in runoff.

decay curves (Table III). [All reported regression results are based on the means for treatment or day. The generalized least squares (GLS) method (20) was used for the regressions of Table III. This table reports results for one model for all years, starting with all parameters different for each year and simplifying based on testing corresponding parameters for equality. The resultant weight models were generally consistent with a Poisson or lognormal error structure.] These equations predict that at t = 0, the initial concentration is (a + b) μg/L; k is the first order rate constant; and as t increases, these equations predict that C_{ro} approaches (a) μg/L. The equations allow calculation of DT_{50}s (50% disappearance times) for the chemical in runoff by solving for t when $C_{ro} = 0.5(C_{ro, t=0})$. Runoff DT_{50} values in Table III correspond to soil DT_{50}s of 35 (16, 1987), 18 (8, 1995), and 15 (21, 1996) days. In our field work, we routinely observe that DT_{50} (runoff conc.) < DT_{50} (soil conc.). This observation is reasonably due to rapid leaching of the runoff-available residue to just below the runoff extraction zone (22-24) but not as quickly below the zone removed in soil sampling procedures (in our case, usually the top 2.5 cm soil layer). The analyses reported in Table III yield identical rate constants k for 1987 and 1995; consequently, the DT_{50}s are the same for these two study seasons. The plot size differences between these two studies did not influence the observed persistence of atrazine in runoff.

Table III. Atrazine Concentration in Runoff*
Model: C_{ro} = a + bexp(-kt)
Parameters, (T Values), and [95% Confidence Intervals]

Year	a	b	k	R^2	DT_{50}, days†
			4.4 ha		
1987	1.00 (4.41) [0.50, 1.50]	923 (55.1) [705, 1210]	0.111 (22.6) [0.101, 0.122]	0.986	6.23 [5.68, 6.89]
			0.21 ha		
1995	1.00 (4.41) [0.50, 1.50]	206 (50.7) [164, 259]	0.111 (22.6) [0.101, 0.122]	0.986	6.23 [5.68, 6.89]
1996	1.00 (4.41) [0.50, 1.50]	397 (77.5) [336, 470]	0.193 (23.8) [0.175, 0.211]	0.986	3.59 [3.29, 3.95]

* C_{ro}, conc., μg/L; t, days after application.
† 50% Disappearance time.

Metolachlor. As with atrazine, our field work has also shown rapid decreases of metolachlor runoff concentration with time after application (Figure 2). The

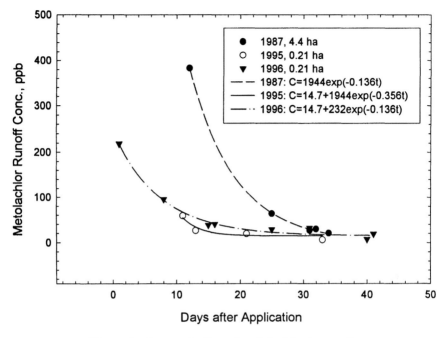

Figure 2. Concentration of metolachlor in runoff.

disappearance curves of Figure 2 fit modified first order equations (Table IV) that provide DT_{50}s that are shorter than the respective soil DT_{50} values. [The results of Table IV were obtained by the statistical procedures of Table III.] The corresponding metolachlor soil DT_{50}s were 20 (*16*, 1987), 29 (*8*, 1995), and 17 (*21*, 1996) days. As with the results for atrazine in Table III, the metolachlor results of Table IV show a similarity in rate constant k (and therefore in persistence in runoff) across plot size (1987 and 1996).

Concentrations of Atrazine and Metolachlor in Runoff as Functions of Soil Concentrations

Leonard et al. (*19*) developed a power equation, $Y = 0.05X^{1.2}$, $R^2 = 0.86$, to describe the relation between runoff concentrations (Y) of various herbicides transported in the water phase and their respective soil surface concentrations (X). These investigators

Table IV. Metolachlor Concentration in Runoff[*]
Model: $C_{ro} = a + b\exp(-kt)$
Parameters, (T Values), and [95% Confidence Intervals]

Year	a	b	k	R^2	DT_{50}, days[†]
		4.4 ha			
1987	0	1944 (84.2)	0.136 (22.3)	0.988	5.12
		[1598, 2364]	[0.122, 0.149]		[4.66, 5.67]
		0.21 ha			
1995	14.7 (5.74)	1944 (84.2)	0.356 (17.2)	0.988	1.95
	[9.14, 20.32]	[1598, 2364]	[0.311, 0.400]		[1.73, 2.23]
1996	14.7 (5.74)	232 (94.4)	0.136 (22.3)	0.988	5.12
	[9.14, 20.32]	[205, 263]	[0.122, 0.149]		[4.66, 5.67]

[*] C_{ro}, conc., µg/L; t, days after application.
[†] 50% Disappearance time.

Table V. Atrazine and Metolachlor Concentration in Soil and Runoff[*]
Model: $C_{ro} = a + bC_s^3$
Parameters, (T Values), and [95% Confidence Intervals]

Year	ha[†]	a	b	R^2
			Atrazine	
1987	4.4	0	0.0626 (8.69)	0.986
			[0.045, 0.080]	
1995	0.21	0	0.0874 (8.15)	0.985
			[0.0532, 0.122]	
1996	0.21	0	0.0369 (17.1)	0.990
			[0.0317, 0.0422]	
			Metolachlor	
1987	4.4	-0.0326 (5.34)	0.00650 (18.7)	0.987
		[-0.0521, -0.0132]	[0.0054, 0.0076]	
1995, 1996	0.21	0.0159 (2.83)	0.00260 (19.6)	0.990
		[0.0034, 0.0284]	[0.00230, 0.00289]	

[*] C_{ro}, runoff conc., µg/L; C_s, soil conc., µg/kg. [†] Plot size, ha.

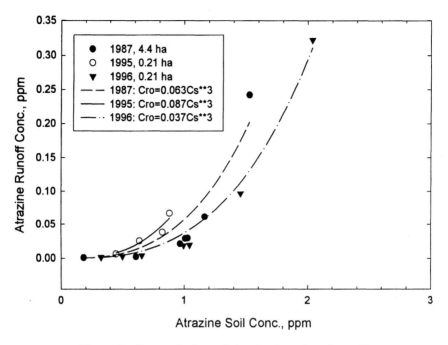

Figure 3. Concentration of atrazine in soil and runoff.

considered the coefficient to be an "extraction coefficient" and suggested that greater distance from linearity shown by the exponent to reflect lower extraction efficiency with increasing time after application. That is, runoff extraction was more efficient early after application when soil concentrations were at their highest. If aged soil residues become more tightly bound or degraded, extraction becomes less efficient and hysteresis is observed (25, pp. 65-66; 26). From our field data sets we have related the runoff concentrations of atrazine (Figure 3) and metolachlor (Figure 4) to their respective soil surface concentrations and have developed power regression equations (Table V) to describe these relationships.

The analyses of Table V reveal a trend toward statistically significant differences (the 1995 atrazine data are not consistent here) between the extraction coefficients b with respect to plot size and/or year. [Table V results were based on linear measurement error regression (27).] For both herbicides the extraction coefficient is lower with the smaller plot size. For atrazine, $b_{0.21} = 0.59 b_{4.4}$ (1996 results compared to 1987); for metolachlor, $b_{0.21} = 0.40 b_{4.4}$ (1995 and 1996 data compared to 1987). If these trends are due to the plot size differences, the longer runs of the larger plots before the samples passed through the sampling flumes may have led to increased extraction efficiency from the runoff-active zone of the soil. On the larger plots there is greater time for the herbicides to desorb from the soil. Higher runoff concentrations of atrazine and metolachlor from the larger plots as a function of time after application are indicated in Figures 1 and 2. Additionally, the extraction coefficients are consistently higher for atrazine than for metolachlor: the 10-fold

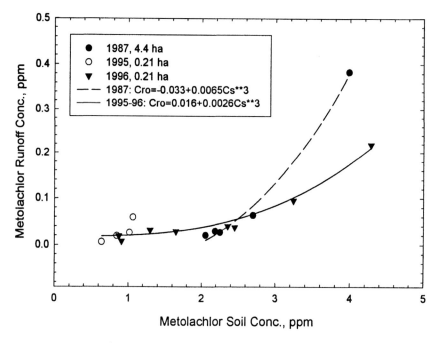

Figure 4. Concentration of metolachlor in soil and runoff.

differences (within same year, atrazine > metolachlor) seen in Table V are consistent with the two-fold differences in K_{oc}s (Table II, metolachlor > atrazine).

Trifluralin and Pendimethalin Yield in Runoff

In 1992, we investigated runoff of the dinitroaniline herbicide, trifluralin. Our 1996 work included runoff of another dinitroaniline, pendimethalin. Since both water solubilities (0.3 ppm) and K_{oc}s (8000 mL/g, trifluralin; 5000 mL/g, pendimethalin) are the same or similar for the two compounds (Table II), we have compared the yield in runoff as a function of flow and have developed regression equations for these observed relationships (Table VI). [The statistical treatment for Table VI was similar to that for Table V.] Yield is expressed as percent of application since the dinitroaniline application rates for the two years were slightly different (Table I).

Table VI. Yields of Trifluralin (1992) and Pendimethalin (1996) in Runoff*
Model: $Y = a + bV^{1/2}$
Parameters, (T Values), and [95% Confidence Intervals]

Year	ha	a	b	R^2
1992	4.4	-0.0104 (7.12) [-0.0150, -0.0058]	0.00412 (15.7) [0.00329, 0.00495]	0.987
1996	0.21	-0.0140 (2.97) [-0.141, -0.0139]	0.00765 (7.72) [0.00522, 0.0101]	0.826

*Y, herbicide yield, % of appl.; V, runoff flow, mm.

The flow coefficient b (Table VI) is statistically higher for the 1996 (0.21 ha) data than for the 1992 (4.4 ha) results ($b_{0.21\ ha} = 1.86 b_{4.4\ ha}$). This higher coefficient from the smaller plot size is consistent with the application difference between the two years--incorporated in 1992 but not in 1996 (Table I). Trifluralin and pendimethalin show water solubilities and K_{oc}s (Table II) that encourage association with sediment in runoff (22, Fig. 9-8; 24); the greater coefficient b from the smaller plots is also consistent with the larger sediment yields from these plots (see below).

Sediment Yield in Runoff

From our 4.4 ha plots we have nine years of monthly sediment yields in runoff. For a 9-year period, 1988-1996, we have calculated means and standard errors of the means ($\bar{Y}$, yields normalized with respect to flow, kg/ha/mm) for each of the months April, May, June (Figure 5). We have also calculated the respective means and standard errors of sediment yields from the 0.21 ha plots for the 1995 season (mostly in May) and the 1996 season (mostly in April). Comparative box plots (Figure 5) show the differences in the sediment yield/runoff flow data via a graphical method of exploratory data analysis (28). Figure 5 is based on Friendly's adaptation (29, pp. 285-317) of McGill et al. (30).

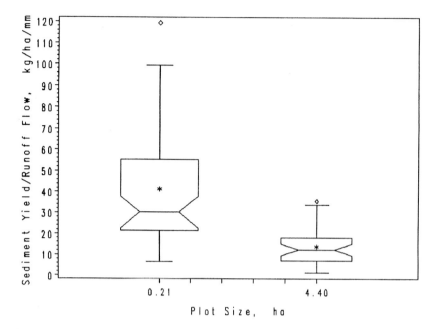

Figure 5. Comparative variable width notched box plots of the sediment yield/runoff for the two plot sizes. Each box represents the middle 50% of the data. The width of each box is proportional to the square root of the number of observations. The asterisks are the respective arithmetic means, the lines within the notches are the medians, and the notches are approximate 95% confidence intervals for the medians. The upper and lower whiskers represent the range of the data for each plot size distribution with the maximum extent not exceeding the median value ±1.5 interquartile range (the length of the box). Points that exceed that limit, the diamonds above the top whiskers, can be considered outliers.

To formally test for a plot size difference in the Y data, an ANOVA was performed. In order to match the method of collection for the larger plot data, the smaller plot yield and runoff data were totaled 30-day periods in just 1995 and 1996 because these were the only years with matching months during the period of record. That is, the appropriate data of 1995 and 1996 from both studies were matched. A traditional completely randomized model was used; the treatment structure was one-way with just two levels, the plot size. The matched values were samples in time resulting in eight data points, four for each plot size. The plot size means were 13.0 kg/ha/mm for the 4.4 ha plots and 41.3 kg/ha/mm for the 0.21 ha plots. The two means are significantly different ($P \leq 0.016$) and give the ratio of mean values:

$$Y_{0.21\,ha} / Y_{4.4\,ha} = 3.2.$$

Summary and Conclusion

In this assessment of our field data, with respect to plot size (0.21 and 4.4 ha), of runoff losses of herbicides (both of moderate and low water solubility) we have observed that this behavior shows trends variably assignable to plot size differences. From both plot areas of our studies, the water soluble compounds atrazine and metolachlor demonstrated rapid concentration declines with increasing time between application and the runoff event. When these runoff concentration data are regressed with time after application (Tables III and IV), the resulting modified first order equations produce the same DT_{50}s across plot size. Regression of atrazine and metolachlor runoff concentrations against soil concentrations (Table V) revealed more effective desorption from the soil in the larger plots.

The herbicides of low water solubility (trifluralin and pendimethalin) also show trends that are consistent with a plot size influence. The probability of this effect is enhanced since the observed trend (Table VI) is in the same direction as that of sediment. But the issue is clouded by the fact that application differences (year effect) are also in line with the trends of Table VI.

Sediment yield in runoff shows a definite trend attributable to plot size. In the analysis of these sediment runoff data, the three 4.4-ha data sets, each extending over a 9-year period, demonstrate a consistently lower runoff sediment yield compared to the two 0.21-ha studies (Figure 5). This difference stands when the 1995 0.21-ha data (collected mostly in May) are compared with the May, 1995 4.4-ha data and when the 1996 smaller plot data (mostly from April) are compared with the larger plot data of April, 1996 (ANOVA).

An elemental agricultural area (*11*, pp. 176-181) will show the following characteristics: uniform soil type, constant vegetation cover, constant overall slope, and uniform distribution of precipitation. In our studies, these conditions are generally met at both plot size levels. The lack of a consistent trend attributable to plot size in our runoff data for atrazine and metolachlor points to a general similarity of the above plot characteristics across the areas of our studies, as these characteristics affect runoff of these compounds of high water solubility. The data indicate that from soil of similar atrazine and metolachlor concentrations, larger concentrations result in a runoff event from the larger plot size. A general breakdown in elemental area across the plot sizes in the work described in this paper shows up in our sediment data and is suggested by the results for the herbicides of low water solubility.

Acknowledgment

The authors are grateful for the statistical assistance of Deborah L. Boykin, USDA, ARS, Stoneville, MS.

Literature Cited

1. Bailey, G. W.; Swank, R. R., Jr. In *Agricultural Management and Water Quality*; Schaller, F. W.; Bailey, G. W., Eds.; Iowa State University Press: Ames, Iowa, **1983**, 27-47.

2. Smith, C. N.; Brown, D. S.; Dean, J. D.; Parrish, R. S.; Carsel, R. F.; Donigian, A. S. Jr. *Field Agricultural Runoff Monitoring (FARM) Manual* (EPA/600/3-85/043); U. S. Environmental Protection Agency: Athens, Georgia, **1985**, 1.
3. Richards, R. P.; Baker, D. B. *Environ. Toxicol. Chem.* **1993**, *12*, 13-26.
4. Johnson, H. P.; Moldenhauer, W. C. In *Agricultural Practices and Water Quality*; Willrich, T. L.; Smith, G. E., Eds; Iowa State University Press: Ames, Iowa, **1970**, 3-20.
5. Gottschalk, L. C. In *Handbook of Applied Hydrology*; Chow, V. T., Ed; McGraw-Hill: New York, New York, **1964**, 17-1 to 17-34.
6. Southwick, L. M.; Willis, G. H.; Bengtson, R. L.; Lormand, T. J. *Bull. Environ. Contam. Toxicol.* **1990**, *45*, 113-119.
7. Willis, G. H.; Fouss, J. L.; Rogers, C. E.; Southwick, L. M. In *Groundwater Residue Sampling Design*; Nash, R. G.; Leslie, A. R., Eds; ACS Symposium Series 465, American Chemical Society: Washington, DC, **1991**, 195-211.
8. Southwick, L. M.; Willis, G. H.; Fouss, J. L.; Rogers, J. S.; Carter, C. E. In *Proceedings of the Twenty-seventh Mississippi Water Resources Conference*; Daniel, B. J., Ed; Water Resources Research Inst.: Mississippi State, Mississippi, **1997**, 239-246.
9. Southwick, L. M.; Willis, G. H.; Mercado, O. A.; Bengtson, R. L. *Arch. Environ. Contam. Toxicol.* **1997**, *32*, 106-109.
10. Bengtson, R. L.; Carter, C. E.; Morris, H. F. In *Drainage in the 21st Century: Food Production and the Environment*; Brown, L. C. Ed; American Society of Agricultural Engineers: St. Joseph, Michigan, **1998**, 530-537.
11. Huggins, L. F.; Burney, J. R. In *Hydrologic Modeling of Small Watersheds*; Haan, C. T.; Johnson, H. P.; Brakensiek, D. L., Eds; American Society of Agricultural Engineers, St. Joseph, Michigan, **1982**, 169-225.
12. Southwick, L. M.; Willis, G. H.; Reagan, T. E.; Rodriguez, L. M. *Environ. Entomol.* **1995**, *24*, 1013-1017.
13. Baldwin, F. L.; Santelmann, P. W.; Davidson, J. M. *Weed Science.* **1975**, *23*, 285-288.
14. Baldwin, F. L.; Santelmann, P. W.; Davidson, J. M. *J. Environ. Qual.* **1975**, *4*, 191-194.
15. Hornsby, A. G.; Wauchope, R. D.; Herner, A. E. *Pesticide Properties in the Environment.* Springer-Verlag, Inc.: New York, New York, **1996**.
16. Southwick, L. M.; Willis, G. H.; Bengtson, R. L.; Lormand, T. J. *J. Irr. Drain. Engin.* **1990**, *116*, 16-23.
17. Guenzi, W. D., Ed. *Pesticides in Soil and Water.* Soil Science Society of America, Inc.: Madison, Wisconsin, **1974**.
18. Sawhney, B. L.; Brown, K., Eds. *Reactions and Movement of Organic Chemicals in Soils.* Soil Science Society of America, Inc.: Madison, Wisconsin, **1989**.
19. Leonard, R. A.; Langdale, G. W.; Fleming, W. G. *J. Environ. Qual.* **1979**, *8*, 223-229.
20. Carroll, R. J., Ruppert, D. *Transformations and Weighting in Regression:*

Monographs on Statistics and Applied Probability. Chapman and Hall: New York, New York, **1988**.

21. Southwick, L. M.; Willis, G. H.; Fouss, J. L. In *Drainage in the 21st Century: Food Production and the Environment.* Brown, L. C., Ed; American Society of Agricultural Engineers: St. Joseph, Michigan, **1998**, 668-675.
22. Leonard, R. A. In *Pesticides in the Soil Environment: Processes, Impacts, and Modeling.* Cheng, H. H., Ed; Soil Science Society of America, Inc.: Madison, Wisconsin, **1990**, 303-349.
23. Leonard, R. A.; Wauchope, R. D. In *CREAMS: A Field-Scale Model for Chemicals, Runoff, and Erosion from Agricultural Management Systems.* Knisel, W. G., Ed; U. S. Dept. of Agriculture: Washington, DC, **1980**, 88-112.
24. Wauchope, R. D. *J. Environ. Qual.* **1978**, *7*, 459-472.
25. Koskinen, W. C.; Harper, S. S. In *Pesticides in the Soil Environment: Processes, Impacts, and Modeling.* Cheng, H. H., Ed; Soil Science Society of America, Inc.: Madison, Wisconsin, **1990**, 51-77.
26. Ma, L., Southwick, L. M., Willis, G. H., Selim, H. M. *Weed Sci.* **1993**, *41*, 627-633.
27. Fuller, W. A. *Measurement Error Models.* John Wiley and Sons, New York, New York, **1987**.
28. Hoagland, D. C; Mosteller, F.; Tukey, J. W. (Eds.). *Understanding Robust and Exploratory Data Analysis.* New York, New York: John Wiley and Sons, Inc., **1983**.
29. Friendly, M. *SAS System for Statistical Graphics.* SAS Institute: Cary, North Carolina, **1996**.
30. McGill, R.; Tukey, J. W.; Larsen, W. *Amer. Stat.* **1978**. *32*, 12-16.

Chapter 12

Estimation of Potential Loss of Two Pesticides in Runoff in Fillmore County, Minnesota Using a Field-Scale Process-Based Model and a Geographic Information System

Paul D. Capel[1] and Hua Zhang[2]

[1]Water Resources Division, U.S. Geological Survey, Minneapolis, MN 55455
[2]Department of Civil Engineering,
University of Minnesota, Minneapolis, MN 55455

In assessing the occurrence, behavior, and effects of agricultural chemicals in surface water, the scales of study (i.e., watershed, county, state, and regional areas) are usually much larger than the scale of agricultural fields, where much of the understanding of processes has been developed. Field-scale areas are characterized by relatively homogeneous conditions. The combination of process-based simulation models and geographic information system technology can be used to help extend our understanding of field processes to water-quality concerns at larger scales. To demonstrate this, the model "Groundwater Loading Effects of Agricultural Management Systems" was used to estimate the potential loss of two pesticides (atrazine and permethrin) in runoff to surface water in Fillmore County in southeastern Minnesota. The county was divided into field-scale areas on the basis of a 100 m by 100 m grid, and the influences of soil type and surface topography on the potential losses of the two pesticides in runoff was evaluated for each individual grid cell. The results could be used for guidance for agricultural management and regulatory decisions, for planning environmental monitoring programs, and as an educational tool for the public.

The movement of agricultural chemicals from their point of application to the broader environment is an area of concern. One approach to assessing the potential for agricultural chemicals to contaminate surface- and ground-water systems is to use process-based simulation models that incorporate the coupled physical, chemical, and biological processes that govern chemical transport (1-6). One of the advantages of process-based simulation models is their capability to predict contaminant transport and concentration in both space and time. The application of these models generally is limited to site-specific studies because they require many specific parameter inputs. For example, the GLEAMS (Groundwater Loading

Effects of Agricultural Management Systems; 4,5) model is designed for an agricultural field-scale area with homogeneous soils, uniform rainfall, and a single crop type and agricultural management practice.

In assessing the occurrence, behavior, and effects of agricultural chemicals in surface and ground waters, the scale of study is usually much larger than the scale of single agricultural fields. At these larger scales (i.e., watershed, county, state, and regional areas), soil types, rainfall amounts, crop types, and management practices can and do vary greatly. Because of this nonhomogeneity, the traditional process-based models cannot be applied legitimately to large areas, but the information provided by process-based models can be useful in the interpretation of chemical observations made at these large scales. The extension of process-based models to larger scales can be accomplished by dividing the large area into many small areas (field-scale) so that the homogeneous conditions required by the process-based models are met, and then evaluating the larger area piece by piece with the process-based models. This method would not be possible by manual techniques, but in recent years, geographic information systems (GIS) have emerged as powerful tools to manipulate, analyze, and display many forms of geographically referenced information. The combination of process-based simulation models and GIS technology can be used to extend our understanding of field processes to help guide agricultural management decisions, plan environmental monitoring programs, and interpret field observations at various scales. Corwin *et al.* (7) have reviewed how the combination of process-based models and GIS have been used in vadose zone and ground-water environments. There have been a number of efforts to combine process-based models with GIS for transport of nutrients (8-10) and sediment (10-13) from agricultural fields to surface waters.

In this chapter, the GLEAMS model (5) was used to estimate the potential loss in runoff of two pesticides (atrazine and permethrin) with contrasting chemical properties by evaluating the spatially distributed influence of soil type and surface topography. All other agricultural and environmental parameters (crop type, management practice, application rate, rainfall) were held constant in order to make the results comparable across the entire county The GIS technology was used to provide site-specific information on soil type and slope needed for the GLEAMS model, and to assign the estimated loss of the pesticides in runoff for agricultural areas of the county.

Methods

Description of the study area. Fillmore County is in southeastern Minnesota. The western part of the county is a gently rolling plain, whereas the eastern part is deeply dissected by streams. The county is drained almost entirely by the eastward-flowing Root River system, except in the southern border areas, where surface water flows to the south (Figure 1). The subsurface is predominately karst. Fillmore County has a typical continental climate characterized by a wide variation in temperature, scanty winter precipitation, and normally ample summer rainfall.

The average annual temperature is 4.5°C, with the coldest month in January (-10.8°C) and warmest month in July (22°C). The average annual rainfall is 80 cm (14).

Fillmore County relies mainly on agriculture for its economy. In 1992, out of a total area of about 2225 km², 650 km² were used for the production of corn, 190 km² acres for soybeans, and 90 km² for oats (15). About 82.4% of the county was classified as agricultural land, 16.6% as forested land, and 0.7% as urbanized land. The agricultural land was targeted in this study.

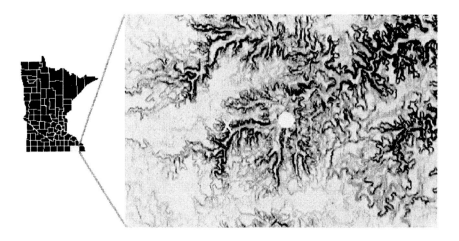

Figure 1. Location and topography of Fillmore County, Minnesota. The darker shades on the map indicate steeper slopes. The white dot in the center shows the location of Preston, Minnesota. The white "X" is the sampling location of Forestville Creek.

Sources of environmental data. Three geographic databases were used in the study: State Soil Geographic Database (STATSGO; 16), Digital Elevation Model data (DEM; 17), and Land Use and Land Cover data (LCLU; 17).

The STATSGO database (1:250000 scale) is the digital form of soil maps that are linked to a database. It was developed by the Soil Conservation Service (SCS, currently called the Natural Resources Conservation Service) and was designed for regional and statewide natural-resource monitoring, planning, and management (16). The STATSGO map units represent polygons of soil association. Two major portions of the STATSGO database -- soil components and soil layers -- were used in this study. The soil-component table describes 60 soil properties of the soil phase. The soil-layer table includes 28 soil properties for each soil layer at a specific depth.

The DEM database was derived from the U.S. Geological Survey's digital topographic maps (1:250000 scale). The DEM grid of Fillmore County was used to derive the slope value for each cell.

The LULC database describes the vegetation, water, and natural and cultural features on the land surface (1:250000 scale). This land-use coverage is based on the Anderson classification of nine major land-use categories. Each major land-use category is further divided into secondary categories (17).

Daily temperature and precipitation values from 1992 to 1994 at Preston, Minnesota (Figure 1), were used (State Climatology Office, Minnesota Department of Natural Resources, personal communication). A 3-year period was selected to help assure representative results for average annual water runoff, as well as pesticide losses. Monthly mean radiation was selected from the GLEAMS users manual for Minneapolis, Minnesota (5).

Data preparation and model calculations. The GLEAMS model was developed by scientists from the U.S. Department of Agriculture, Agricultural Research Service during the mid-1980s (4,5). The GLEAMS model was designed to include the major processes that potentially affect water, soil, nutrient, and pesticide transport through the plant root zone into the subsurface and through surface runoff from a field-size management unit. GLEAMS consists of four separate model components: hydrology, erosion, nutrients, and pesticides.

The hydrology component of GLEAMS uses daily climatic data to calculate the water balance. Precipitation is partitioned between surface runoff and infiltration into the soil surface. The SCS curve-number method is used to estimate runoff. A storage-routing technique is used to simulate redistribution of infiltrated water within seven computational layers in the plant-root zone. Water evaporation and plant transpiration are estimated with a modified Penman equation. The soil erosion component of the GLEAMS model uses a modified version of the Universal Soil Loss Equation for storm-by-storm estimations of rill and interill erosion in overland flow areas. Eroded soil is routed with runoff according to particle size. The sediment yield in runoff is used to estimate the transport of sorbed pesticides.

The pesticide component of GLEAMS uses the sorption characteristics of pesticides to calculate their distributions between the solution and sorbed phases. The model assumes an instantaneous equilibration between solution and sorbed phases. The model also considers pesticide transformation and upward movement by evaporation and plant uptake.

The crop type selected in this study was corn, which is the major crop in Fillmore County. The planting season was assumed to begin on May 1 of each year. The initial fraction of plant-available water in soils was assumed to be 0.50 for all simulations. The effective rooting depth was set at 56 cm. Atrazine and permethrin were selected as example chemicals because of their contrasting properties of water solubility and soil sorption. Soil dissipation half-lives differ by about a factor of two. Atrazine is relatively soluble and more persistent in soil, whereas permethrin strongly sorbs to soil particles and is less persistent. The pesticide application rates were assumed to be 2.0 kg/ha. The application of the chemicals was assumed to occur on July 1 of each year.

The GLEAMS model is designed for field-size areas. Therefore, to apply the model to Fillmore County, the county was divided into small areas (cells) by the GIS system. Each cell is a square with a length of 100 m. The cell grid contained a total of 222,336 cells for the whole county.

The slope, expressed in rise over run, was derived from the DEM database using the GIS program (18). About 1393 km^2 of the county have a slope that ranges from 0 to 0.1 (rise/run, 0 to 6 degrees). Only 10.4 km^2 of the county have a slope greater than 0.70 (35 degrees). On the basis of these results, nine representative slopes (0.0001, 0.001, 0.01, 0.05, 0.1, 0.25, 0.5, 0.75, and 1.0) were chosen for evaluation by GLEAMS.

The spatial distribution of the soils in Fillmore County was derived from the STATSGO database (16). There were 26 STATSGO mapping units present in Fillmore County. The first soil layer (top soil) of each STATSGO soil component was used for the GLEAMS model simulation. This soil layer is up to 61 cm in depth, corresponding to the plant-root zone. There are 19 soil types in Fillmore County, but only 7 major types: silty clay loam, silt loam, loam, clay loam, sand, sandy loam, and fine sandy loam. The minor soil types were grouped with the most similar of the seven major soil types. Soil organic matter was taken from the STATSGO database for the top soil layer. The model input of the other soil parameters for each specific soil type was taken from the GLEAMS manual (5). These other parameters included the SCS curve number, saturated conductivity, porosity, field capacity, wilting point, evaporation constant, and soil particle size distribution.

The GLEAMS model was run for the 63 slope and soil combinations (7 major soil types and 9 representative slopes) to estimate the annual potential losses in runoff for atrazine and permethrin as a percentage of the amount applied. Since each soil mapping unit had more than one soil type, there was more than one soil type in each grid cell. The potential losses in runoff from each grid cell were calculated by a weighted average based on the distribution of soil types in that soil mapping unit. An ARC/INFO Macro Language program was written to delineate the appropriate choices of slopes and soil types for Fillmore County and to display the results of the GLEAMS model (19). The program converted the surface topography data into a slope grid and the soil-polygon coverage into a soil grid. The program then combined the slope and soil grids into a single composite grid. Finally, the program displayed the values for the potential losses in runoff of the two pesticides calculated by GLEAMS model in map form.

General GLEAMS model results. Because of its water solubility, atrazine is predicted to be predominately in the dissolved phase in runoff. Therefore, soil particle size, which strongly influences the volume of runoff, is the most important control on the amount of atrazine lost in runoff (Figure 2). It is also predicted that there is only a minimal effect of slope on the annual percent losses of atrazine (Figure 2). The slope, according to the GLEAMS manual (5), is used mainly to estimate the peak-flow rate for daily water runoff. Sediment yield. however, is sensitive to slope because the peak-flow rate is used in the erosion component of the model to calculate the sediment-transport capacity. There is a small increase in the predicted atrazine loss for silty clay loam soil as the slope changes from 0.1 to 1 because of the contribution of increased loss of atrazine associated with soil particles (Figure 2). Permethrin was predicted to be more strongly associated with soil particles. Therefore, a larger part of the predicted losses were due to transport in the sorbed phase instead of the dissolved phase. The loss of permethrin was controlled jointly by the soil type and the slope of the landscape.

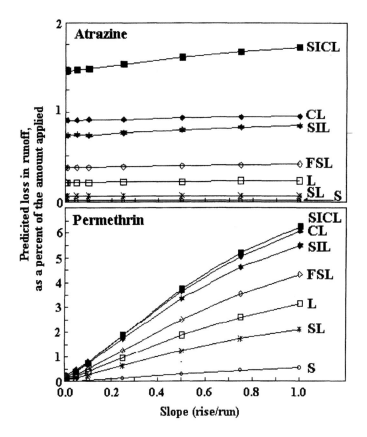

Figure 2. Predicted loss of atrazine and permethrin in runoff to surface water as a function of soil type and slope. The soil types are abbreviated as: S, sandy; SL, sandy loam; L, loam; FSL, fine sandy loam; SIL, silt loam, CL, clay loam; SICL, silty clay loam.

Mapping the GLEAMS model results. The mean annual losses of atrazine and permethrin were estimated for each grid cell for the entire county. The results were regrouped into categories (Table I) that are mapped in Figures 3 and 4.

For all the agricultural cells in the county, the range of potential loss of atrazine in runoff is 0.21 to 1.2% of the amount applied. Areas that have the greatest predicted losses of atrazine ($\geq 1.0\%$) are along the main branch of the Root River system (black area on Figure 3). This narrow band along the river valley has mostly silty clay loam and silty loam soils. As Table I indicates, about 3.2% of the county area falls into this highest group of predicted atrazine loss. Areas predicted to yield intermediate levels of atrazine loss (0.6 - 0.8%) occur extensively (about 44% of the county). These areas are characterized by low slope and finer soil textures. About 10% (excluding forested and urbanized areas) of the county has estimated annual atrazine losses of <0.6%. Sandy loam is the common soil type in these areas of lower potential runoff because water tends to pass through its soil instead of running off. Again, the spatial distribution of predicted atrazine loss is controlled more by soil type than by slope.

Table I. Statistical distribution of predicted potential loss of atrazine and permethrin in runoff to surface-water for Fillmore County, Minnesota

Atrazine			Permethrin		
Loss Category (percent)	Area in Loss Category km^2	(percent)	Loss Category (percent)	Area in Loss Category km^2	(percent)
---[a]	400.0	18.0	---[a]	400.0	18.0
< 0.4 %	56.3	2.5	< 0.5 %	1003.5	45.2
0.4 to < 0.6 %	159.2	7.2	0.5 to < 1.0%	258.6	11.6
0.6 to < 0.8 %	979.6	44.0	1.0 to < 1.5%	311.5	14.0
0.8 to < 1.0 %	558.1	25.1	1.5 to < 2.0%	34.4	1.5
≥ 1.0 %	71.6	3.2	$\geq 2.0\%$	216.9	9.7
All	2224.8	100.0	All	2224.8	100.0

[a] nonagricultural areas (forested and urbanized)

For all the agricultural cells in the county, the range of potential loss of permethrin in runoff is 0.1 to 6.2% of the amount applied. In contrast to atrazine, the spatial pattern of predicted permethrin loss is largely controlled by slope. The regions of highest loss potentials ($\geq 2.0\%$) occur mainly along the Root River Valley (black area in Figure 4), where the slopes are relatively steep and account for a little less than 10% of the total county area (Table I). More than half of the county falls into the lower permethrin loss categories (<1%) (Table I).

A simple field validation of GLEAMS combined with GIS. The accuracy of the estimated atrazine losses derived from GLEAMS combined with GIS can be partly validated with one set of field observations. Within Fillmore County, the Forestville Creek springshed (38.6 km^2, Figure 5) has been monitored for atrazine

Figure 3. Potential loss in runoff for atrazine, as a percentage of the amount applied, for Fillmore County based on the GLEAMS (Groundwater Loading Effects of Agricultural Management Systems model).

Figure 4. Potential loss in runoff for permethrin, as a percentage of the amount applied, for Fillmore County based on the GLEAMS (Groundwater Loading Effects of Agricultural Management Systems model).

(20). A springshed includes both the surface area and the subsurface rock and sediment that contribute water to a spring (21). A springshed can have autogenic and allogenic parts. The autogenic part of a springshed is in karst areas where the surface drainage is much less substantial than the subsurface drainage. Precipitation falling on the karsted land surface can quickly drain into sinkholes or infiltrate directly downward to the subsurface drainage system. The allogenic part of a springshed coincides with the surface drainage-basin boundary and overlies relatively impermeable materials. The runoff on the allogenic part of a springshed would be retained on the land surface until the water moves into the adjacent autogenic part of the springshed.

The western part of Forestville Creek springshed is allogenic and runoff flows eastward on the land surface; the eastern part is characterized by underground flow. The runoff that disappeared from the western part of the springshed reappear at the base of steep cliffs, where impermeable shale is present and where the water table and the land surface intersect. Springs from this area form the headwaters of Forestville Creek.

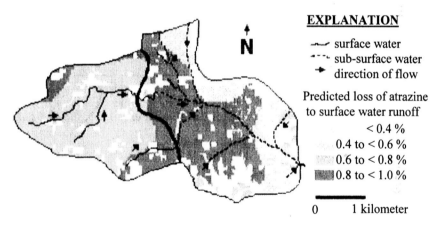

Figure 5. Potential loss in runoff for atrazine, as a percentage of the amount applied, for Forestville Creek springshed, Fillmore County, Minnesota based on the GLEAMS (Groundwater Loading Effects of Agricultural Management Systems model). The western portion of the springshed is allogenic and the eastern portion is autogenic. The thick black line running approximately north and south is the dividing line between these two areas.

The atrazine load (which is based on weekly atrazine concentrations and daily discharge measurements) in Forestville Creek from May 1983 to May 1984 was estimated to be between 4.2 and 5.6 kg/yr. This was 0.16 to 0.22% of the estimated amount applied in the springshed in 1983 (20). The potential losses of atrazine in runoff shown in Figure 6, as predicted by the GLEAMS model, are based on the assumption that all agricultural land was planted in corn and all agricultural land received an application of atrazine. When the assumptions are adjusted with the actual estimated areas of corn (50% of total row crop area) and use of atrazine (75% of corn crop) in the springshed for 1983, the mean annual loss of atrazine in runoff to surface water was predicted to be 0.22% of the amount applied. The predictions are in good agreement with field observations.

Discussion

In this example, the combined use of GLEAMS and GIS provided predictions of the potential losses of atrazine and permethrin in runoff for a large, nonhomogeneous area. The results are expressed as maps and could be used in several ways, such as guidance for agricultural management decisions (for example, decisions on crop type, chemical use, and tillage practice), guidance for county or regional regulatory decisions, as a framework for planning environmental monitoring programs, and as an educational tool to increase public awareness of potential surface-water contamination problems.

There are current limitations in combining process-based models with GIS technology from aspects of both tools. The predictions using the combined tools are only as good as the predictions of the stand-alone, process-based model itself. As the understanding of the behavior, transport, and fate of agricultural chemicals in the environment grows and is incorporated into process-based models, the accuracy of model predictions should increase. On the GIS side, many of the most important databases that could be utilized by process-based models are available only at relatively coarse scales, generally 1:250000. At these coarser scales, the results of field-scale modeling efforts will be somewhat ambiguous because of the problem of the precise boundary delimitation. As these databases become available in finer scales, this limitation will be reduced. When the two technologies are combined to model transport of chemicals in surface runoff to streams and, then, subsequent transport in the streams, additional problems of routing the surface water flow between and through the grid cells arise.

Beyond the generation of vulnerability-types maps, as are produced in this study, the combined use of process-based models and GIS may be of particular value in the interpretation of the data generated in surface and ground water monitoring programs. Environmental monitoring programs are almost always undertaken on large scales (watershed to regional) and their interpretation is too often limited to basic descriptive statistical summaries because the results are difficult to interpret at the level of process-based mechanisms for the conditions of individual locations. The occurrence of an agricultural chemical in surface or ground water, however, is the result of the multiple environmental processes and conditions to which it has been exposed since it was applied. Therefore, the use of process-models combined with GIS can be a valuable tool for extending our field-based understanding to large areas and enabling process-based interpretations of environmental monitoring data.

Acknowledgments.

Partial funding for this project was provided by the U.S. Geological Survey, National Water-Quality Assessment Program's National Pesticide Synthesis project. Any use of trade, product, or firm names is for descriptive purposes only and does not imply endorsement by the U.S. Government.

Literature Cited.

1. Carsel, R.F.; Mulkey, L.A.; Lorber, M.N.; Baskin, L.B. *Ecolog. Model.*, **1985**, *30*, 49-69.

2. Wagenet, R.J.; Hutson, J.L. *J. Environ. Qual.*, **1986**, *15*, 315-322.

3. Dean, J.D.; Huyakorn, P.S.; Donigian, A.S.; Voos, Jr., K.A.; Schanz, R.W.; Meeks, Y.J.; Carsel, R.F. *Risk of unsaturated/saturated transport and transformation of chemical concentrations (RUSTIC);* U. S. Environmental Protection Agency, Athens, GA, EPA/600/3-89/048a, **1987**, 202 p.

4. Leonard, R.A.; Knisel, W.G; Still, D.A. *Trans. Am. Soc. Agric. Eng.,* **1987**, *30*, 1403-1418.

5. Knisel, W.G.; Davis, F.M., Leonard, R.A.;. *GLEAMS Versoin 2.03, User Manual*, U.S. Department of Agriculture, Agricultural Research Service, Tifton, GA, 1992; 199 p.

6. Fong, F.K,; Mulkey, L.A. *Water Res. Res.* **1990**, *26*, 291-303.

7. Corwin, D.L.; Vaughan, P.J.; Loague, K. *Environ. Sci. Technol.*, **1997**, *31*, 2157-2175.

8. Geleta, S.; Sabbagh, G.J.; Stone, J.F.; Elliott, R. L.; Mapp, H. P.; Bernardo, D.J.; Watkins, K.B. *J. Environ. Qual.*, **1994**, *23*, 36-42.

9. He, C.; Riggs, J.F.; Kang, Y. T. Water Resour. Bull. **1993**, *29*, 891-900.

10. Engel, B.A.; Srinivasan, R.; Arnold, J.; Rewerts, C.; Brown, S.J. Water Sci. Technol., **1993**, *28*, 685-690.

11. Mitchell, J.K.; Engel, B.A.; Srinivasan, R.; Wang, S.S.Y. Water Resour. Bull. **1993**, *29*, 833-842.

11. Mitchell, J.K.; Engel, B.A.; Srinivasan, R.; Wang, S.S.Y. Water Resour. Bull. **1993**, *29*, 833-842.

12. Mellerowica, K.T.; Rees, H.W.; Chow, T.L.; Ghanem, I.J. Soil Water Conserv. **1994**,.*49*, 194-200.

13. Tim, U.S.; Jolly, R. *J. Environ. Qual.*, **1994**, *23*, 25-35.

14. Soil Conservation Service, *Soil Survey of Fillmore County, Minnesota*, Series 1954, No. 1; U.S. Department of Agriculture, Washington, DC, 1958; 23 p.

15. Minnesota Agriculture Statistics Service. *Minnesota Agriculture Statistics*; U.S. Department of Agriculture, Washington, DC, 1994, 4 p.

16. Soil Conservation Service, *State Soil Geographic Data Base (STATSGO) Data Users Guide*, Misc. Publication No. 1492; U.S. Department of Agriculture, Washington, DC, 1993. Web site: http://www.ncg.nrcs.usda.gov/statsgo.html.

17. Thelin, G.P.; Gilliom, R.J. *Classification and Mapping of Agricultural Land for National Water-Quality Assessment*, U.S. Geological Survey, Reston, VA, Circular 1131, 1997. Web site: http://water.wr.usgs.gov/pnsp/circ1131.

18. ESRI, *Understanding GIS, The ARC/INFO Method.* Rev. 6; ESRI, Redlands, CA, 1992. 2 pp.

19. Zhang, H. MS Thesis, Univ. of Minnesota: Minneapolis, MN, 1997, 139 p.

20. Grow, S.R. *Water Quality in the Forestville Creek Karst Basin of Southeastern Minnesota*; MS Thesis, University of Minnesota: Minneapolis, MN, 1986, 147 p.

21. Alexander, C.A., Jr., Green, J.A., Alexander, S.C., Spong, R.C. *Springshed, Plate 8, Part B of Geologic Atlas, Fillmore County, Minnesota.* Minnesota Department of Natural Resources, St. Paul, MN, 1996.

Chapter 13

Herbicide Transport in Subsurface Drainage Water Leaving Corn and Soybean Production Systems

T. B. Moorman[1], R. S. Kanwar[2], and D. L. Karlen[1]

[1]National Soil Tilth Laboratory, Agricultural Research Service,
U.S. Department of Agriculture, Ames, IA 50011
[2]Department of Agricultural and Biosystems Engineering,
Iowa State University, Ames, IA 50011

> Herbicides are transported through subsurface drainage to surface waters from corn-growing areas of the USA and Canada. Herbicide losses are highly variable, ranging between 0.01 to 10 g/ha. The magnitude of herbicide loss results from precipitation patterns, herbicide-soil interactions, and farming practices. This report reviews existing literature and presents new research concerning effects of farming practices on herbicide losses in drainage water. Conservation tillage practices which increase infiltration tend to increase herbicide losses. Increasing intensity of drainage and increased frequency and rate of herbicide use also increase herbicide losses. Banding lowers the application rate and reduces annual losses and average concentrations of atrazine compared to broadcast applications. Metolachlor losses were reduced by banding, but the effect was only statistically significant in continuous corn systems.

Pesticides are a component of the highly productive farming systems in the American Midwest and elsewhere. Sixty percent of pesticides used in the USA are applied to cropland in 12 Midwestern states. Gains in agricultural production from weed and insect control have not come without a cost. Pesticide use in agriculture has contaminated surface and ground water, leading to concerns over the impact of long-term, low-dose exposure to pesticides on human health (*1, 2, 3*). In some areas of the United States, subsurface drainage diverts shallow groundwater beneath the soil profile into surface waters. Monitoring data show that waters leaving subsurface drainage systems contribute substantial amounts of pesticides and nutrients to surface waters.

The purpose of this paper is to examine the magnitude of herbicide losses in subsurface drainage waters leaving corn and soybean cropping systems in the north

0.090% from the plots (*9*). The pattern of metolachlor losses was similar at these two scales, but greater total losses were observed at the plot scale than at the watershed scale. Although several factors contributed to the discrepancy in herbicide losses between the watershed and plot scale, the drainage density appears to be a strong contributing factor. Other studies have experimentally assessed the effects of drainage tube density. Losses of pesticides were nearly 4 times greater when drains were spaced 5 m apart than at 20 m spacing in southeast Indiana (*10*). In production fields the drainage network may be unevenly spaced and in some cases the network is unknown. In addition, herbicides can be applied at precise rates on known dates at the plot scale, while usage estimates and application windows are known with less precision at the watershed scale.

Jaynes et al. (*11*) compared atrazine losses at different scales by monitoring at sites receiving only subsurface drainage, both subsurface and surface drainage, and streamwater in the 5130 ha Walnut Creek watershed in central Iowa. The concentrations of atrazine exiting the subsurface drain (Figure 1a) increase as drainage increases during periods of precipitation. The discharge-atrazine concentration relationship is less distinct at the site monitoring a larger county drain (Figure 1b), which receives water and herbicides in runoff through surface inlets and subsurface drainage from both herbicide-treated and untreated fields. The discharge and atrazine concentration responses for the entire watershed are also shown at a site further downstream on Walnut Creek (Figure 1c). The temporal pattern of losses within the growing season were similar at the field and watershed scales (Figure 1). At the watershed scale, the distribution of corn and atrazine use remains fairly constant from year to year, with corn averaging 43% of the watershed land area (*12*). Jaynes et al. (*11*) used the data in Figure 1 and similar monitoring data to show that herbicide losses (expressed as kg loss/kg applied) from subsurface drainage alone were generally less than 10% of the total herbicide losses in the watershed. Similar results were obtained in the Sugar Creek watershed in central Indiana (*13*). Atrazine losses were similar in concentration and timing between individual field subsurface drains and Sugar Creek. Total amounts of herbicide loss were not obtained from the subsurface drainage systems, therefore, the contribution of subsurface drainage to total losses could not be estimated.

Quantification of the subsurface drainage contribution to atrazine loads in large watersheds has not been accomplished at this time. Richards and Baker (*14*) concluded that the subsurface drainage was not a key contributor to several watersheds of 11 to 16,095 km^2 in size. This conclusion was based on the lack of correlation between the pesticide and nitrate concentrations and the generally lower pesticide concentrations in subsurface drainage compared to stream and river waters. However, several studies have shown that there is little linkage between nitrate and herbicide in subsurface drainage (*11, 15*). Elsewhere, researchers concluded that atrazine and it's metabolite, DEA (deethylatrazine), were transported through subsurface drains into the Minnesota River watershed (38,583 km^2), but they were unable to differentiate between herbicides transported by runoff directly into surface water, by runoff into the subsurface drainage network, or by leaching through soil into

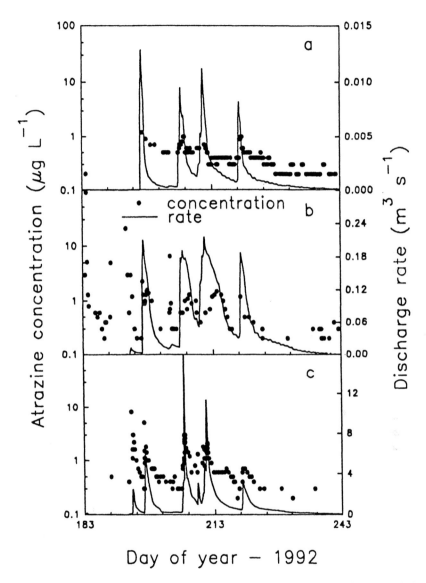

Figure 1. Atrazine concentrations and water flow rate from (a) subsurface drain beneath an 18 ha field treated with atrazine, (b) surface and subsurface drainage water from a 366 ha area, and (c) Walnut Creek stream water (5100 ha area). Source: Reprinted with permission from ref. 11.

central part of the USA and southern Canada. This region combines intensive agriculture with high herbicide use with extensive areas where subsurface drainage networks have been installed. Fausey et al. (*4*) reports that between 10 and 50% of all cropland in the 8 state north-central part of the U.S. has subsurface drainage. In addition, we report effects of management practices (tillage, rotations, swine manure application and banded herbicide application) on drainage water quality at a field site in northeast Iowa and discuss the effect of herbicide sorption and degradation processes on herbicide movement of under these management practices. A broader geographic perspective on pesticide leaching processes was presented by Flury (*5*).

Methodology and Scale

Most of our knowledge of pesticide losses in subsurface drainage is from measurements made using plots with individual subsurface drains. Such drainage plots are ideal for evaluating the effect of management practices on water quality. This methodology has the advantage of precision application of herbicides and the replicated measurement of drainage and herbicides. In most cases, plots are hydrologically isolated by various means to prevent runoff and lateral water flow between plots. The research site at Iowa State University's Northeast Research Farm near Nashua, Iowa is one example of a drainage plot research facility. The site contains 36 plots of 0.4 ha area. In 1979, subsurface drains were installed at 29 m spacings, approximately 1.2 m deep. Each 58 by 67 m plot has a subsurface drain through the center and along the north-south borders. Center drains were routed to sumps for monitoring the drainage water while border drains isolated plots on the north and south sides. Soils at the site are the Floyd loam (fine-loamy, mixed, mesic Aquic Hapludoll), Kenyon loam (fine-loamy, mixed, mesic Typic Hapludoll) and Readlyn loam (fine-loamy, mixed, mesic Aquic Hapludoll), which are described previously (*6*). These soils are moderately well to poorly drained and lie over loamy glacial till. In contrast to the Nashua site, are the smaller 0.04 ha plots with drains at a 13.3 m spacing on a poorly drained Hoytville silty clay in Wood Co., Ohio (*7*).

Are results obtained on drainage plots representative of drainage losses from production systems at the field or watershed scale? Concerns have been expressed as to the representativeness of herbicide losses from drainage plots compared to field or watershed scales. The need for hydrologic isolation of drainage plots may prevent the use of sites where topography increases lateral flow and runoff to the subsurface drains, such as at the 18 ha production field site described by Moorman et al. (*8*). At this site, surface runoff transports herbicides to areas of the field directly over the subsurface drains, where maximum concentrations of 313 µg/L of atrazine and 38 µg/L of metribuzin were measured in the accumulated runoff. Concern has also been expressed over the greater intensity of drainage tubes in plots compared to typical farmers fields. Ng et al. (*9*) estimated the subsurface drain density at 833 m drain line/ha in the Nissouri Creek watershed compared to 1298 m/ha in small plots near Woodslee, Ontario. Losses of atrazine (expressed as a percentage of the applied amount) in runoff and subsurface drainage were 0.039% from the watershed and subsurface drains (*15*). However, the DAR ratios (ratio of DEA to atrazine concentrations) in river water increased by two orders of magnitude in the 5 month period following atrazine application. Since low DAR ratios can be produced through direct transport of atrazine into water, this could indicate the predominance of transport in runoff early in the growing season when atrazine concentrations in river water are maximal and before atrazine is degraded into DEA. DAR ratios measured in drainage plot studies in Iowa (*16*) also increased over time, but the DAR ratios were much larger those in the Minnesota River and the magnitude of increase over the growing season was much less.. This would tend to support the interpretation of the DAR ratios in the Minnesota River by Schottler et al. (*15*). However, the overall difficulty in interpreting DAR ratios (or other metabolite ratios) preclude their use as a quantitative tool for estimation of the subsurface drainage component of herbicide transport.

The use of relatively small field sites or drainage plots to estimate the contribution of subsurface drainage to herbicide loads in larger watersheds remains the sole practical method of measuring pesticides in subsurface drainage, although scaling results obtained from plots or fields to the large watershed is not straightforward. Ng et al. (*9*) described an approach to this problem in their study of the Nissouri Creek watershed. In addition to measurement of herbicide losses in runoff and subsurface drains, soil properties and distributions, herbicide application (area and quantity), subsurface drain density, rainfall and hydrologic data were identified as necessary for scaling estimates from the plot scale to the watershed scale.

Magnitude of Herbicide Loss

The movement of herbicides into the drainage water are affected by the rate and timing of application, rainfall pattern and amount, and the soils above the drainage tubes. Annual losses from several studies in the American and Canadian Midwest (Table I) show substantial variation in atrazine losses due to these interacting factors. There is clearly no simple relationship between application rate and losses. Atrazine and metolachlor losses in subsurface drainage ranged between 13 to 71% of the total herbicide lost in runoff and drainage under a variety of soil and tillage practices (*5*). Figure 2 shows the relationship between annual losses of atrazine and precipitation or drainage using data from the sources given in Table I. Annual atrazine losses above 0.1% occur over a range of annual precipitation and drainage amounts. High rainfall shortly after herbicide application probably account for atrazine losses exceeding 0.1% of applied in years with lower annual precipitation. Losses below the 0.1% level are almost always in years with less than 1000 mm precipitation. Similarly, drainage exceeding 200 mm usually results in losses exceeding 0.1%. Linear regression shows that for each 500 mm of annual precipitation the fractional (%) losses of atrazine increase about 10 fold. The relationship between annual precipitation and herbicide loss in drainage is not as strong as the relationship between drainage and herbicide loss (Figure 2). For each 120 mm of drainage, atrazine losses increase by approximately 0.1% of the applied amount. Atrazine concentrations in subsurface drainage have exceeded the MCL for drinking water established by USEPA of 3 µg/L, but the duration of these exceedences is generally short (7, 8, 16, 17, 18). For purposes of comparison, subsurface drainage of 100 mm water would require transport of 3 g/ha atrazine (or 0.3% of a 1.0 kg/ha application) to reach an average concentration equivalent to the 3 µg/L MCL. Although data on losses of other herbicides is less extensive than that for atrazine, the long-term trends in losses appear to be similar to atrazine in both their variability and relationship to precipitation and drainage.

Table I. Annual losses of atrazine in subsurface drainage waters in different corn and corn-soybean production systems under field conditions

Crop Rotation	Atrazine (g/ha)[a]	Annual Loss (range)		Site (Ref.)
		(g/ha)	(% applied)	
Cont. corn 1990-1991	0 - 1680	0.69 - 5.40	0.04-0.32	Ames, Iowa (16)
Cont. corn 1987-1990	0 - 2400	0.77 - 6.49	0.039 - 0.27	Ottawa, Ontario (19)
Cont. corn 1971-1974	1200	2.51 - 15.43	0.083 - 1.28	Woodslee, Ontario (20)
Cont. corn 1987-1990	1700	9 - 32	0.12 - 1.88	Woodslee, Ontario (17)
Cont. corn 1990-1993	1500 - 2200	0.29 - 0.98	0.02 - 0.12	Ottawa, Ontario (18)
Cont. corn 1986-1988	4000	0.4 - 3.7	0.01 - 0.09	Waseca, Minnesota (21)
Corn-soybean 1992-1995	0 - 1121	0.02 - 2.16	0.004 - 0.47	Kelly, Iowa (8)
Corn-soybean 1987-1990	0 - 2200	0 - 31.3	0.41[b]	Wood Co., Ohio (7)
Corn-soybean 1990-1996	541 - 2800	0.2 - 7.3	0.007 - 0.26	Nashua, Iowa

[a]Atrazine application rates of zero indicate years when atrazine was not applied.
[b]Averages over different tillage treatments.

Soils and Herbicide Behavior

Although water movement to subsurface drains exerts a strong influence on herbicide movement, sorption and persistence in the soil also affect movement. Both of these processes are determined by the specific physical and chemical properties of the herbicide and the soil. Herbicides that persist in the surface soil and subsoil are more likely to be transported into drainage waters. The results obtained in a comparison of

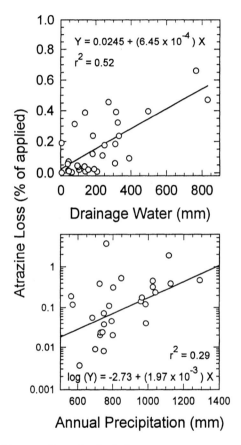

Figure 2. Annual losses of atrazine in drainage waters, expressed as a percentage of applied herbicide, in relation to drainage water or annual precipitation, based on reports cited in Table 1. In some instances, atrazine losses from individual sites are averaged over tillage types or other factors in order to obtain an average annual loss from a single site.

herbicide losses under different tillage systems at site near Nashua, Iowa (Table II) illustrate this relationship. Greater amounts of atrazine were lost than the other herbicides relative to their application rates. The half-life of atrazine at this field location was 40 days, whereas the alachlor half-life was only 24 days (22). Other studies also observed greater losses of atrazine with subsurface drainage relative to other compounds, including significant losses of atrazine in years when it was not applied (7, 8, 16, 21).

Table II. Effects of tillage on mean annual losses of herbicides (g/ha) in subsurface drainage from plots near Nashua, Iowa.

Year	Herbicide[a]	Tillage System			
		Chisel plow	Mold. Plow	Ridge Till	No Till
1990	Atrazine	3.2	0.9	4.5	7.3
1991	(CC)	2.6	0.7	4.0	4.5
1992		0.3	0.2	0.5	0.6
Mean		2.03	0.6	3.0	4.1
1990	Alachlor	0.12	0.02	0.13	0.13
1991	(SR)	0.31	0.61	0.30	0.24
1992		0.001	0.001	0.004	nd[b]
Mean		0.14	0.21	0.15	0.12
1990	Cyanazine	0.9	0.10	0.37	1.33
1991	(CR)	0.02	0.05	1.82	0.61
1992		0.02	0.002	0.04	0.01
Mean		0.31	0.05	0.74	0.65

[a]Atrazine was applied at 2.8 kg/ha to continuous corn (CC). Alachlor at 2.24 kg/ha and cyanazine at 2.8 kg/ha were applied to corn (CR) -soybean (SR) rotation.
[b]Not determined.

Sorption of most nonionic herbicides is governed by soil organic C content. Losses of herbicides and insecticides in subsurface drainage was inversely related to their K_{oc} (10). In contrast, Logan et al. (7) determined that the relative losses of several pesticides in subsurface drainage water corresponded to the pesticides soil half-life, but not K_{oc} or water solubility. The sorption of pesticides may vary in different parts of the landscape and high organic C contents in areas over subsurface drains may help prevent losses to drainage waters (23). Many newer herbicides are weakly ionic at neutral to alkaline pH levels, and clay content may be important in retention of these compounds in the soil profile (24).

Soil physical conditions can also affect herbicide movement into subsurface drains. The partitioning of rainfall between infiltration and runoff is dependent upon rainfall intensity and duration, soil structure, and crop residue cover. The rapid movement of herbicides into subsurface drains following precipitation events has been attributed to macropore flow (*10, 25*). Despite the apparent importance of macropore flow, few studies exist where macropore flow is documented with respect to tillage or other management changes and herbicide loss into tile drainage.

Tillage Effects on Herbicide Loss

Tillage systems affect the organic C content of soil, soil physical conditions and the infiltration and movement of water. Fawcett et al. (*26*) examined a large number of studies and concluded that herbicide losses in runoff from conservation tillage were reduced by 42 to 70% compared to losses from moldboard plow systems. Changes in soil properties following adoption of conservation tillage may have contradictory effects with respect to herbicide movement, as illustrated by the data from four tillage systems at the Nashua site (Table II). Subsurface drain water samples were collected on weekly basis for NO_3-N and herbicide analyses, as described previously (*22, 27*). Annual herbicide losses during the three years of the study vary by as much as an order of magnitude within tillage systems. Losses of atrazine and cyanazine tended to be greater from the ridge till and no till systems, while alachlor losses were generally similar among the tillage types. The maximum annual average concentrations were 2.5 µg/L alachlor in 1992, 5.7 µg/L atrazine in 1991, and 4.3 µg/L cyanazine in 1991. The volume of drainage water was reduced by moldboard plowing in the continuous corn system relative to the other tillage treatments, but in the corn-soybean system with no-till, ridge-till and moldboard tillage drainage quantities were generally equivalent and slightly lower than chisel plow. The greater losses of atrazine and cyanazine in the ridge-till and no-till systems are in contrast to the greater organic C contents in soil under these conservation tillage systems. The organic C levels were 23.7 (moldboard tillage), 29.1 (chisel plow), 32.9 (ridge-till) and 37.3 (no-till) g C/kg in 1992 (*28*). The greater herbicide movement in no-till soil, despite the increase in organic matter, suggest that decreased tillage has increased herbicide transport in macropore flow at the Nashua site.

The results obtained at Nashua are generally similar to atrazine and metolachlor losses in drainage from no-till and conventional till systems obtained at Ottawa, Ontario (*18*). Metolachlor losses were less than atrazine losses and herbicide losses from no-till plots were larger than losses from the conventional till system. Similar results were also obtained in plot studies in Ohio (*7*). In another study, losses of metolachlor and atrazine in three different tillage systems varied among the four years of study (*17*). In the year of greatest herbicide loss (1989), the conventional tillage system lost more of these herbicides in drainage water than the ridge-till or no-till systems. To the extent that conservation tillage systems increase water infiltration, they increase the potential for herbicide loss in subsurface drainage. However, this trend may not be evident with herbicides that have short to intermediate persistence.

Application Rates and Herbicide Losses

Little information is available comparing losses in drainage water as a direct effect of application rate, although several studies have examined banding as an application technique. Banding refers to the application of herbicide in strips that usually cover the crop row. This results in a net reduction in herbicide application on a field area basis. This technique is combined with cultivation in the non-banded area to control weeds. In Ontario, banding reduced maximum concentrations in drainage water by over 50% (29). Figure 3 shows the amounts of atrazine and metolachlor lost from the Nashua plots in subsurface drainage. Six plots in a corn-soybean rotation with chisel plow tillage were given a treatment of herbicide banding (only one third of the area on the rows was applied with herbicides and area between the rows was cultivated) and six plots received broadcast herbicides. Differences resulting from the broadcast and banding methods of application for both herbicides (atrazine and metolachlor) were most apparent in 1993 and 1995. Banding reduces herbicide inputs and results in a significant ($P < 0.07$) decrease of 88% in mean (1993 - 1996) leaching losses for atrazine. The four-year average atrazine concentration in water from plots receiving the banded atrazine was 0.12 µg/L was also significantly lower than the 0.45 µg/L concentration for the broadcast treatment. However, the 1993 - 1996 mean losses for metolachlor were not significantly different, although banding reduced losses by 39%. Annual volumes of drainage were similar in the two systems.

Manure Effects on Herbicide Loss

Manure application to soils can result in effects on both the persistence and transport of herbicides and insecticides. Different animals produce manures with varying composition, but all contain large amounts of decomposable carbon and bioavailable nitrogen. Swine manure production probably exceeds 14×10^6 Mg in the USA (30). Solid swine manure contains 3 to 4.9 g N/kg manure and liquid swine manure can contain up to 6.6 g N/L (31). Most of the N in swine manure is present as NH_4^+ or organic N. Long-term application of manures generally increases soil organic C, while soluble C increases were noted in soil shortly after manure applications (32). Indirectly, the increase in soil organic C often causes increases soil aggregation and tilth, which may increase water infiltration and transport. Manure contains considerable amounts of both dissolved and particulate organic matter, which have been reported to increase atrazine desorption (33), or increase alachlor (34) and atrazine sorption (35). Barriuso et al. (35) concluded that effects of organic matter amendments, including manure, differed according to whether they were added prior to the herbicide or afterwards. Businelli (33) also reached the same conclusion; atrazine desorption with a pig slurry extract was approximately twice as great as that desorbed by calcium chloride solution, suggesting that manure applied closely after atrazine would increase atrazine in the soil solution. However, pretreatment of soil with pig slurry before atrazine addition increased atrazine sorption by about fourfold.

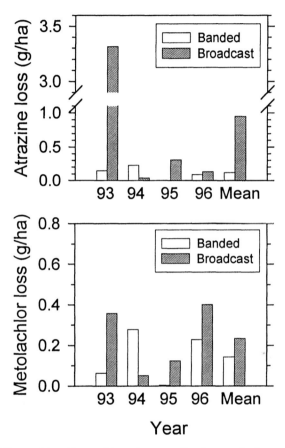

Figure 3. Effects of banding and broadcast application methods on mean annual losses of atrazine and metolachlor in subsurface drainage water from corn plots in a corn soybean rotation near Nashua, Iowa. Differences in the 1993-1996 means are significant for atrazine at the $P < 0.05$ level, but not significant for metolachlor, by the Kruskal-Wallis test.

Figure 4 shows indirect evidence of manure effects on herbicide loss in drainage water at the Nashua site. Atrazine (formulated as Extrazine) was broadcast applied at 0.54 kg/ha with metolachlor at 2.8 kg/ha to previously manured soil or band applied to soil treated with pre-plant urea-ammonium nitrate (UAN). The fall swine manure application was equivalent in N content to the UAN application of 135 kg N/ha. The plots were cropped to continuous corn. The average loss of atrazine applied by banding over 5 yr was 0.89 g/ha in continuous corn (Figure 4) compared to 0.12 g/ha loss of banded atrazine in the corn-soybean system (Figure 3). This difference is attributed to the greater frequency of atrazine application in the continuous corn. However, when atrazine was broadcast onto previously manured soil the average loss was not different from the banded herbicide applied to non-manured soil (Figure 4), in contrast to the reduction in loss obtained by banding in the rotated corn (Figure 3). While this is not a direct comparison of manure effects, it appears that the manure reduced atrazine leaching.

Manure does not appear to affect metolachlor leaching as significantly greater ($P < 0.05$) amounts of herbicide are lost after broadcast to the manured soil than after banding (no manure) (Figure 4). The average concentration was also reduced by banding to 28% of the broadcast concentration (1.48 µg/L). The average metolachlor loss from the broadcast treatments in continuous corn systems was 0.56 g/ha (Figure 4) compared to the 0.23g/ha loss from corn-soybean rotation (Figure 3). However, the average losses of band applied metolachlor from continuous and rotated corn are nearly identical.

Previous studies on the effects of manure on herbicide persistence are contradictory. Application of slurried swine manure or cattle manure at rates of 40 tons/ha two months before atrazine application increased the half-lives of atrazine from 60 days to 96 and 91 days, respectively (36). In the same study, metolachlor half-lives were increased from 44 days to 83 and 76 days for swine slurry and cattle manure, respectively, and a subsequent study produced similar results (37). Different results were obtained by Topp et al. (38), who compared atrazine dissipation in slurries prepared from soils with or without 4 previous annual applications of dairy manure. In all slurries with the history of manure application atrazine degradation was more rapid than in soil slurries without the previous use of manure. In another study, dairy manure increased the degradation and mineralization of high concentrations of atrazine, but ammonium addition inhibited mineralization. (39). Some triazine-degrading bacteria are inhibited by mineral N (40), whereas others are not (41, 42), and triazine degradation by bacteria are stimulated by additions of carbon sources other than atrazine (39, 42). The decrease in ^{14}C-atrazine mineralization in the Floyd loam soil may be attributable to inhibition of microbial degradation by the additional N in the manure.

The fate of ^{14}C-atrazine in manured and untreated Floyd soil from the Nashua site is shown in Table III. More atrazine or atrazine metabolites were bound in the manured soils, with correspondingly less mineralized by microorganisms. Periodic measurements of atrazine by extraction and HPLC showed an average half-life of 55 days and indicated no effect of manure on atrazine disappearance (data not shown).

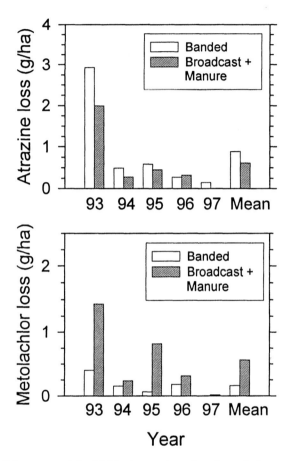

Figure 4. Mean annual herbicide losses in subsurface drainage water from continuous corn plots treated with manure and broadcast herbicides or UAN fertilizer and banded herbicides near Nashua, Iowa. Differences in the 1993-1996 means are significant at the $P < 0.05$ level for metolachlor, but not significant for atrazine by the Kruskal-Wallis test.

The increase in bound residues would be consistent with increased atrazine sorption in the manured soils relative to the non-manured soils. Further experiments will be needed to determine if increased sorption results from manure applications in these soils. We have not performed similar studies with metolachlor.

Table III. Distribution of ^{14}C at 171 days following addition of manure and 1 mg/kg of [^{14}C-*ring*]atrazine to Floyd loam soil from the Nashua site.

Manure (kg N/ha)[a]	Atrazine (% of applied ^{14}C)	$^{14}CO_2$ (%)	Non-extractable (%)[b]
0	25	51	30
224	25	19	52
448	24	27	45

[a] Liquefied swine manure was applied at rates which supplied these levels of N to the soil.
[b] Non-extractable residues remaining after two extractions with 80% methanol.

Conclusions

The following conclusions can be reached from our analysis of data from the studies at Nashua and other reports.

1. There is substantial temporal variability both within growing seasons and across years in the amounts of herbicides lost in subsurface drainage. This variability is primarily driven by trends in precipitation, which are closely linked to drainage losses.

2. Herbicides which are weakly sorbed and have longer persistence in surface and subsurface soils are more prone to losses in subsurface drainage.

3. Banding of herbicides resulted in lower atrazine and metolachlor losses to shallow groundwater, but this effect is largely attributable to the decreased application rate. This result suggests that other rate reduction practices will result in decreased herbicide loads in drainage waters. Weed control needs and soil erosion potential will be important considerations in adoption of herbicide banding.

4. Conservation tillage practices which tend to increase infiltration will tend to increase herbicide losses in subsurface drainage. The increased loss in subsurface drainage may by accompanied by decreased herbicide loss in surface runoff.

5. Manure decreased atrazine loss by subsurface drainage, but did not alter the pattern of metolachlor loss. The mechanisms responsible for this effects are not

clearly understood, but our results and those of others indicate that manure can alter the persistence and leaching of herbicides. Large amounts of land may be manured and these interactions could result in unforeseen consequences.

Acknowledgments

The authors thank Carl Pederson and Ken Pecinovsky for technical support at the Nashua site and Alissara Reungsang for data on the fate of atrazine in manure-treated soils. We also thank the USDA MSEA Program and the Leopold Center for Sustainable Agriculture for supporting research described in this report.

Literature Cited

1. Fairchild, D. M. In *Groundwater Quality and Agricultural Practices*, Fairchild, D. M., Ed.; Lewis Publishers, Inc.: Chelsea, MI., 1987; pp 273-294.
2. Hallberg, G. R. *Agric. Ecosys. Environ.* **1989**, *26*, 299-367.
3. Johnson, C. M.; Kross, B. C. *Am. J. Ind. Med.* **1990**, *18*, 449-456.
4. Fausey, N. R.; Brown, L. C.; Belcher, H. W.; Kanwar, R. S. *J. Irrig. Drainage Eng.* **1995**, *121*, 283-288.
5. Flury, M. *J. Environ. Qual.*, **1996**, *25*, 25-45.
6. Karlen, D. L.; Berry, E. C.; Colvin, T. S.; Kanwar, R. S. *Commun. Soil Sci. Plant Anal.* **1991**, *22*, 1985-2003.
7. Logan, T. J.; Eckert, D. J.; Beak, D. G. *Soil Tillage Res.* **1994**, *30*, 75-103.
8. Moorman, T. B.; Jaynes, D. B.; Cambardella, C. A.; Hatfield, J. L.; Pfeiffer, R. L.; Morrow, A. J. *J. Environ. Qual.* **1999**, *28*, 35-45.
9. Ng, H. Y.; Gaynor, J. D.; Tan, C. S.; Drury, C. F. *Wat. Res.*, **1995**, *29*, 2309-2317.
10. Kladivko, E. J.; Van Scoyoc, G. E.; Monke, E. J.; Oates, K. M.; Pask, W. *J. Environ. Qual.* **1991**, *20*, 264-270.
11. Jaynes, D. B.; Hatfield, J. L.; Meek, D. W. *J. Environ. Qual.* **1999**, *28*, 45-59.
12. Hatfield, J. L.; Jaynes, D. B.; Burkart, M. R.; Cambardella, C. A.; Moorman, T. B.; Prueger, J. H.; Smith, M. A. *J. Environ. Qual.*, **1999**, *28*, 11-24.
13. Fenelon, J. M.; Moore, R. C. *J. Environ. Qual.* **1998**, *27*, 884-894.
14. Richards, R. P.; Baker, D. B. *Environ. Toxicol. Chem.*, **1993**, *12*, 13-26.
15. Schottler, S. P.; Eisenreich, S. J.; Capel, P. D. *Environ. Sci. Technol.*, **1994**, *28*, 1079-1089.
16. Jayachandran, K.; Steinheimer, T. R.; Somasundaram, L.; Moorman, T. B.; Kanwar, R. S.; and Coats, J. R. *J. Environ. Qual.* **1994**, *23*, 311-319.
17. Gaynor J. D.; McTavish, D. C.; Findlay, W. I. *J. Environ. Qual.* **1995**, *24*, 246-256.
18. Masse, L.; Patni, N. K.; Jui, P. Y.; Clegg, B. S. *Trans. ASAE* **1996**, *39*, 1673-1679.
19. Frank, R.; Clegg, B. S.; Patni, N. K. *Arch. Environ. Contam. Toxicol.* **1991**, *21*, 41-50.

20. Von Stryk, F. G.; Bolton, E. F. *Can. J. Soil Sci.* **1977**, *57*, 249-253.
21. Buhler, D. D.; Randall, G. W.; Koskinen, W. C.; Wyse, D. L. *J. Environ. Qual.* **1993**, *22*, 583-588.
22. Weed, D. A. J.; Kanwar, R. S.; Stoltenberg, D. E.; Pfeiffer, R. L. *J. Environ. Qual.* **1995**, *24*, 68-79.
23. Novak, J. M.; Moorman, T. B.; Cambardella, C. A. *J. Environ. Qual.* **1997**, *26*, 1271-1277.
24. Moorman, T. B.; Keller, K. E. In *Herbicide-Resistant Crops: Agricultural, Environmental, Economic, Regulatory, and Technical Aspects*; Duke, S. O. Ed.; CRC Press: Boca Raton, FL, 1996; pp 283-302.
25. Czapar, G. F.; Kanwar, R. S.; Fawcett, R. S. *Soil Tillage Res.* **1994**, *30*, 2489-2498.
26. Fawcett, R. S.; Christensen, B. R.; Teirney, D. P. *J. Soil Water Conserv.* **1994**, *49*, 126-135.
27. Kanwar, R. S.; Colvin, T. S.; Karlen, D. L. *J. Prod. Agric.* **1997**, *10*, 227-234.
28. Karlen, D. L.; Kumar, A.; Kanwar, R. S.; Cambardella, C. A.; Colvin, T. S. *Soil Tillage Res.* **1998**, *48*, 155-165.
29. Gaynor, J. D.; Tan, C. S.; Drury, C. F.; Van Wesenbeeck, I. J.; Welacky, T.W. *Water Qual. Res. J. Canada* **1995**, *30*, 513-531.
30. Lal, R.; Kimble, J. M.; Follett, R. F.; Cole, C. V. *The Potential of US Cropland to Sequester Carbon and Mitigate the Greenhouse Effect*; Ann Arbor Press: Chelsea, MI, 1998; pp 74-76.
31. Hatfield, J. L.; Brumm, M. C.; Melvin, S. W. In *Agricultural Uses of Municipal, Animal, and Industrial Byproducts, Conserv. Res. Rept. 44*; Wright, R. J.; Kemper, W. D.; Milner, P. D.; Power, J. F.; Korcak, R. F., Eds.; USDA, Agric. Res. Serv.: Washington, DC, 1998, pp 78-90.
32. Gregorich, E. G.; Rochette, P.; McGuire, S.; Liang, B. C.; Lessard, R. *J. Environ. Qual.* **1998**, *27*, 209-214.
33. Businelli, D. *J. Environ. Qual.* **1997**, *26*:102-108.
34. Guo, L.; Bicki, T. J.; Felsot, A. S.; Hinesly, T. D. *J. Environ. Qual.* **1993**, *22*, 186-194.
35. Barriuso, E.; Baer, U.; Calvet, R. *J. Environ. Qual.* **1992**, *21*, 359-367.
36. Rouchaud, J.; Gustin, F.; Cappelen, O.; Mouraux, D. *Bull. Environ. Contam. Toxicol.* **1994**, *52*, 568-573.
37. Rouchaud, J.; Gustin, F.; Callens, D.; Bulcke, R. *Weed Res.* **1996**, *36*, 105-112.
38. Topp, E.; Tessier, L.; Gregorich, E. G. *Can. J. Soil Sci.* **1996**, *76*, 403-409.
39. Gan, J.; Becker, R. L.; Koskinen, W. C.; Buhler, D. D. *J. Environ. Qual.* **1996**, *25*, 1064-1072.
40. Mandelbaum, R.; Wackett, L. P.; Allan, D. L. *Appl. Environ. Microbiol.*, **1993**, *59*, 1695-1701.
41. Assaf, N. A.; Turco, R. F. *Biodegradation*, **1994**, *5*, 29-35.
42. Struthers, J. K.; Jayachandran, K.; Moorman, T. B. *Appl. Environ. Microbiol.* **1998**, *64*, 3368-3375.

Chapter 14

Metolachlor Volatilization Estimates in Central Iowa

J. H. Prueger[1], J. L. Hatfield[1], and T. J. Sauer[2]

[1]National Soil Tilth Laboratory, Agricultural Research Service,
U.S. Department of Agriculture, Ames, IA 50011
[2]Biomass Research Center, Agricultural Research Service,
U.S. Department of Agriculture, University of Arkansas, Fayetteville, AR 72704

> Volatilization of pesticides has been considered to be a large part of the loss from fields after application; however, few studies have quantified the amount lost to the atmosphere. Volatilization rates of a pre-emergent herbicide, Dual (active ingredient (a.i.) metolachlor 2.24 kg ha^{-1}, (2-chloro-N-(2-ethyl-6-methylphenyl)-N-(2-methoxy-1-methylethyl)acetamide)), were estimated from a 175 ha field in central Iowa over a period of 4 years. A micrometeorological approach for metolachlor was developed and refined. Flux profile concentrations of metolachlor were found to be large during the first 12-24 hours after application and quickly declined for the duration of the study. Precipitation generally contributed to large but short in duration flux peaks. In 1995 results showed that over a 10-day period, 22% of the applied metolachlor volatilized to the atmosphere.

Studies on the fate and impact of agricultural pesticides on environmental quality are necessary to better manage pesticide use in agricultural operations. Improved pesticide use requires an understanding of the processes that can cause potential environmental problems. Pesticide volatilization from soils and plant surfaces represents a source of off-site movement and a major dissipation pathway to the atmosphere for a variety of pesticides used in agriculture (*1-6*). Other modes of pesticide transport to the atmosphere include vapor and droplet drift during application, and wind erosion of soil particulates onto which pesticides are adsorbed.

Pesticide volatilization can be affected by a variety of chemical and environmental factors. Volatility of a pesticide is related to its vapor pressure, while actual volatilization rates are dependent upon microclimatic parameters that modify the effective vapor pressure of pesticides (*7*). Parochetti (*8*) found that chlorpropham volatilization from wet soils increased with soil temperature but declined to almost zero in dry soils. Spencer (9) reported that vapor densities of dieldrin (1a, 2, 2a, 3, 6, 6a, 7, 7a)-3,4,5,6,-9,9-hexachloro-1a, 2, 2a, 3, 6, 6a, 7, 7a-octahydro-2,7:3,6-dimethanonaphth[2,3-b]oxirene;1,2,3,4,10,10-hexachlor-6,7-epoxy-1,4,4a,5,6,7,8,8a-octahydro-endo,exo-1,4:5,8-dimethanonaphthalene in dieldrin-soil mixtures increased with temperature and dieldrin concentration, but were not affected by soil water content.

Relative humidity was reported to have an effect on dieldrin volatilization (10). At very low relative humidities, it was found that dieldrin accumulated at the soil surface, which resulted in rapid volatilization when the surface was re-moistened by exposure to 100% relative humidities. Results from field experiments showed that pesticides applied to the surface of fallow soil initially volatilize at rates proportional to the vapor density of the pure chemical (11). These findings suggested that field pesticide volatilization over an extended period of time is affected by soil water content and incident solar radiation.

Dual, a pre-emergence herbicide (a.i. metolachlor) has become one of the herbicides of choice among many farmers in the Midwest in recent years. (Use of trade names does not imply preference or endorsement by the USDA but are provided to aid the reader.) Metolachlor is a relatively nonvolatile (vapor pressure 0.00173 Pa at 20°C) clear to amber color odorless liquid. Parochetti (12) found that while 0.1% of metolachlor was lost from the soil surface after eight days, approximately 50% was lost from a glass surface, and from 11.5 to 36.6% volatilized from plant residue surfaces. Burkhard (13) calculated that metolachlor at a concentration of 80 µg g^{-1} on a wet soil weight basis volatilizes from moist soil at 20°C at a rate of between 1.5 to 4.5 ng cm^{-2} hr^{-1} when the air flow rate is 30 l hr^{-1}. Actual field volatilization losses of metolachlor have not been adequately quantified.

Nations (14) reported detection of metolachlor along with thirteen other pesticides in rainfall samples in northeastern Iowa over a three year period beginning in October 1987 through September 1990. Hatfield (15) sampled for metolachlor in rainfall samples from April 1991 through August 1994 and found that differences among years were related to rainfall patterns from April through July. Deposition amounts ranged between 20 and 140 µg m^{-2} during the 4 years of the study. Goolsby (16) measured concentrations of atrazine (6-chloro-N-ethyl-N'-(1-methylethyl)-1,3,5-triazine-2,4-diamine; 2-chloro-4-ethylamino-6-iso-propylamine-s-triazine) and alachlor (2-chloro-N-(2,6-diethylphenyl)-N-(methoxymethyl) acetamide; 2-chloro-2',6'-diethyl-N-(methoxymethyl) acetanilide) in rainfall across the Midwest and found similar deposition rates to those reported by Hatfield (15). The mode of metolachlor transport into the atmosphere is not entirely understood but is most likely a combination of drift during application and volatilization after application.

The objective of this paper was to evaluate estimates of metolachlor volatilization in central Iowa from 1992 to 1995 using a flux-gradient based micrometeorological technique.

Theoretical Considerations

Estimating metolachlor volatilization from a field requires reliable estimates of metolachlor vapor leaving a surface. An important consideration is the type of trapping media to be used. A general description would include a range of trapping media from physical adsorption using polyurethane foam plugs (PUF) to chemically based adsorption using activated charcoal. Two types of trapping media were used in Iowa, chemical adsorption using macroporus polymeric beads (XAD-8) that have a high surface area and a chemically homogenous nonionic structure and by physical adsorption with PUF. One method used in sampling pesticide vapor from a field involves use of a chamber with a known volume placed over a soil surface. The chamber is attached to a vacuum pump that draws air from the chamber at constant rate for a given time period. Based on the volume of the chamber, flow rate of the pump, and time of sampling period, flux rates of metolachlor volatilization can be estimated. A considerable drawback to this method is that the volume of air under the chamber is effectively decoupled from the atmosphere above the surface. Increased air temperature, relative humidity and decoupled wind speeds under the chamber can result in an artificial environment affecting the volatilization rates of the pesticide, thus introducing bias into the flux estimates. Another method involves using principles derived from micrometeorology to estimate evaporation and sensible heat flux from a surface. This method allows for pesticide flux estimates to be made *in situ* under natural micrometeorological conditions.

In 1992 the flux-gradient theory was employed using an aerodynamic profile for metolachlor. Bowen-ratio estimates of latent and sensible heat flux densities were used to approximate turbulent-transfer coefficients for metolachlor volatilization. The flux-gradient theory is based on the assumption that turbulent transfer of scalars is analogous to molecular diffusion and can thus be determined as the product of the mean vertical mixing ratio gradient and a turbulent-transfer coefficient (*17*). Expressed in general form, the gradient profile for metolachlor can be defined as:

$$M = K(z)\frac{\partial c}{\partial z} \qquad (1)$$

where M is the metolachlor flux, $K(z)$ is the turbulent-transfer coefficient ($m^2\ s^{-1}$), and $\partial c/\partial z$ (ng m^{-3}) is the concentration gradient of metolachlor as a function of height z above a surface. Volatilized metolachlor flux estimates are based on the assumption that similarity exists in the transport of metolachlor vapor and scalar properties of sensible heat and water vapor. This is reasonable since only the vapor phase of metolachlor is of concern. An alternative method involves an aerodynamic expression from Thornthwaite (*18*) to estimate evaporation in the turbulent layer near a surface and modified by Parmele *(19)* to estimate pesticide flux, and is expressed as:

$$P = \frac{k^2(\overline{c_2}-\overline{c_1})(\overline{u_2}-\overline{u_1})}{\phi_m \phi_h [\ln(z_2/z_1)]^2} \quad (2)$$

where k is the von Karman coefficient ($\approx$ 0.40, dimensionless), c_1 and c_2 are metolachlor concentrations at heights z_1 and z_2 (m), respectively, u_2 is wind velocity (m s^{-1}) at z_2, and ϕ_m and ϕ_h are diabatic stability correction functions (dimensionless) for momentum and sensible heat flux, respectively.

The functions ϕ_m and ϕ_h have been described by several empirical expressions *(20-23)* and represent stability corrections in the boundary layer near the surface. Majewski *(24-25)* successfully used this form of the equation to estimate the volatilization of four different pesticides. In this study the most recent review of field data by Hogstrom *(26)* was used for ϕ_m and ϕ_h. For the unstable case (z/L<0):

$$\phi_m = (1 - 19 \frac{z}{L})^{-0.25} \quad (3)$$

$$\phi_h = 0.95 (1 - 11.6 \frac{z}{L})^{-0.5} \quad (4)$$

while for the stable case (z/L>0):

$$\phi_m = (1 + 5.3 \frac{z}{L}) \quad (5)$$

$$\phi_h = (1 + 8.0 \frac{z}{L}) \quad (6)$$

where z is height (m) and L is the Monin-Obukov length expressed as:

$$L = -\frac{u_*^3}{k\ g/T\ H/\rho c_p} \tag{7}$$

here u_* is the friction velocity (m s^{-1}), k and H have been previously defined, g is the acceleration due to gravity (m s^{-2}), T is air temperature (K), ρ is the density of air (kg m^{-3}), and c_p is the specific heat of air (J kg^{-1} K^{-1}).

In the lower 2-3 m of the boundary layer, z/L can be approximated by the bulk Richardson number (dimensionless) expressed as:

$$Ri = g\frac{(\frac{\partial T}{\partial z})}{T(\frac{\partial u}{\partial z})^2} \tag{8}$$

where all terms have been previously defined. Metolachlor fluxes were estimated from measurements of metolachlor concentration profiles and then corrected for atmospheric stability effects using data from wind speed and temperature profiles.

The micrometeorological requirements for the Thornthwaite-Holzman technique (*19*) must include a large uniformly surfaced area with an upwind distance (fetch) of at least 100 times the height of the instruments to ensure that the boundary layer is in equilibrium with the underlying surface (*27*) as well as sampling periods of sufficient duration to approximate steady state conditions.

Materials and Methods

Beginning in 1992 and continuing through 1995, a series of field scale volatilization studies were conducted in the Walnut Creek watershed located approximately 10 kilometers south of Ames, Iowa (lat. 41° 5730'; long. 93° 3730'). These studies coincided with normal planting operations for corn (*Zea mays* L.), which ranged from late April to early June, depending on local weather conditions. The length of each experiment averaged between 7-10 days after the pre-emergent herbicide was applied. A 175 hectare field was selected to ensure adequate size and fetch requirements necessary for a micrometeorological based method to estimate metolachlor volatilization. In this field two major soil types were represented, a Webster and Clarion series (fine-loamy, mixed, mesic Typic Haplaquolls) with an average soil organic carbon content ranging between 22-34 g kg^{-1} and a 0-2% slope. A complete description of the watershed is provided by Hatfield (*28*).

A preliminary study was first conducted in 1992 using the flux-gradient technique involving equations 1 and 2. Turbulent diffusivity coefficients (K) in equation 1 were derived from Bowen-ratio estimates of sensible heat flux. The trapping media used in 1992 was XAD-8. In this study the field was treated with a pre-emergent herbicide (tradename Dual, a.i. metolachlor) using a 0.25 m banded strip over a 0.76 m wide row at a rate of 2.24 kg ha^{-1}. Fields surrounding the study site were not treated with metolachlor thus ensuring a buffer of approximately 1 km that were not treated with metolachlor. Air sampling for metolachlor vapor typically began within one hour after the herbicide application was completed.

Micrometeorological instruments were deployed near the center of the field adjacent to the metolachlor sampling mast to measure surface energy balance components of net radiation (R_n), soil heat flux (G), and sensible (H) and latent heat flux (E) densities (W m^{-2}). Net radiation and soil heat flux were measured using a REBS Q*7 net radiometer and a REBS HFT-1 soil heat flux plate. Sensible and latent heat fluxes were estimated using Bowen-ratio and eddy covariance techniques. Bowen-ratio instrumentation included two aspirated psychrometers positioned at 0.25 and 1.25 m above the surface to measure air temperature and relative humidity gradients.

A Campbell Scientific 1-d sonic anemometer and krypton hygrometer were used for the eddy covariance measurements. In addition, a 10 m wind speed and aspirated psychrometer profile tower was erected near the experimental site to characterize boundary-layer stability conditions from wind speed and temperature profiles near the surface. Anemometers and psychrometers were positioned on the 10 m tower at 1.10, 1.57, 2.27, 3.26, 4.62, 6.62, and 8.74 m above the soil surface. The data from the 10-m tower were used to estimate parameters for calculating L and Ri in equations 3-8. Meteorological data were sampled on a 10 second interval and averaged for 30 min. Eddy covariance measurements were sampled at 10 Hz and internally averaged in the datalogger every 10 min, and then averaged and output every 30 min. In subsequent studies following 1992 in the same field, micrometeorological and pesticide sampling equipment were used. The only differences were (1) PUF plugs were used in place of XAD-8, and (2) different microclimate conditions as dictated by changing weather patterns from year to year.

Pesticide Sampling Mast. Pesticide sampling masts were constructed of 3 m length by 0.0254 m i.d. galvanized steel pipes vertically erected in the field adjacent to the micrometeorological instruments. Clamps were attached along the length of the pipes to hold sampling canisters made of glass tubes that were 0.0254 m i.d. by 0.15 m long. The canisters were tapered at one end to a stem of 0.0085 m diameter onto which tygon tubing was attached and connected to a Staplex (Model TFIA) high volume rotary motor calibrated to a flow rate of 40 l min^{-1} through each sampling canister. Vertical placement of the canisters was 0.15, 0.27, 0.51, 0.69, 1.25 and 3.00 m above the soil surface. In 1992 the trapping media used was XAD-8, after which in subsequent studies PUF plugs were employed. This was done for two reasons: (1) PUF plugs were a fraction of the cost of XAD-8, and (2) the trapping efficiencies for metolachlor were greater at higher flow rates with the PUF plugs than with XAD-8. Each sampling canister contained PUF plugs with dimensions of 0.0254 m in diameter by 0.075 m in length and placed end-to-end in the sampling canister. The second PUF plug served as a trap for breakthrough flow from the first PUF plug. The individual canisters were wrapped with aluminum reflective tape to protect against photodegradation. The PUF plugs were cut from standard carpet polyurethane foam sheets and pre-cleaned using separate methanol and hexane washes and allowed to air dry. Twenty-five (out of 1500) PUF plugs were randomly selected and analyzed for general chromatogram cleanliness and particularly for interferences in the region of metolachlor absorbance using gas chromatography (GC) and confirmed with mass spectrometry (MS). Within one hour after each sampling exchange, the PUF plugs were labeled and stored in a freezer at a temperature of -10°C.

Metolachlor Extraction. Extraction of XAD-8 for metolachlor was done by rinsing the polymeric beads with 100 ml of ethyl acetate for a period of 30 minutes. The eluate was evaporated to 1.5 ml and prepared for analysis with GC/MS. The extraction process of the PUF plugs was accomplished using a Hewlett Packard (HP) 7620 Super Critical Extraction (SFE) method with carbon dioxide as the extracting fluid at a density of 0.70 g ml^{-1} and a pressure of 11.5 MPa. The plugs were eluted for 45 minutes trapping the metolachlor on a diol cartridge that was eluted with ethyl acetate. The extraction efficiency averaged greater than 90%. Analysis of the extract was accomplished using GC/MS in the selected ion mode (SIM). The GC was equipped with an HP Ultra-1 column (12 m x 0.2 mm x 0.33 μm). A temperature program of 50-200°C at 6°C min^{-1} was used. The minimum detection limit was 30 ng per PUF plug. The MS was run at 70 eV with an ion source temperature of 250°C.

Results

Ambient Micrometeorological Conditions. Typical ambient air temperatures during the month of May in central Iowa averaged 25-30°C during mid-day and afternoon periods while nighttime temperatures ranged between 6-9°C. Average wind speeds were between 2-5 m s^{-1} during the day and 1-2 m s^{-1} at night. Wind directions during daytime and evening hours typically were from the south-southwest. On occasion the wind direction would shift to the north-northwest after midnight for approximately 6-8 hours and then slowly rotate back to the south-southwest. Since the meteorological and pesticide sampling instrumentation were positioned in the center of the fields, wind directions from either the north or south easily provided the minimum fetch to height ratios as suggested by Panofsky (27).

Preliminary Results from 1992. In 1992 using XAD-8 as the trapping media, metolachlor fluxes were estimated using equations 1 and 2. Bulk K values (equation 1) were estimated from the Bowen-ratio estimates of sensible heat. The ϕ functions *(26)* in equation 2 were derived using wind speed and temperature measurements from the 10 m tower located at the site. Sampling did not begin until approximately 36 hours after the metolachlor was applied due to a malfunctioning vacuum pump. The results in Figure 1 show daytime averages (periods when $R_n > 0$) beginning on DOY 126 and ending on DOY 162. Good agreement was evident between the two methods. The two peaks observed near DOY 132 and 139 represent periods after a light precipitation event (≈ 10 and 20 mm, respectively). Increased soil water content at the surface contributed to greater dissociation of metolachlor from the soil particles thus increasing the source concentration for volatilization. Shortly after DOY 140 metolachlor fluxes quickly decreased and remained relatively constant at near non-detectable levels for the duration of the study. We speculate that the decrease was related to the second precipitation event near DOY 139 (≈ 20 mm) when the remaining available metolachlor source had been dissociated from the surface soil particles and transported downward into the soil profile by infiltrating water and by the increased (burst) volatilization shortly after the precipitation event. From these preliminary results we learned that through careful monitoring of the sampling and micrometeorological instruments it is possible to obtain reliable field estimates of metolachlor volatilization and thus laid the foundation for continued metolachlor volatilization research in central Iowa. In subsequent studies we adopted use of the PUF plugs into our sampling protocol as well as changing our sampling interval. In 1992 only the daytime period was sampled, while in 1993-1995 sampling was done on a 24-hour basis for the duration of each study. Data collection continued regardless of weather conditions. During the first 96 hours after application, the sampling interval was every 2 hours, that is, every 2 hours the PUF plugs were changed. Because of continuous decreasing metolachlor concentration source at the surface, after the first 96 hours, the frequency of the PUF plug exchange was changed to every 4 hours until the end of each experiment. We also used equation 2 exclusively, as Bowen-ratio estimates of latent and sensible heat fluxes during neutral and nighttime periods are generally unreliable, thus rendering estimates of K unreliable. Using this information we can implement the protocols for field volatilization studies of metolachlor.

Metolachlor Concentration Profiles. In order for the micrometeorological approach to be valid, metolachlor vapor concentration profiles need to be generally log-linear with height. This is a fundamental micrometeorological requirement based on the assumption of a constant flux layer near the surface. A log-linear distribution of metolachlor or of any atmospheric constituent (heat, water vapor, CO_2) ensures that the air is well mixed and in equilibrium with the underlying surface.

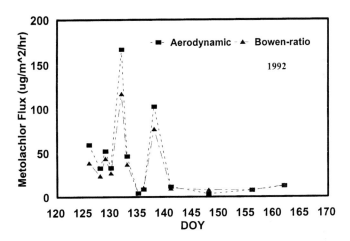

Figure 1. Daily average metolachlor flux estimates using equations 1 and 2 in 1992.

An example of metolachlor vapor concentration profiles during the first two hours after application in 1993 and 1995 are shown in Figures 2 and 3. Profiles were clearly log-linear with height shortly after application with the highest concentration nearest the surface and the lowest furthest away from the surface. The log-linear profiles indicate a well-mixed boundary layer at the surface ensuring that the assumption of an existing constant flux layer near the surface is valid. Figure 4 shows the metolachlor concentration profile for 1995 approximately 7 days after application. The profile can be observed to continue to be log-linear despite the fact that concentration values were significantly lower than at the beginning of the study as observed in Figure 3.

Log-linear profile distributions of metolachlor concentrations were observed each year of the study. Vapor concentrations varied at each height during the course of the study in response to varying concentration source areas and diurnal micrometeorological conditions. Satisfied that the profiles were valid and assuming transport similarity to sensible heat and momentum profiles, metolachlor vapor concentration gradients were related to stability corrections for sensible heat and momentum to calculate metolachlor flux densities (equation 2) from the surface.

Metolachlor Flux Estimates. Metolachlor concentration gradients and meteorological data were used to calculate metolachlor fluxes using equation 2. Figures 5 and 6 show estimated metolachlor fluxes in 1994 for DOY 126 and 127 (May 6-7). This is an example of a period shortly after application (≈ 30 hours) during what was characterized as a dry spring. No precipitation had occurred for 12 days prior to DOY 126. Peak flux estimates are observed to follow a diurnal trend associated with clear sunny conditions with warm temperatures. Metolachlor fluxes were low in the early morning hours and increased rapidly with positive available energy (R_n-G), then decreased in the afternoon hours and into the nighttime period. During the early morning hours of DOY 127 there was a brief precipitation event (≈ 10 mm) followed by patchy cloudy conditions during the daylight period of DOY 127. Significant fluctuations in metolachlor fluxes can be observed that are related to changes in available energy at the surface as periods of passing clouds changed to periods of sunny conditions. Peak fluxes were 2.5 times greater than the previous day (Figure 5) as a result of wet soil surface conditions from the precipitation. Figures 5 and 6 clearly show metolachlor fluxes responding to local micrometeorological conditions. Continuous results of metolachlor volatilization for approximately 11 days in 1995 are shown in Figure 7. In this particular study the planting practice for the field site was a ridge-till technique where most of the previous year's corn residue remained on the soil surface over undisturbed ridges. The soil surface area covered by the corn residue was approximately 80%. In order to display volatilization changes over the course of the study, the results in Figure 7 are expressed in log-scale. This was done because of the disproportionate flux magnitude shortly after application relative to that several days after application. The greatest volatilization occurred during the

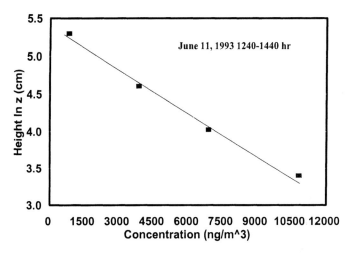

Figure 2. Metolachlor concentration profiles with height 2 hours after application on June 11, 1993.

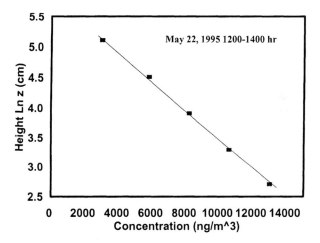

Figure 3. Metolachlor concentration profiles with height 2 hours after application on May 22, 1995.

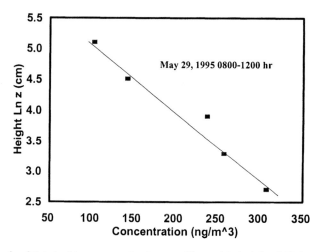

Figure 4. Metolachlor concentration profiles with height 168 hours after application on May 29, 1995.

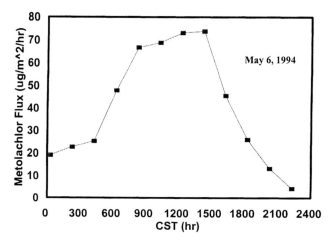

Figure 5. Metolachlor flux estimates for a 24-hour period on May 6, 1994 under clear sky conditions.

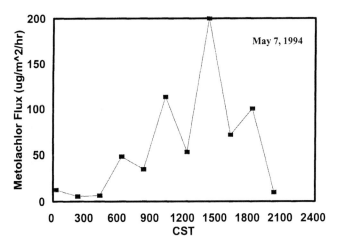

Figure 6. Metolachlor flux estimates for a 24-hour period on May 7, 1994 under patchy-cloud conditions.

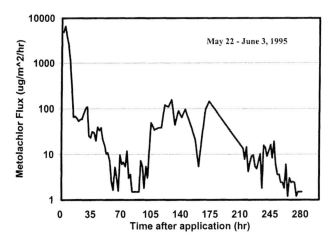

Figure 7. Continuous metolachlor flux estimates beginning on May 22 and ending on June 3, 1995.

first 36 hours after application. Peak fluxes occurred during the first 6 hours after application (7000 μg m^{-2} hr^{-1}) and continued to be large during the first 24 hours after application during which volatilization rates can be observed to diminish sharply. A precipitation event occurred after 100 hours after application, which resulted in a large volatilization flux response for a period of almost 70 hours (Figure 7). The last significant precipitation event occurred from DOY 127 (May 7) through DOY 129 (May 9) totaling 66 mm of precipitation approximately 9 days prior to the application of metolachlor. Precipitation may have promoted dissociation of metolachlor from the soil matrix and particularly from the corn residue in the field, thus enabling the metolachlor to easily volatilize from the different surfaces. The corn residue in the interrows of the field was multi-layered and had an average depth ranging from 0.10-0.15 m. Corn residue covered approximately 80% of the surface so that most of the metolachlor was applied onto a combination of crop residue and soil surfaces. Parochetti (12) had shown that metolachlor adsorption to plant residue is considerably weaker than to soil, thus during the precipitation event much of the metolachlor source was redistributed from the surface to various corn residue layers below before it reached the soil surface. Residue covered soils tend to remain wetter longer than non-residue covered soils due in part to the increased resistance to evaporation from the residue. As the corn residue dried layer by layer from evaporation, metolachlor became readily available for volatilization, thus resulting in an extended period of large volatilization fluxes following the precipitation event. Volatilization losses diminished to near non-detectable levels after 240 hours.

The percent loss of applied metolachlor from volatilization was calculated as the ratio of the total volatilized for the study period to the total of active ingredient metolachlor applied per hectare. Based on an assumption of no losses due to drift during application and that subsequent volatilization losses after the study period were negligible, approximately 22% of the applied metolachlor had volatilized from the field in 1995. This represents a significant loading of metolachlor to the atmosphere.

Conclusions

Metolachlor volatilization was estimated using a flux profile gradient technique. Data were collected over a period of years from the same field during which the technique was continually refined and evaluated. Measurements of volatilized metolachlor were found to be consistently log-linear with height above the surface throughout several years of field volatilization research. Combining metolachlor concentration gradients with stability corrections for heat and momentum calculated from the local micrometeorological data resulted in estimates of metolachlor fluxes. Volatilization data from the treated fields show that 95% of the metolachlor flux can occur during the first 12-24 hours after application. Precipitation occurring shortly after application can result in increased rates of volatilization. Total volatilization losses from the 1995 study were 22% of the total metolachlor applied assuming no loss to drift during application. Depending on wind and temperature conditions during application, drift losses can range between 5-30%, resulting in an even larger percent of total applied losses *(29-30)*. Metolachlor loss from volatilization can be significant and appears to be a function of residue surface cover, timing of precipitation relative to application, and local microclimate parameters. Micrometeorological based techniques can provide a reliable method of estimating field scale volatilization of pesticides.

Literature Cited

1. Caro, J.H.; Taylor, A.W. *J. Agric. Food Chem.* **1971**, *19*, 379-384.
2. Cliath, M.M.; Spencer, W.F. *Soil Sci. Soc. Am. J.* **1971**, *35*, 791-795.
3. Farmer, W.J.; Igue, K.; Spencer, W.F.; Martin, J.P. *Soil Sci. Soc. Am. J.* **1972**, *36*, 443-447.
4. Farmer, W.J.; Spencer, W.F.; Shepherd, R.A.; Cliath, M.M. *J. Environ. Qual.* **1974**, *3*, 343-346.
5. Harper, L.A.; White, Jr., A.W.; Bruce, R.R.; Thomas, A.W.; Leonard, R.A. *J. Environ. Qual.* **1976**, *5*, 236-242.
6. Spencer, W.F.; Farmer, W.J. In *Dynamics, Exposure, and Hazardous Assessments of Toxic Chemicals in the Environment.* Hague, R., Ed.; Assessment of the vapor behavior of toxic organic chemicals. Ann Arbor Sci.: Ann Arbor, MI, 1980; pp 142-162.
7. Basile, M.; Senesi, N.; Lamberti, F. *Agric., Eco. and Environ.* **1986**, *17*, 269-279.
8. Parochetti, J.V.; Warren, G.F. *Weeds,* **1966**, *14*, 281-285.
9. Spencer, W.F.; Cliath, M.M.; Farmer, W.J. *Soil Sci. Soc. Amer. Proc.* **1969**, *33*, 509-511.
10. Spencer, W.F.; Cliath, M.M. *J. Environ. Qual.* **1973**, *2*, 284-289.
11. Glotfelty, E.D.; Taylor, A.W.; Turner, B.C.; Zoller, W.H. *J. Agric. Food Chem.* **1984**, *32*, 639-643.
12. Parochetti, J.V. *Weed Sci. Soc. Am.* **1978**, *Abstr. #17.*
13. Burkhard, N.; Guth, J.A. *Pestic. Sci.* **1981**, *12*, 37-44.
14. Nations, B.K.; Hallberg, G.R. *J. Environ. Qual.* **1992**, *21*, 486-492.
15. Hatfield, J.L.; Wesley, C.K.; Prueger, J.H.; Pfeiffer, R.L. *J. Environ. Qual.* **1996**, *25*, 259-264.
16. Goolsby, D.A.; Thurman, E.M.; Pomes, M.L.; Meyer, M.T.; Battaglin, W.A. *Environ. Sci. Tech.* **1997**, *31*, 1325-1333.
17. Lumley, J.L.; Panofsky, H.A. *The structure of atmospheric turbulence*; Interscience: New York, 1964; 239 p.
18. Thornthwaite, C.W.; Holzman, B. *Mon. Weather Rev.* **1939**, *67*, 4-11.
19. Parmele, L.H.; Lemon, E.R.; Taylor, A.W. *Water, Air and Soil Poll.* **1972,** *1*, 433-451.

20. Swinbank, W.C. *Quart. J. Roy. Meteorol. Soc.* **1968**, *94*, 460.
21. Dyer, A.J.; Hicks, B.B. *Quart. J. Roy. Meteor. Soc.* **1970**, *96*, 715-721.
22. Pruitt, W.O.; Morgan, D.L.; Lourence, T.J. *Quart. J. Roy. Meteorol. Soc.* **1973**, *99*, 370-386.
23. Dyer, A.J.; Bradley, E.F. *Boundary-Layer Meteorol.* **1982**, *22*, 39.
24. Majewski, M.S.; Glotfelty, D.E.; Seiber, J.N. *Atmos. Environ.* **1989**, *23*, 929-938.
25. Majewski, M.S.; Glotfelty, D.E.; Paw U, K.T.; Seiber, J.N. *Environ. Sci. Technol.* **1990**, *24*, 1490-1502.
26. Hogstrom, U. *Boundary-Layer Meteorol.* **1996**, *78*, 215-246.
27. Panofsky, H.A.; Townsend, A.A. *Quart. J. Roy. Meteor. Soc.* **1963**, *90*, 147-155.
28. Hatfield, J.L.; Jaynes, D.B.; Burkart, M.R.; Cambardella, C.A.; Moorman, T.B.; Prueger, J.H.; Smith, M.A. *J. Env. Qual.* **1999**, *28*, 11-24.
29. Nordbo, E.; Kristensen, K.; Kirknel, E. *Pestic. Sci.* **1993**, *38*, 33-41.
30. Arvidsson, T. Spray drift as influenced by meteorological and technical factors; Doctoral dissertation, Swedish University of Agricultural Sciences, Uppsala, Sweden, 1997; pp 9-144.

Chapter 15

Pesticides in Ambient Air and Precipitation in Rural, Urban, and Isolated Areas of Eastern Iowa

M. E. Hochstedler, D. Larabee-Zierath, and G. R. Hallberg[1]

University of Iowa Hygienic Laboratory, Iowa City, IA 52242

Atmospheric transport and deposition of pesticides are significant issues in better understanding human and environmental exposure. Pesticide concentrations in air and precipitation were measured over a one-year-period at four sites in Iowa chosen to characterize rural, urban and regional effects. Twenty-eight pesticides including twenty-one herbicides and seven insecticides were detected in precipitation during the sampling period, October 1996 through September 1997. Pesticide concentrations were greatest during the planting-growing season, April through August. Peak concentrations reached 0.96 ug/L for acetochlor, 1.1 ug/L for 2,4-D, and 3.5 ug/L for atrazine, all commonly used herbicides in Iowa. Concentrations were generally higher at the farm site but were present at all sites indicating distant transport. Measurable air concentrations were occasionally seen, but most were below detection limit for the volume of air sampled.

Iowa is noted for its extensive row-crop agriculture; 94% of the state's area is farmland. Most years over two-thirds of the state's land receives pesticide applications. It is estimated that Iowa applies more herbicide active ingredients than any other state in the USA (1). Preliminary studies of pesticides in precipitation in Iowa suggest that losses to the atmosphere may account for greater environmental losses than those measured in runoff or leached into groundwater. These studies indicated concentrations in rainfall near farm fields ranged as high as 40 ug/L for atrazine. Methyl parathion, malathion, and diazinon were detected in rainfall in Iowa City, but not in farm/rural samples, suggesting an urban input. These compounds are commonly used for lawn and garden insect control (1,2). Prior observations suggest that "scavenging" of pesticides, the incorporation of dissolved gases and particulates

[1]Current address: The Cadmus Group, Inc., Watham, MA 02154

into precipitation, is responsible for their occurrence (*1,3,4,5,6*). While high concentrations have been observed in rainfall, air concentrations may be even greater, a consequence of incomplete scavenging.

The regional occurrence of pesticide residues in precipitation would suggest that their occurrence in air is likely an important route of exposure to the general public, as well as to farmers. Pesticides in ambient air, dust and precipitation do contribute to inhalation and dermal exposures. To non-applicators, such as the general public, the data suggest that inhalation, as well as dermal contact, are routes of routine low-concentration exposure, similar to drinking water in parts of Iowa. Both air and precipitation are important sources of pesticide deposition in non-target areas.

Our study sought to enhance the understanding of the route of human and environmental exposure to pesticides by measuring concentrations in air and precipitation over a one-year-period at four site locations in Iowa. We evaluated the significance of location for sample collection and explored the difference between wet and dry precipitation. Samples were analyzed for a large number of pesticides to extend the database on pesticide exposure.

Materials and Methods

Four site locations within Johnson County, Iowa, were chosen for our study. Figure 1 shows Johnson County in relationship to the rest of Iowa and the sampling site locations within the county. The Farm collection site, located three miles west of Iowa City, was used to characterize local transport effects. This site is surrounded by 260 acres of farmland. The Urban collection site was chosen to measure urban effects and distant transport from agriculture. It is located in an older residential neighborhood in central Iowa City. The Macbride collection site was chosen as a control in an attempt to find a site isolated from localized effects. This site is located ten miles outside of Iowa City in a recreational park which is undeveloped for agricultural or residential use. The Oakdale collection site was chosen for continued monitoring after the initial study. This site has characteristics common to all three and is located at the University of Iowa Hygienic Laboratory. It is near cropland, urban areas, uncultivated fields (pasture), and wooded park areas.

A total of 222 precipitation samples, 124 dry deposition samples, and 288 air samples were collected and analyzed during the one year sampling period, October 1996 - September 1997. All samples were analyzed at the University of Iowa Hygienic Laboratory. Seventy-eight compounds were included in the protocols (See Table I). This list is based on those analytes routinely determined by gas chromatography in the Pesticide Residue Section of the University Hygienic Laboratory.

Air Samples. Air samples were collected and measured using an adaptation of U.S. EPA Method IP-8 (*7*). Approximately 1.5 m^3 of air was sampled over a 24-hour-period using a programmable, battery-operated pump operating at 3 L/min. The pump cycled

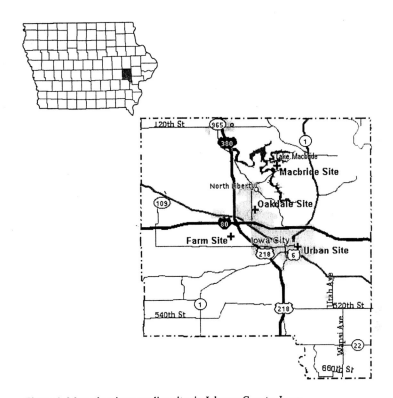

Figure 1. Maps showing sampling sites in Johnson County, Iowa.

Table I. Pesticides and related compounds included in the study. Number of detections in 634 samples from all matrices are given in parenthesis.

Nitrogen Containing Compounds	Organophosphate Insecticides	Chlorinated Herbicides and Related Compounds
EPTC	Isofenphos	2,4-D(14)
Butylate	Phorate	Dicamba(6)
Propachlor	Dimethoate	Silvex(4)
Atrazine(59)	Dyfonate	Bentazon(4)
Trifluralin(1)	Disulfoton	Picloram(4)
Metribuzin	Methyl parathion	Dichlorprop(15)
Cyanazine(16)	Parathion	Pentachlorophenol(16)
Alachlor(2)	Ethoprophos	2,4,5-T(2)
Metolachlor(29)	Terbufos	Chloramben(2)
Pendimethalin(11)	Diazinon	Acifluorfen(1)
Prometon	Malathion	Dinoseb
Propazine	Chlorpyrifos(3)	Bromoxynil(2)
Triallate	Dichlorvos	2,4-DB(1)
Acetochlor(20)	**Metabolites**	Dacthal
Bromacil	Desethyl	MCPP
Butachlor	atrazine(11)	MCPA
Simazine	Desisopropyl	Triclopyr(1)
Carbaryl	atrazine(5)	
Carbofuran(1)		

Chlorinated Insecticides and Related Compounds

alpha-BHC	Heptachlor(6)	Chlordane
beta-BHC	Heptachlor epoxide	Toxaphene
Lindane(4)	Endrin	Aroclor 1016
delta-BHC	Endrin aldehyde(7)	Aroclor 1221
Aldrin	Endrin ketone(1)	Aroclor 1232
Dieldrin(3)	Endosulfan I(1)	Aroclor 1242
DDT(3)	Endosulfan II	Aroclor 1248
DDD	Endosulfan sulfate	Aroclor 1254
DDE(7)	Methoxychlor	Aroclor 1260

on for one minute and off for two, giving rise to a 1.5 m³ air sample per 24 hrs. Air samples were passed through a collection device consisting of an open glass tube, containing a specially cleaned polyurethane foam (PUF) plug. Airborne particulate matter was retained on a glass fiber pre-filter.

PUF plugs were extracted in a Soxhlet extractor for 16 hours with 5% diethyl ether in n-hexane and the extracts concentrated to 3 mL in a Kuderna-Danish concentrator. They were then analyzed by gas chromatography using either electron capture or nitrogen-phosphorus detection. Because the extracts were relatively free of artifacts, protocols for drinking water methods were used. The glass pre-filters were analyzed in the same manner to assess particulate loading.

Precipitation and Dry Deposition Samples. Precipitation samples were collected using automated, wet/dry samplers of the design used for the National Atmospheric Precipitation Assessment Program (Aerochem Metrics Model 301, Bushnell, FL). This collector consisted of two buckets, each containing a large glass beaker. An automated cover was connected to a motor and sensor activated by water. Whenever the sensor detected rainwater, the motor moved the cover from one bucket to the other. Thus wet and dry precipitation were collected sequentially. The rainwater in the wet bucket remained covered to minimize evaporation loss. Samples were also collected in galvanized steel basins (hog pans) to provide adequate sample size during small rainfall events. These basins collected both wet and dry deposition, also known as bulk precipitation. The wet/dry collectors generally operated well through the winter months during snow events. Some difficulty was encountered when the collected snow melted, then refroze, breaking the container resulting in complete loss.

Protocols defined in EPA Methods 507, 508 and 515.1 (*8*) were selected because low detection limits were a priority and samples were generally free of interferences. The precipitation samples were transferred to 1-L glass jars with Teflon lined polypropylene lids specifically cleaned for pesticide analysis. Precipitation collectors were rinsed with distilled water after sample removal and returned to the collector. Samples were refrigerated and extracted by solvent extraction within a 7-day holding period after collection. After concentration and exchange into n-hexane, the extracts were analyzed by gas chromatography with either electron capture or with nitrogen-phosphorus detectors.

Samples were collected from the dry bucket beakers by rinsing each with 200 mL of acetone in the field. The acetone was transferred to 1-L jars for transport to the laboratory. The acetone extract was concentrated and exchanged into n-hexane for analysis by gas chromatography as above.

Method detection limits (MDL) for all compounds generally ranged between 0.01-0.2 ug/L for rain and snow, depending on the analyte and the volume of sample collected. For air the range of detection limits for all compounds was 0.1-0.2 ug/m³. Dry sampler detection limits were in the range of 0.05-0.2 ug/sample, also dependent on the analyte. Compounds not classified as herbicides or insecticides were included in the analytical protocol since they are target compounds in the methods used for this study. Polychlorinated biphenyl's (PCB's) have traditionally been determined by chlorinated hydrocarbon insecticide methods and are included in EPA Method 508. Pentachlorophenol, 4-nitrophenol and 3,5-dichlorobenzoic acid are included in EPA Method 515.1.

QA/QC. The University Hygienic Laboratory follows very strict quality assurance and quality control (QA/QC) guidelines to maintain highest degrees of precision and accuracy. These procedures include, but are not limited to: daily instrument calibration verification, interference checks (blanks), standards verification, and assessment of extraction efficiencies as well as PUF plug capture efficiencies.

Duplicate air samples were taken every other week at one site, rotating the four site locations. Not enough results from air samples were positive to give meaningful statistics for sample duplicates. One reagent blank was prepared for each set of air samples, to test for false positive results. 4-Nitrophenol was often tentatively identified by dual-column gas chromatography; however this result could not be confirmed by mass spectrometry at the concentrations observed. A field exposed spike, prepared by spiking 100 uL of known concentrations of analytes in acetone to the inlet side of the polyurethane foam plug, was taken monthly at each site for air sampling. These spikes generally averaged near 100% for chlorinated hydrocarbon insecticides with standard deviations between 8-14%. Field exposed spikes for acid herbicides gave lower recoveries with 2,4-D averaging 66 % and silvex averaging 80%. Recoveries of nitrogen containing herbicides were generally very good, typically averaging 100%, with standard deviations of 10%. Some pesticides were not recovered from field exposed spikes. Comparisons with spikes prepared in the laboratory suggests that butylate volatilizes from the collector, while trifluralin is poorly extracted from the PUF plug.

Laboratory fortified blanks were prepared with each set of wet precipitation and dry deposition samples. Recoveries fell in the ranges specified by the EPA drinking water methods 507, 508, or 515.1. Blanks prepared with each set of samples showed no contamination.

Results

Figure 2 shows rainfall data from the Iowa Institute of Hydraulic Research, University of Iowa, during the sampling period with corresponding long-term averages. Notably, May was above normal and June and July were below normal.

Pesticides were detected in all sample types: air, wet precipitation, dry deposition and bulk precipitation, at all four locations during the one year study. May and June were months of the highest frequency of pesticide detections. Thirty-four of the 78 compounds in the screening protocol were detected. Of the total of 634 samples collected, the most frequently detected pesticides and related compounds were atrazine (8.7%), metolachlor (4.6%), acetochlor (3.1%), cyanazine (2.5%), pentachlorophenol

(2.5%), dichlorprop (2.4%), and 2,4-D (2.2%). These pesticides were found at all four sampling site locations. Figure 3 shows a comparison of atrazine in bulk precipitation and wet precipitation. The distinctions between the two do not seem to be significant.

Precipitation. Atrazine occurrence in rainwater at each of the four sites is shown in Figure 4. As the most frequently detected pesticide, it was quantified at a maximum concentration of 3.5 ug/L at the Farm site in bulk precipitation. This is not surprising, atrazine is widely used in Iowa, and is the most frequently detected herbicide in most other water quality studies within the state (*1*). There were 8 million lb. of atrazine used on corn in Iowa in 1996 (*9*). Detected only during the spring and summer months of April through August, atrazine occurrence shows a definite seasonal trend. It is noteworthy that atrazine was detected more frequently at the Urban site than any of the others, including the Farm site.

Figure 5 shows data for metolachlor, the second most frequently detected pesticide in this study. It is also widely used in Iowa. Its occurrence shows the same trend as atrazine, being detected only during the spring and summer months and at all four collection site. In contrast to the atrazine case, the concentrations of metolachlor were at or near the limit of detection. Consequently, the sensitivity of this measurement was very sample volume dependent. Figure 4 also illustrates the dependence of the determination upon sample volume. Some very low level detections may have been missed using this protocol. Acetochlor occurrence in rainwater at each of the four sites is shown in Figure 6. As the third most frequently detected pesticide, it was quantified at a maximum concentration of 1.1 ug/L at the Farm and Macbride sites. While the Macbride site was chosen to represent a non-agricultural source term, it is interesting to note that the same concentration was detected at the Macbride site as at the Farm site, where the pesticide was likely applied. Unlike atrazine and metolachlor, acetochlor does not have a long use-history on Iowa row crops. Figure 7 shows the levels of 2,4-D detected in rainwater at all four sites. It follows the same trends as atrazine, metolachlor, and acetochlor. It is noteworthy to point out that 2,4-D is the only herbicide of these 4 detected in rainwater which is currently registered for both agricultural and non-agricultural (commercial and residential) use.

Table II shows chlorinated insecticides in all types of precipitation for all four sites. These insecticides, most of which are no longer registered for agricultural usage, were present in ultratrace amounts, and detected throughout the year, showing no seasonal trends.

Air Samples. Table III shows results for pesticides in air samples collected using the PUF plug. Most detections were below the quantitation limit of 0.5 ug/sample. At the farm site 2,4-D was detected in air at a maximum concentration of 0.34 ug/sample. Both pentachlorophenol and endrin aldehyde were detected more frequently than any other pesticides and they were detected at all four sampling sites, and more frequently than any other pesticides. Other pesticides found in air include atrazine, cyanazine, dicamba and lindane. These were found in air samples collected in the fall (October and November) and also in the spring/summer (May through September). The

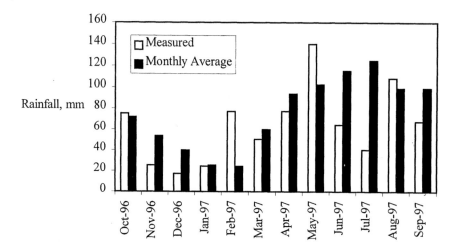

Figure 2. Comparison of measured rainfall versus average amounts during the time of the study.

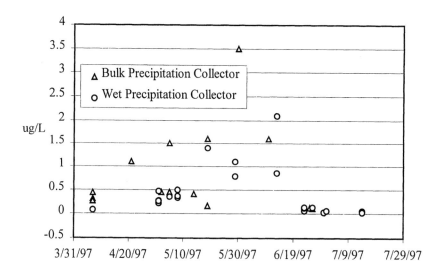

Figure 3. Atrazine concentrations measured in rainfall events collected in the wet sampler versus that collected in the bulk precipitation collector.

225

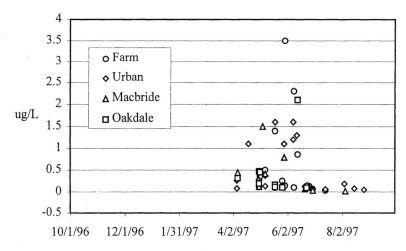

Figure 4. Atrazine found in rainfall events at each of the four sampling sites.

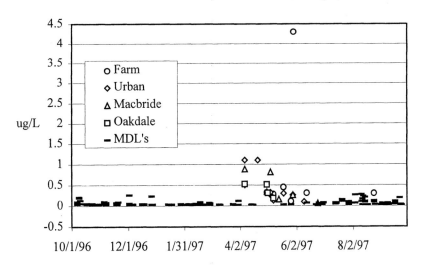

Figure 5. Metolachlor found in rainfall events at each of the four sampling sites. The MDL values shown are dependent on sample volume and illustrate how close the results are to the method sensitivity.

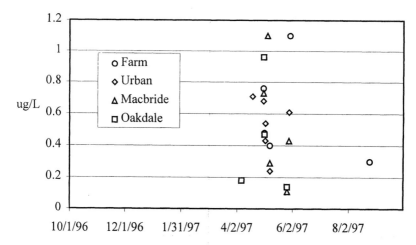

Figure 6. Acetochlor found in rainfall events at each of the four sampling sites.

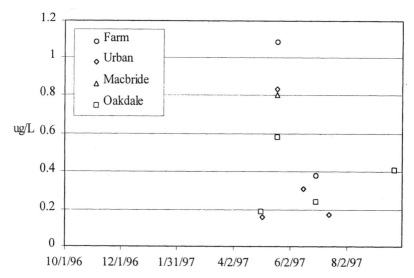

Figure 7. 2,4-D found in rainfall events at each of the four sampling sites.

Table II. Rain and bulk precipitation data for the chlorinated insecticides detected from each site for each sample. The value is the concentration found in each rainfall or snowfall event.

Month	Farm Analyte	ug/L	Urban Analyte	ug/L	Macbride Analyte	ug/L	Oakdale Analyte	ug/L
Oct-96								
Nov-96								
Dec-96								
Jan-97	Heptachlor Chlordane	0.016 0.037 0.038	Dieldrin Endosulfan I Heptachlor Lindane	0.0093 0.0081 0.011 0.035 0.0099 0.018			Heptachlor Lindane Chlordane	0.0073 0.016 0.021 0.010 0.038
Feb-97								
Mar-97								
Apr-97			EK[1]	0.0094				
May-97								
Jun-97	Dieldrin	0.0085	Dieldrin	0.0036				
Jul-97								
Aug-97								
Sep-97								

[1] EK is endrin ketone

Table III. Air sampler data for all analytes from each site. The value is the concentration in the air at each sampling.

Month	Farm Analyte	ug/m³	Urban Analyte	ug/m³	Macbride Analyte	ug/m³	Oakdale Analyte	ug/m³
Oct-96	PCP[1]	0.07	PCP[1]	0.032	Dicamba	0.018	PCP[1]	0.072
					PCP[1]	0.1		0.058
								0.076
Nov-96					PCP[1]	0.025	Cyanazine	0.11
Dec-96								
Jan-97								
Feb-97								
Mar-97					EA[2]	0.0097		
Apr-97								
May-97			EA[2]	0.031			EA[2]	0.021
Jun-97	2,4-D	0.34						
	Atrazine	0.14						
Jul-97	Lindane	0.47	EA[2]	0.0089			2,4-D	0.22
Aug-97	EA[2]	0.0093			Atrazine	0.014		
Sep-97								

[1] PCP is pentachlorophenol
[2] EA is endrin aldehyde

relatively high concentration of lindane determined at the Farm site in July is fully consistent with the volatility of this insecticide, as inferred from its vapor pressure and its Henry's Law constant.

Dry Samples. Table IV summarizes the dry sampler data. Generally, frequency of detection for all pesticides is less than observed with the wet sampler. However, analysis of the dry samples reveals that more different chemical classes of pesticides are deposited from the atmosphere in association with particulate matter than those observed in wet precipitation. Consequently, fundamentally different physicochemical processes are controlling deposition in the dry phase and the wet phase. The fewest detections were observed at the Macbride site. Approximately the same number of detections were observed for the Oakdale and the Farm sites. More pesticides were detected at the urban site than at any of the others. These detects represent a combination of herbicides and insecticides used in row crops together with several insecticides which are either banned in the U.S. or sold specifically for non-agricultural uses. These include DDT, along with its primary environmental degradate, DDE.

Conclusions

Seasonal variations in frequency of occurrence of pesticides in precipitation is observed for many currently used pesticides. These would include: triazine herbicides, chloroacetanilide herbicides, chlorophenoxy acid herbicides, organochlorine and organophosphate insecticides. These results are consistent with those reported by Hatfield, et al. (*10*) in which nearly 90% of the detections of atrazine, alachlor and metolachlor occurred between mid-April and early July. In addition, several pesticides which are banned from commerce have also been detected. These chemicals, no longer in use, do not show a seasonal trend and appear throughout the year in ultratrace concentrations. This would suggest that the major source of contamination is from recently applied pesticides.

Since detections occurred at the Macbride site, a recreational area removed from cropland areas, as well as at the Urban site, these studies reaffirm the hypothesis that transport over long distances does occur. More detections occurred at the Urban site than any of the other sites. While it is likely that urban use may contribute somewhat to the frequency of detections, the most frequently detected compounds must originate from agricultural landscapes.

Systematic differences in wet precipitation and bulk precipitation were not evident from the data collected. The number of detections in air was too few to make meaningful conclusions. Efforts must be made to lower detection limits by modifying collection techniques to permit larger sample volume.

Table IV. Detections from the dry deposition sampler data from each site. The value is the total mass recovered from the collector.

Month	Farm Analyte	μg	Urban Analyte	μg	Macbride Analyte	μg	Oakdale Analyte	μg
Oct-96			Chlorpyrifos	0.31				
Nov-96			Picloram	0.032				
Dec-96			DDE	0.004				
Jan-97								
Feb-97								
Mar-97								
Apr-97								
May-97	Picloram	0.27	PCP[1]	0.019	Atrazine	0.12	EA[2]	0.0095
		0.5		0.031		0.11	Atrazine	0.19
	Atrazine	0.11	Silvex	0.084	Cyanazine	0.12		0.11
		0.1	Atrazine	0.28				0.1
		0.25		0.13			Cyanazine	0.15
		0.14		0.15				0.12
	Cyanazine	0.04	Cyanazine	0.1				
	Metolachlor	0.25	Pendimethalin	0.1				
		0.2						
Jun-97	Atrazine	0.1	Chlorpyrifos	0.055				
Jul-97								
Aug-97			PCP[1]	0.043				
Sep-97			DDE	0.0039			Dicamba	0.071
			DDT	0.01				
			Atrazine	0.047				
			Chlorpyrifos	0.039				

[1] PCP is pentachlorophenol
[2] EA is endrin aldehyde

Acknowledgment

The authors wish to thank Ed Engroff, Leonard Marine, Carlos Rodriguez, Scott Hinz, and Mark Hurt for field sampling and field record analysis. Our thanks to Mary Jo Kline, Greg Jacobs, Dick Sweeting, Peter Ma, Rick Derrig, Bernie Kirby, Ryan Carter, and Karla Benninghoven for sample extraction. Thanks to Paul Beney, Pam Mollenhauer, Wayne Patton, and Vicki Reedy for sample analysis. Thank you to Matt Nonnenmann for assisting with data reduction. Our thanks also to Dr. Mary J.R. Gilchrist, Director of the University Hygienic Laboratory, for the support and time necessary to complete this work. A special thanks to the Center for Health Effects of Environmental Contamination (CHEEC) for their part in sponsoring the project.

Literature Cited

1. Nations, B.K.; Hallberg, G.R. *J. Environ. Qual.* **1992**, *21*, 486-492.
2. Nations, B.K.; Hallberg, G.R.; Libra, R.D.; Kanwar, R.S. In *Agricultural Research to Protect Water Quality*; Soil and Water Conservation Society: Ankeny, IA, **1993**, pp 142-145.
3. Ligocki, M.P.; Leuenberger, C.; Pankow, J.F. *Atmos. Environ.* **1985**, *19*, 1609-1617.
4. Atlas, E.: Giam, C.S. *Science* **1981**, *211*, 163-165.
5. Glotfelty, D.E.; Williams, G.H.; Freeman, H.P; Leech, M.M. In *Long Range Transport of Pesticides*; Kurtz, D.L.,ed.; Lewis Publishers: Chelsea, MI, **1990**; pp 199-221.
6. *Pesticides in the Atmosphere: Distribution, Trends, and Governing Factors*, Majewski, M.S.; Capel, P.D., ed.; Ann Arbor Press, Inc.: Chelsea, MI, **1995**; 214 pp.
7. EPA Method IP-8: Organochlorine and Other Pesticides, EPA Compendium of Methods for the Determination of Air Pollutants in Indoor Air. Draft published March **1989**.
8. Methods for the Determination of Organic Compounds in Drinking Water - EPA/600/4-88/-039 - December 1988 (Revised July **1991**).
9. Hallberg, G.R. Derived from *National Agricultural Statistical Service and Iowa State University Extension Service Surveys of Agrichemical Use and Practices.* And updated from Mayerfeld, D.B., Hallberg, G.R., Miller, G.A., Wintersteen, W.K., Hartzler, R.G., Brown, S.S., Duffy, M.D., and DeWitt, J.R. *Pest Management in Iowa: Planning for the Future*; IFM 17, Iowa State University Extension, Ames, IA, **1996**.
10. Hatfield, J.L.; Wesley, C.K.; Prueger, J.H.; Pfeiffer, R.L. *J. Environ. Qual.* **1996**, *25*, 259-264.

Chapter 16

The Midwest Water Quality Initiative: Research Experiences at Multiple Scales

J. L. Hatfield[1], D. A. Bucks[2], and M. L. Horton[3]

[1]National Soil Tilth Laboratory, Agricultural Research Service, U.S. Department of Agriculture, Ames, IA 50011
[2]Natural Resources and Sustainable Agricultural Systems, Agricultural Research Service, U.S. Department of Agriculture, Beltsville, MD 20705
[3]CSREES, U.S. Department of Agriculture, Washington, DC 20250

Increasing concern about the role of farming practices on water quality provided the impetus for a research, education, and extension program in the Midwest. This multi-agency and multi-disciplinary research program was directed to evaluate the effect of farming practices on water quality and to develop farming practices that could be adopted by producers. Projects were conducted in Iowa, Minnesota, Missouri, Nebraska, North Dakota, Ohio, South Dakota, and Wisconsin. Research studies addressed issues on degradation processes to water flow through the soil as a pesticide and nutrient transport mechanism. This combined effort has integrated many components, and the results have been successfully transferred to users across the Midwest. The results of these efforts demonstrate that it is possible to develop an integrated program that addresses issues on nonpoint source pollution.

In the mid-80's there began to be increasing concern about the potential impact of farming practices on ground and surface water quality. These concerns led to the development of the Presidential Initiative on Water Quality in 1989. The principles of this Initiative were described by Swader (1) as:

"The nation's ground water resources should be protected from contamination by fertilizer and pesticides without jeopardizing the economic viability of U.S. agriculture;
Both the immediate needs to halt contamination and the future needs to alter farm production practices should be addressed;
Farmers ultimately must be responsible for changing production practices to avoid contaminating ground and surface waters."

In developing this program there were several challenges for the overall program within the U.S. Department of Agriculture (USDA). These challenges were to 1) conduct biological, physical, and chemical research; 2) address the management of chemicals for crop production; 3) develop alternative cropping systems, in order to 4) educate, demonstrate, and assist farmers in making appropriate changes in production practices; and 5) monitor implementation of improved management practices and systems.

These goals and challenges were the foundation to the formation of the Management Systems Evaluation Areas (MSEA) Program that began in 1990. This program was described by Onstad (2).

Development of the MSEA research and education program revolved around the goal to "identify and evaluate agricultural management systems that can protect water quality for the Midwest." To achieve this goal there were six general objectives:

1) Measure the impact of prevailing and modified farming systems on the content of nutrients and pesticides in ground and surface waters;

2) Identify and increase understanding of the factors and processes that control the fate and transport of agricultural chemicals;

3) Assess the impact of agricultural chemicals and practices on ecosystems associated with agriculture;

4) Assess the projected benefits of implementing modified farming systems in the Midwest;

5) Evaluate the social and economic impacts of modified management systems; and,

6) Transfer appropriate technology to farmers for use on the land.

These goals and objectives were encompassed in the five projects located in Iowa, Minnesota, Missouri, Nebraska, and Ohio. The overall structure of the MSEA program has been described by Hatfield (3); however, for this paper some of the structure will be described to help report the successes. The MSEA program involved over 150 researchers from the USDA-Agricultural Research Service, USDA-Cooperative State Research, Education and Extension Service, U.S. Department of Interior-United States Geological Survey, U.S. Environmental Protection Agency, and participating State Agricultural Experiment Stations and Cooperative Extension Services in Iowa, Minnesota, Missouri, Nebraska, Ohio, North Dakota, South Dakota, and Wisconsin. Within each project there were numerous state and local agencies that were involved in the project development and implementation.

The objective of this paper is to describe the MSEA program and the structure that was used to obtain the success for the original objectives of the project.

Structure of the MSEA Program

Communication and coordination of the MSEA program was foremost in the initial meetings of the project investigators. Facilitation of the communication and coordination management of this program was delegated to a Steering Committee comprised of a Principal Investigator (PI) from each project who was typically an Agricultural Research Service (ARS) and a State Agricultural Experiment Station (SAES) investigator and a representative from the United States Geological Survey (USGS), the U.S. Environmental Protection Agency (EPA), and the USDA Cooperative Extension Service Water Quality Coordinators. Each of the PI's represented a different agency within USDA so that a broad representation of the USDA agencies was possible with a small number of individuals. The primary responsibility of the Steering Committee was to coordinate the technical details, research, and scientific progress of the MSEA program. Scientific management of the individual projects remained with the Principal Investigators within each project. It was felt that the Steering Committee was needed to ensure that the overall goals of the Water Quality Initiative were being addressed through the scientific accomplishments of the projects selected to address these objectives within the Midwest Water Quality Initiative. During the period from 1996-98, the composition of the Steering Committee expanded to include three representatives from each project. These were the ARS Principal Investigator, the SAES Principal Investigator, and the Cooperative Extension Specialist. This evolution in the committee structure has allowed for a broader diversity of representation from each project and more focus on the integration of the research, education, and technology transfer goals of the MSEA program. The overall administrative structure of the MSEA program has been fluid to accommodate the program needs rather than a dictate from any of the agencies. This has allowed the project investigators to adopt a structure that has facilitated the original goals to be focused and evaluated as rapidly as possible.

Within the Steering Committee there were a number of Technical Subcommittees and Working Groups. The Technical Subcommittees were considered as key overarching components of the MSEA program critical to the success of the program while the Working Groups were considered to be groups that would have a changing function throughout the life cycle of the program. The Technical Subcommittees were: Data Base Development and Management; Quality Assurance and Quality Control; and Technology Transfer. The Technical Working Groups were Cropping Systems Evaluation, Process Modeling, and Socio-Economic Evaluations. Each of these committees was chaired by representatives from the Steering Committee but comprised of investigators from individual projects in order to incorporate the greatest amount of scientific and educational expertise into the project as possible.

The Steering Committee was responsible for the organization of program meetings, preparing program summary reports, and working with USDA administration to develop and execute workshops or meetings to describe MSEA program results. Communication was critical to the program goals, and during the initial phase of the project, monthly teleconferences were held with the Steering Committee along with quarterly meetings of the program Principal Investigators. There were also periodic planning meetings where all the Technical Subcommittees and Working Groups met along with many of the researchers and extension specialists to discuss accomplishments and products. The objectives of these meetings were to ensure that the program planning and implementation phases were on schedule and that all individual projects were developing as detailed in the original workplan. Communication of the project goals and the timelines was critical to ensure that all investigators understood the linkage among all of the program components. Each of these meetings rotated among projects in order to have site visits to the individual projects. This was not totally achieved because of the program need for detailed analysis of results that were not dependent upon site visits.

Data Base Development and Management. There were common parameters observed in the different MSEA sites. These included a meteorological station at each site with a wet/dry precipitation sampler. Soil and water samples were analyzed for a standard suite of agricultural herbicides, atrazine, alachlor, metribuzin, and metolachlor, and nitrate-nitrogen. Plant and soil parameters were standardized among sites as much as possible in order to be able to share information among sites and to evaluate different models across a number of sites. These data were placed into data bases for each site to provide as much sharing of data as possible and to lead toward the development of common data bases at the end of the project. The overall format of the original data set was described in Ward (4) as an internal MSEA report, and the concepts are reported in USDA (5).

Data bases being developed will be used to archive the data from the various sites for the purpose of evaluating models of agricultural chemical movement and dissipation under different farming practices. These collective data bases have been used to evaluate the Root Zone Water Quality Model (RZWQM) across the different MSEA sites. These efforts are described by Watts (6) and show the utility of a project like MSEA to provide an evaluation of models across a range of climates, soils, and farming practices.

Quality Assurance/Quality Control. One of the major components of the MSEA program was the analytical process to quantify the concentrations of the herbicides and nitrate-nitrogen in soil and water. Detection limits for the herbicides were established at 0.2 µg/L and for nitrate-nitrogen at 1.0 mg/L. These levels were adequate to assess the presence of the analytes, atrazine, alachlor, metribuzin, and metolachlor, in the water and soil samples and to meet the project goals. Water samples were analyzed at each project using individual project laboratories while soil samples were analyzed at the National Soil Tilth Laboratory. The analytical descriptions for the Iowa project are reported in Hatfield (7). Quality Assurance/Quality Control protocols were developed for each project and reviewed by external groups to ensure that the procedures and protocols would produce the data quality required to meet the project objectives.

An external QA/QC laboratory was used as part of the MSEA program plan. The purpose of this laboratory was to analyze split and duplicate samples and to evaluate the performance of each laboratory through the use of reference samples. This part of the process included the distribution of a standard reference mixture each quarter, comparison of the analytical results with the standard, and evaluation of the individual laboratory QA/QC protocols and Standard Operating Procedures (SOP's) to ensure continual adherence to the overall protocols. The reference samples were distributed and analyzed quarterly. Throughout the life of the project all of the laboratories were able to maintain the original QA/QC protocols and the detection limits for all chemicals. This process extended to the soil samples, even though only one laboratory was used for analysis, to ensure continual performance relative to the project objectives.

Technology Transfer. One of the original MSEA program objectives was the transfer of information to various users. A Water Quality Extension Specialist was hired for the overall MSEA program through the USDA-Extension Service. The goals of this effort were to: translate results from research sites into information that could be used by producers, agribusinesses, and consultants; provide information about specific research activities to a wide range of audiences; and assist in the research planning process so that the research would address farming practices that could be adopted or modified by a wide range of producers. There were two scales of information transfer within the MSEA program: within individual states through field days, workshops, and extension training sessions, and across the region through coordinated educational materials that could be used by a wide range of audiences. Both of these outreach efforts are based on the integration of research findings from the various projects into more comprehensive educational materials. Throughout the initial 5 years of the MSEA program, over 2,000,000 people were directly reached with the results from the different projects.

Cropping Systems Evaluation. The charge to the Cropping Systems Evaluation technical working group was to evaluate the different cropping practices at each site to ensure that a comprehensive picture of Midwestern agriculture would be portrayed through these studies. A description of the overall cropping and tillage practices across the projects is shown in Table I. There was a large diversity in the different tillage practices among the study sites. There have been a series of manuscripts prepared on the environmental impacts of ridge tillage for *Soil and Tillage Research* as described by Hatfield (*8*). These papers have provided a summary of the progress on ridge tillage and the potential environmental benefits of adopting a different tillage management system. Ridge tillage is only practiced on 4 percent of the cultivated land in the Midwest, and these papers show the changes that occur in this soil tillage system.

Process Modeling. The Process Modeling technical working group was charged with the evaluation of the Root Zone Water Quality Model (RZWQM). This model has been described in detail by USDA (*9-10*). Each of the MSEA sites assumed a particular responsibility in the evaluation of the model across all of the sites. The goal was to use this model to evaluate the potential impact of changing farming practices across a range of soils, climates, and cropping systems. The overview of this effort is described by Watts (6). The evaluation of the model required the integration of the data among projects, which provided a test of the initial data base system for sharing results among investigators.

Socio-Economic Evaluations. Socio-economic evaluations of producers' responses was one of the original goals of the Water Quality Initiative. One of the critical elements of the MSEA program was the evaluation of the sociological response of producers to potential changes in management practices and the economic implications of these changes. This technical working group tackled several different problems over the course of the MSEA program: surveys of producer response, focus groups, economic evaluations, and linkage of environmental and economic benefits of changing farm management practices. The integration of the socio-economic component into the MSEA program strengthened the overall program because producers were able to identify with the results and their potential application to production systems.

Research and Education Sites

The research and education programs of MSEA were designed to cover a number of different soils, climates, and geological areas. A detailed description of the individual projects and their setting is described for Iowa (*11*), Minnesota-Northern Sand Plains (*12*), Missouri (*13*), Nebraska (*14*), and Ohio (*15*). A general overview of the sites is given in USDA (*5*), and their location is shown on Figure 1. To help understand the various linkages among the projects and the overall program it is necessary to examine the individual project objectives. These are listed below:

Iowa: Evaluation of the Impact of Current and Emerging Farming Systems on Water Quality

1. Quantify the physical, chemical, and biological factors that affect the transport and fate of agricultural chemicals.

2. Determine the effects of crop, tillage, and chemical management practices on the quality of surface runoff, subsurface drainage, and ground water recharge.

3. Integrate information in meeting objectives 1 and 2 with data about soil, atmospheric, geologic, and hydrologic processes to assess the impact of these factors on water quality.

4. Evaluate the socioeconomic effects of current and newly developed management practices.

Table I. Cropping and Tillage Systems Used in the MSEA Program from 1991 Through 1996

Crop Rotation	Tillage Practice	Projects	Number of Sites
Grass and legumes	None	MO	1
Continuous corn	Moldboard plow	IA	2
	Chisel plow (fall)	IA, MN, NE, OH	8
	Disk only (spring)	IA	2
	Ridge-tillage	IA	2
	No-till	IA	2
Corn-soybean	Moldboard plow	IA	1
	Chisel plow (fall)	IA, MO, OH	4
	Ridge-tillage	IA, MN, MO	7
	No-till	IA, MO, OH	3
	Disk & till-plant	NE	1
Soybean-sorghum	Chisel plow (fall)	MO	1
Corn-soybean-wheat	Chisel plow	MO	1
Corn-soybean-wheat-clover	Ridge-tillage	OH	1
Sweet corn-potato	Chisel plow	MN	1

Figure 1. Location of MSEA research sites in the midwest United States.

5. Understand the ecological effects of agrichemicals, distinguishing them from the impacts of other agricultural practices. Evaluate alternative management practices for their long-term effectiveness in preventing ecological degradation, in contributing to restoration of the ecosystem, and in maintaining agricultural productivity.

Minnesota-Northern Sand Plains: Midwest Initiative on Water Quality: Northern Corn Belt Sand Plain
1. Investigate the impact of ridge-till practices in a corn and soybean cropping system on ground water quality and on the transport of nitrate-nitrogen, atrazine, alachlor, and metribuzin in the saturated and unsaturated zones.
2. Determine the effects of nitrogen management by soil tests and plant analysis.
3. Characterize ground water flow through the sand and gravel aquifers and correlate the characteristics of the aquifers to the transport and storage of agrichemicals.
4. Determine the relationship between the rates of ground water recharge and the rates of agricultural chemical loading to ground water.

Missouri: Alternative Management Systems for Enhancing Water of an Aquifer Underlying Claypan Soils
1. Measure the effects of conventional and alternative farming systems on surface and ground water quality.
2. Study the mechanisms responsible for the fate and transport of agrichemicals in soil and water.
3. Determine how information from the plots and fields can be scaled up to watershed and regional levels.
4. Develop and refine models of the physiochemical, economic, and social processes of farming activities that affect soil and water contamination.
5. Develop and evaluate alternative cropping systems and technologies designed to protect water quality through the use of site-specific management techniques.
6. Establish the relative profitability of alternative farming systems and determine farmers' attitudes toward adoption of these systems.
7. Develop education programs to increase farmers' awareness and understanding of the relative profitability and environmental benefits of alternative farming systems.

Nebraska: Management of Irrigated Corn and Soybeans to Minimize Ground Water Contamination
1. Compare the net effects on ground water quality of conventional and alternative management systems for irrigated crop production.

2. Increase knowledge about the fate and transport of agricultural chemicals under conventional and improved irrigated production systems.

3. Develop and evaluate new technologies for managing pesticides, nitrogen, and irrigation to reduce ground water contamination.

4. Develop models and decision-making systems to aid farmers in choosing management strategies that are environmentally sound and profitable.

5. Identify and analyze the social and economic factors that influence the acceptance and use of management options for improving water quality.

6. Evaluate the economic impacts on the farm and estimate the economic impacts on the region of alternative management practices to improve water quality, including household income and aggregate economic output.

7. Develop a "nitrogen budget" for the various management systems to evaluate fertilizer efficiency and the potential for nitrate leaching.

Ohio: The Ohio Buried Valley Aquifer Management Systems Evaluation Area

1. Characterize the baseline hydrogeologic, geochemical, and geomicrobial environments of the buried river valley aquifer at the Ohio MSEA, in the Piketon region, and in each of the research plots.

2. Assess the effects of the different farming systems on the ecological, hydrogeologic, geochemical, and geomicrobial environment of each system.

3. Determine the dynamic and spatial leaching fluxes of applied pesticides and nitrate under different agricultural management systems.

4. Determine crop production responses to the different agricultural management systems.

5. Determine the expected profitability of each commodity produced under alternative agricultural systems and the variability of profits.

6. Identify areas in a region in which to establish the most promising alternative agricultural systems and then to assess the likely benefits of the systems in the locations.

7. Determine socioeconomic factors affecting the adoption of alternative agricultural management systems.

8. Develop practical predictive models and systems for identifying the effects of an agricultural management system on water quality at specific sites, as well as the production levels and profitability of the system.

9. Augment existing agricultural data bases related to water quality.

10. Disseminate MSEA research results and provide technical assistance to farmers who are implementing new farm management systems.

This overall list of objectives provides an understanding of the scope of the project, which covers all aspects of agricultural chemical movement, fate, and dissipation within a wide range of soils, climates, and geologic settings.

Across all of the projects there was an underlying expectation for the development of products to enhance water quality. The expected products from these effects were minor, suggesting that in simulation models, this parameter would be somewhat insensitive. Some of the studies were designed to evaluate the role of preferential movement patterns of water through soil and the potential implications for agricultural chemical movement below the root zone. The laboratory studies provided the foundation for understanding and interpreting the results from plot and field studies. Without this foundation it was not possible to quantify the results obtained in the early part of the MSEA program. Fortunately, there were several component studies that were part of the MSEA program and also in allied research groups that were part of this effort.

Plot and field studies covered the bulk of the research effort within the MSEA program. Every project had their own individual research projects at this scale. These studies were replicated to evaluate a range of hypotheses across the sites. At this scale we began to see the emergence of the factors that interact to determine the response of soil or crop management practices on water quality. Typical of these studies are those reported by Ward (*25*), who described the overall goals of the program at the onset, and the later report by Hatfield (*8*), who summarized the implications of the tillage management practices on water quality. These studies have shown that it is possible to alter the impact of farming practices and have a positive effect on water quality. Many of these practices are being placed into farming systems that are being demonstrated through field trials and demonstration areas.

Watershed scale studies were conducted in Iowa and Missouri. These studies required the linkage of field management practices to observations of the various agricultural chemicals in surface water, subsurface drainage, and ground water. The development of these projects required a different approach than other research scales and were more weather dependent. Observations of subsurface and stream discharge from Walnut Creek in Iowa showed a large interannual and annual variation due to the meteorological conditions among years. Jaynes (*26*) estimated that a minimum of 7 years of records may be necessary to overcome the effects of weather variation on the observed agricultural chemical patterns across multiple fields and management practices. For producers to have confidence in the results from water quality studies will require a long-term commitment to the objectives, and procedures that can capture the different chemicals with accuracy and precision.

Conferences and Workshops

The MSEA program developed an active effort to organize and sponsor conferences on water quality. The first conference was held in 1993 in St. Paul, Minnesota and produced a proceedings entitled *Agricultural Research to Protect Water Quality* published by the Soil and Water Conservation Society of Ankeny, Iowa. The second conference was held in 1995 in Kansas City, Missouri and produced a four volume series entitled *Clean Water Clean Environment: 21st Century* and was published by the American Society of Agricultural Engineers in St. Joseph, Michigan. Each of these conferences attracted over 500 participants. A third conference is scheduled for

studies included: 1) identification of environmentally sound farming systems that are acceptable to farmers; 2) assessments of landscapes and farming systems for their vulnerability to water contamination from farm chemicals; 3) information about the effects of farm chemicals on a region's ecology; 4) information about the suitability of management systems for specific farms in the Midwest; and 5) basic understanding of the behavior of farm chemicals in the environment. These products have been developed through research studies at each site, and there are reports beginning to emerge that represent the integration of information among sites.

Research Studies

Water is the primary transport medium for agricultural chemicals, and to understand the differences among the sites, the hydrologic components evaluated are given in Table II. The individual projects covered a large range of scales from laboratory column studies for individual pesticide fate and transport to fields positioned within large watersheds over a number of years. Laboratory studies were designed to answer specific questions while the field studies were designed to assess the interactions among the various components of the farming system. Investigators within the MSEA program addressed all of these different scales throughout the course of the program, however, to accomplish this task required the cooperation of many different scientists. Examples of this type of research are studies on the movement of water and solutes in the sand plains soils (*16*), the development of techniques to measure water movement in soils (*17*), the evaluation of leaching of organic carbon following anhydrous ammonia application (*18*), *in situ* characterization of sand in relation to solute movement (*19*), and the comparison of tillage and crop effects on infiltration rates (*20*). Each of these studies provided a critical piece of information that was necessary to help interpret the results from a wide range of studies. Project meetings were devoted to sharing results among sites to help researchers understand the techniques being evaluated and the potential application to other soils and climates.

Typical of laboratory studies were those conducted to evaluate the half-life of atrazine, alachlor, and metolachlor, the effect of different microbial populations on the degradation rates of atrazine, and the transformations of nitrogen under different soil conditions, e.g., temperature and organic carbon content. Seybold (*21*) found that atrazine adsorption was higher in forest soils than prairie soils, which explained why lower mobility of atrazine has been observed in these soils. Baluch (*22*) evaluated the degradation products of atrazine and found the highest degradation rates in the surface layer. Observations from the MSEA program have shown that atrazine and metolachlor are present in the upper 50 cm of the soil profile and concentrations below this depth are typically 1 ppb or less. Jayachandran (*23*) examined atrazine degrading microorganisms and concluded that these populations increased with increased frequency of atrazine use. Observations across the Midwest have shown that atrazine continues to be used on 67% of the corn production area (*5*). Novak (*24*) evaluated the effects of aggregate size on atrazine sorption kinetics but found that the

Table II. Hydrologic Components of the MSEA Program Evaluated for Each Project

Site	Unsaturated zone	Saturated zone	Surface water
Iowa			
Walnut Creek	Subsurface drainage Evapotranspiration	Alluvial aquifer	Stream Skunk River Wetland
Deep Loess	Vadose zone	Saturated loess Glacial till aquifer	Stream
Nashua	Subsurface drainage Vadose zone	Glacial till aquifer	
Minnesota-Northern Sand Plains			
Princeton, MN	Direct recharge	Sand Plain aquifer	Stream Wetland
Aurora, SD	Vadose zone	Sand Plain aquifer	
Oakes, ND	Evapotranspiration	Sand Plain aquifer	
Sand Plains, WI	Vadose zone	Sand Plain aquifer	Stream
Missouri			
Goodwater Creek	Vadose zone	Fractured clay pan	Stream
Nebraska			
Platte alluvium	Vadose zone	Alluvial aquifer	Losing stream
Ohio			
Scioto alluvium	Vadose zone	Alluvial aquifer	Stream Wetland

June 1999 in New Orleans, Louisiana on the topic of Best Management Practices to Reduce Loadings to Receiving Waters.

Each of the state projects has hosted at least one workshop and conference on water issues within the state and the research being conducted to address these local problems. These efforts have produced proceedings that serve as information sources and are compilations of information from MSEA investigators and other researchers. There have been other educational and extension workshops on water quality as a direct result of the MSEA program. The MSEA program has had an impact on the knowledge base across the Midwest on the effect of farming practices on water quality.

Overall Accomplishments

The MSEA program was designed to be a regional project to address the goals and objectives of the Presidential Initiative on Water Quality. It is not possible to list all of the different reports that have been generated from the program, and those listed are given to demonstrate the type of research report that has been produced. Most of these reports incorporate a number of different authors, often from multiple locations. Some of the research reports cover aspects of specific studies within a project while others integrate across projects. Over the course of the MSEA program there have been over 780 research reports and over 60 educational bulletins prepared. However, these represent only the beginning because many of the reports are just beginning to emerge from the field studies. At this time there are approximately 48 scientific papers in preparation for various scientific journals.

There has been a positive research experience from the MSEA program. The results of this type of research are having an impact on the development of new management practices that will further protect the water resources of the United States.

The MSEA Program was able to achieve its goals because of the dedication and commitment of all of the participants. Integrative research to address water quality problems requires a structure similar to the MSEA program where individual scientists can incorporate their expertise into a framework of linking laboratory and field/watershed scale research. There have been several improved management practices from the MSEA research effort. These are being evaluated during the second phase of the MSEA program. These efforts have led to the development and evaluation of technology for improved nitrogen placement in soil that reduces the leaching potential (*27*), the capture of subsurface drainage water for reuse as irrigation water (*28*), the evaluation of improved nitrogen and irrigation management methods for irrigated areas (*29-30*), the evaluation of landuse practices, e.g., buffer strips or wetlands (*31-32*), and the evaluation of herbicides for their weed control efficacy and environmental impact across watersheds (*33*). It is not possible to list all of the projects and their potential implications for water quality; however, the incorporation of these findings into practices adopted by producers will further enhance the water quality of the United States.

Future Efforts

Although MSEA can be considered a success in terms of achieving the goal of determining the impact of farming practices on water quality, the program should not be considered as complete. There are many continuing efforts needed to address many of the problems confronting agriculture from a nonpoint pollution viewpoint. Part of this effort is encompassed in the second phase of MSEA, which has been expanded to include a larger scale approach dealing with multiple fields and livestock systems in a watershed context. This program has been named Agricultural Systems for Environmental Quality and is directed toward a systems approach to evaluate ground and surface water and air quality from a range of agricultural systems in the original MSEA sites plus the Delta region of Mississippi, North Carolina, South Carolina, and the Lake Erie basin of Ohio. These projects have been blended into the revised MSEA goals with the same potential for success in showing that farming practices can successfully address the issues of nonpoint source pollution and deliver information to producers that will influence their decision-making.

Literature Cited

1. Swader, F.N. *Agricultural Research: The Challenge in Water Quality*. Proc. Agricultural Research to Protect Water Quality; Soil and Water Conservation Society, Ankeny, IA, 1993; p. 16-20.
2. Onstad, C.A.; Burkart, M.R.; Bubenzer, G.D. *J. Soil and Water Conserv.* **1991**, *46*, 184-188.
3. Hatfield, J.L.; Anderson, J.L.; Alberts, E.E.; Prato, T.; Watts, D.G.; Ward, A.; Delin, G.; Swank, R. *Management Systems Evaluation Areas--An Overview.* Proc. Agricultural Research to Protect Water Quality; Soil and Water Conservation Society, Ankeny, IA, 1993; p. 1-15.
4. Ward, A.D.; Alberts, E.E.; Anderson, J.L.; Hatfield, J.L.; Watts, D.G. *Management Systems Evaluation Areas Program: Data collection, management, and exchange*; Internal MSEA Report, 1991; 35 p.
5. USDA. *Water Quality Research Plan for Management Systems Evaluation Areas (MSEA's): An Ecosystems Management Program*; Agricultural Research Service Bulletin, ARS-123, Washington, D.C., 1994; 45 p.
6. Watts, D.G.; Fausey, N.R.; Bucks, D.A. *Agron. J.* **1998**. (In Press)
7. Hatfield, J.L.; Jaynes, D.B.; Burkart, M.R.; Cambardella, C.A.; Moorman, T.B.; Prueger, J.H.; Smith, M.A. *J. Environ. Qual.* **1999**, *28*, 11-24.
8. Hatfield, J.L.; Allmaras, R.R.; Rehm, G.W.; Lowery, B. *Soil and Tillage Res.* **1998**, *48*, 145-154.

9. USDA. *Root Zone Water Quality Model*; Version 1.0 Technical Documentation; GSPR Technical Report No. 2; USDA-ARS Great Plains System Research Unit, Ft. Collins, CO, 1992; 236 p.
10. USDA. *Root Zone Water Quality Model*; Version 1.0 Users Manual; GSPR Technical Report No. 3; USDA-ARS Great Plains System Research Unit, Ft. Collins, CO, 1992; 103 p.
11. Hatfield, J.L.; Baker, J.L.; Soenksen, P.J.; Swank, R.R. *Combined agriculture (MSEA) and ecology (MASTER) project on water quality in Iowa*. Proc. Agricultural Research to Protect Water Quality; Soil and Water Conservation Society, Ankeny, IA, 1993; p. 48-59.
12. Anderson, J.L.; Dowdy, R.H.; Lamb, J.A.; Delin, G.N.; Knighton, R.; Clay, D.; Lowery, B. *Northern Cornbelt Plains Management System Evaluation Area*. Proc. Agricultural Research to Protect Water Quality; Soil and Water Conservation Society, Ankeny, IA, 1993; p. 39-47.
13. Alberts, E.E.; Prato, T.; Kitchen, N.R.; Blevins, D.W. *Research and education to improve surface and ground water quality of a claypan soil*. Proc. Agricultural Research to Protect Water Quality; Soil and Water Conservation Society, Ankeny, IA, 1993; p. 21-38.
14. Watts, D.G.; Schepers, J.S.; Spalding, R.F.; Peterson, T.A. *The Nebraska MSEA: Management of irrigated corn and soybeans to minimize groundwater contamination*. Proc. Agricultural Research to Protect Water Quality; Soil and Water Conservation Society, Ankeny, IA, 1993; p. 60-68.
15. Ward, A.D.; Nokes, S.E.; Workman, S.R.; Fausey, N.R.; Bair, E.S.; Jagucki, M.L.; Logan, T.; Hindall, S. *Description of the Ohio buried valley aquifer agricultural management systems evaluation area*. Proc. Agricultural Research to Protect Water Quality; Soil and Water Conservation Society, Ankeny, IA, 1993; p. 69-79.
16. Komor, S.C.; Emerson, D.G. *Water Resourc. Res.* **1994**, *30*, 253-267.
17. Kung, K-J.S.; Donohue, S.V. *Soil Sci. Soc. Am. J.* **1991**, *55*, 1543-1545.
18. Clay, D.E.; Clay, S.A.; Liu, Z.; Harper, S.S. *Bio. Fertil. Soils,* **1995**, *19*, 10-14.
19. Hart, G.L.; Lowery, B.; McSweeney, K.; Fermanich, K.J. *Geoderma*, **1994**, *64*, 41-55.
20. Logsdon, S.D.; Jordahl, J.J.; Karlen, D.L. *Soil & Tillage Res.* **1993**, *28*, 179-189.
21. Seybold, C.A.; McSweeney, K.; Lowery, B. *J. Environ. Qual.* **1994**, *23*, 1291-1297.
22. Baluch, H.U.; Somasundaram, L.; Kanwar, R.S.; Coats, J.R. *J. Environ Sci. Health,* **1993**, *B28(2)*, 127-149.
23. Jayachandran, K.; Stolpe, N.B.; Moorman, T.B.; Shea, P.J. *Soil Biol. Biochem.* **1998**, *30*, 523-529.
24. Novak, J.M.; Moorman, T.B.; Karlen, D.L. *J. Agric. Food Chem.* **1994**, *42*, 1809-1812.

25. Ward, A.D.; Hatfield, J.L.; Lamb, J.A.; Alberts, E.E.; Logan, T.J.; Anderson, J.L. *Soil and Tillage Res.* **1994,** *30,* 39-74.
26. Jaynes, D.B.; Hatfield, J.L.; Meek, D.W. *J. Environ. Qual.* **1998.** (In Press)
27. Ressler, D.E.; Horton, R.; Baker, J.L.; Kaspar, T.C. *Applied Eng. in Agric., ASAE.* **1997,** *13,* 345-350.
28. Fausey, N.R. *Subirrigation/drainage systems for water table management in the Midwest.* Irrigation Association Technical Conference Proc., Atlanta, GA, 1995; p. 242-246.
29. Schepers, J.S.; Francis, D.D.; Vigil, M.; Below, F.E. *Commun. Soil Sci. Plant Anal.* **1992,** *23,* 2173-2187.
30. Blackmer, T.M.; Schepers, J.S. *J. Prod. Agric.* **1995,** *8,* 56-60.
31. Baker, J.L.; Mickelson, S.K.; Hatfield, J.L.; Fawcett, R.S.; Franti, D.W.; Peter, C.J.; Tierney, D.P. *Reducing herbicide runoff: Role of best management practices.* Proc. Brighton Crop Protection Conference, Brighton, England, 1995; p. 479-487.
32. Workman, S.R.; Fausey, N.R.; Brown, L.C.; Bierman, P. Couple wetland-agricultural ecosystem for water quality remediation. In *Versatility of Wetlands in the Agricultural Landscape*; Campbell, K.L., Ed.; Tampa, FL, 1995; p. 159-168.
33. Ghidey, F.; Alberts, E.E.; Lerch, R.N. *J. Environ. Qual.* **1997,** *26,* 1555-1563.

Chapter 17

Reconnaissance Survey of Sulfonamide, Sulfonylurea, and Imidazolinone Herbicides in Surface Streams and Groundwater of the Midwestern United States

T. R. Steinheimer[1], R. L. Pfeiffer[1], K. D. Scoggin[1], and W. A. Battaglin[2]

[1]National Soil Tilth Laboratory, Agricultural Research Service, U.S. Department of Agriculture, 2150 Pammel Drive, Ames, IA 50011
[2]WRD, U.S. Geological Survey, Box 25046, Mail Stop 406, Lakewood, CO 80225

The study objective was to conduct a small scale synoptic survey of representative water resources draining agricultural land for occurrence of several herbicide residues. These new classes of herbicides are commonly applied pre-emergence or post-emergence in conservation tillage systems to control grasses and broadleaf weeds in cropped and noncropped areas. Both surface water and groundwater samples were collected from 44 midwestern locations during the summer of 1997, and analyzed for herbicide residues of 15 sulfonylurea and imidazolinone chemicals, and one sulfonamide. Each site was sampled between mid-June and late-October with several stream sites sampled twice. The method, developed jointly by the chemical manufacturer's and the U.S. Environmental Protection Agency, provides a 100 ng/L limit of quantitation in surface water for all analytes. Analytes were detected and identity confirmed in surface water at six sites and in ground water at two sites. The most frequently detected herbicides were imazaquin, imazethapyr, and nicosulfuron. For field studies in which the source of surface and ground water associated with the farming system on the agricultural landscape is known, the sensitivity of the method can be improved with only minor modifications in detection criteria.

Introduction

Development of sustainable grain production practices are necessary in order to preserve our natural resource base while assuring an adequate food supply. Central to these newly adopted agronomic practices is greatly reduced tillage, with the practice of no-till becoming more common across the Midwest (*1-2*). One consequence of this is an increased reliance on chemical herbicides for weed control. A corn-soybean rotation in these conservation tillage systems typically involves a herbicide applied pre-plant for burndown of all existing weeds, followed by a second herbicide application at planting or shortly after crop emergence. Typically, the post-emergent application occurs 10-21 days after planting. Among the newer herbicide classes developed primarily for use in conservation tillage grain production are the sulfonylurea (SU) and the imidazolinone (IMI) families and one sulfonamide (SA). Molecular structures, common chemical names, and formulated current product names are given in Figure 1 for chemicals representative of these new herbicide classes. They are characterized by high activity against weed species, very low application rates in comparison to earlier classes of herbicides, and extremely low toxicity to humans and other mammalian species (3). However, some sensitivity to other plants grown for fruit, grain, or fiber has been reported (4). Although nearly identical in their mode of action, they bear little similarity in structure. Thus, the biogeochemical reactions controlling their degradation in the soil dictate uniquely different chemical pathways of environmental-fate for each (5-7). While the soil half-life for each is relatively short, both may produce stable degradates which can persist or degrade very slowly, especially in anaerobic soil zones. Table 1 lists some physical and chemical properties of each compound. The recommended application rates listed for these post-emergent herbicides are maximum rates for grass and broadleaf control in croplands. Frequently, lower rates are actually applied either as a consequence of tank-mix custom applications which include other active ingredients, or as the result of higher than expected efficacy on specific target weeds. The somewhat higher rates for imazapyr (Arsenal^R) and sulfometuron methyl (Oust^R) result from their primary usage on noncropland areas.

Because of their structure, the acid imidazolinones possess unique acid-base properties. Any of five distinct chemical species may dominate in aqueous solution at various pH's. In the pH range most critical for environmental issues (pH 5-9) the species with the imidazolinone ring unionized but with the carboxylic acid group largely dissociated becomes dominant. As the pH drops to 5, the undissociated neutral species predominates. In contrast, nearly all sulfonylurea herbicides are weak acids, with acidities similar to acetic acid. Therefore, at pH's normally encountered in soils, pH 5-7, the SU's are largely dissociated, with the dominant species being the anion formed by dissociation of the more acidic urea-hydrogen adjacent to the sulfonyl group on the SU bridge. Depending on the buffering capacity of the water sample, the proportion of SU-anion increases as the pH rises. Also, as soil pH rises above 7, persistence appears to increase due to a decrease in the rate of hydrolysis. Another consequence of this weakly-acidic behavior is a pH-dependence on water solubility. For many SU herbicides, water solubility increases by more than an order of magnitude when the pH is raised from 5 to 7. Values listed in Table 1 are given for the near-neutrality pH's common to most upper midwestern soils.

Products registered for weed control on cropland and containing SU's and IMI's as active ingredients first appeared in the early 1980's and continue today. In the grain belt of the midwest herbicides were applied to 97% of the corn acreage and 97% of the soybean acreage in 1996 (8). While atrazine continues to be the most widely used herbicide followed by metolachlor and dicamba, the use of newer post-emergent chemistry offered by the SU's and IMI's has steadily increased. For example, nicosulfuron usage on corn increased from 25,000 lbs in 1992 to 54,000 lbs in 1995 in Iowa, and, in Illinois, from 8,000 lbs in 1992 to 38,000 lbs in 1995 (8-11). Similarly, imazethapyr usage on soybean increased from 228,000 lbs in 1992 to 393,000 lbs in 1995 in Iowa, and, in Illinois, from 148,000 lbs in 1992 to 342,000 lbs in 1995 (9-12). In addition, imazethapyr usage on imidazolinone-tolerant corn hybrids has grown in recent years.

The midwest is the largest and most intensively cultivated agricultural region in the U.S. Nearly 80% of the corn and soybeans grown in the country are produced in this area which includes much of the Mississippi River drainage basin. In producing these

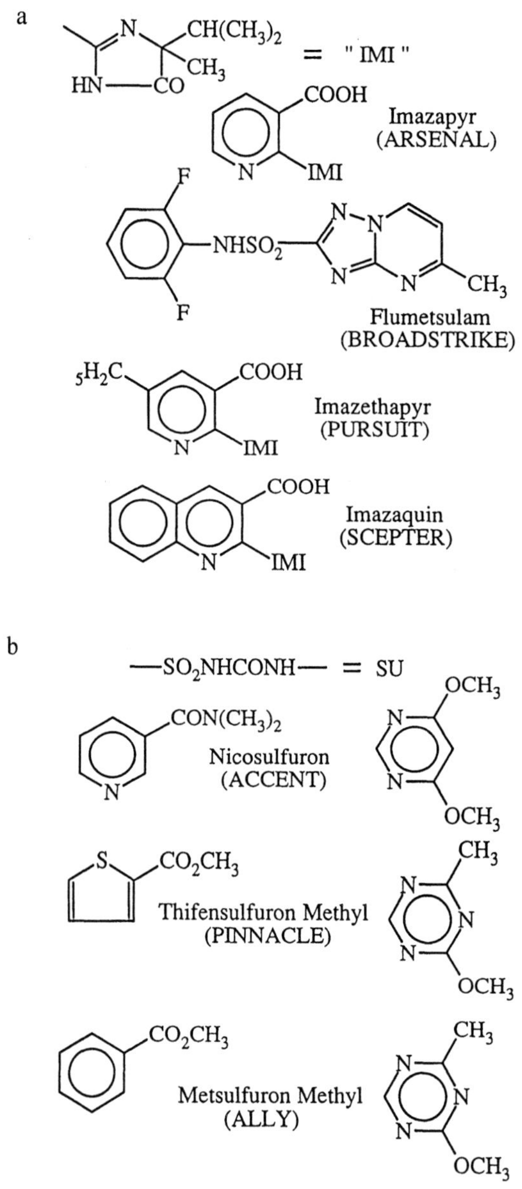

Figure 1. Molecular Structure, Common Chemical Name, and Formulated Product Name for Sulfonamide, Sulfonylurea, and Imidazolinone Herbicides.

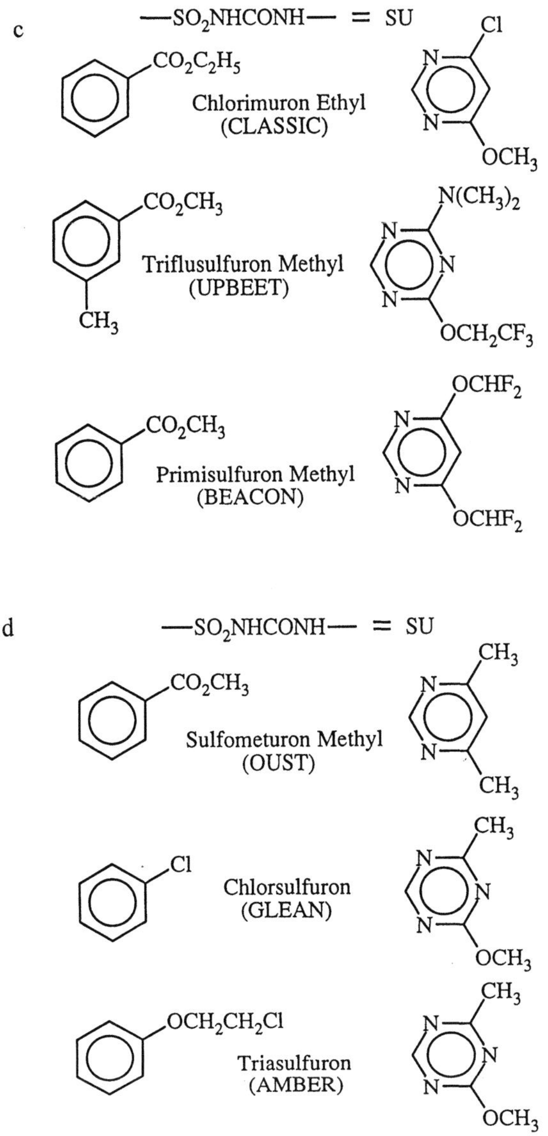
c
—SO2NHCONH— = SU
CO2C2H5
Cl
Chlorimuron Ethyl
(CLASSIC)
OCH3
CO2CH3
N(CH3)2
Triflusulfuron Methyl
(UPBEET)
CH3
OCH2CF3
OCHF2
CO2CH3
Primisulfuron Methyl
(BEACON)
OCHF2
d
—SO2NHCONH— = SU
CH3
CO2CH3
Sulfometuron Methyl
(OUST)
CH3
CH3
Cl
Chlorsulfuron
(GLEAN)
OCH3
CH3
OCH2CH2Cl
Triasulfuron
(AMBER)
OCH3

Figure 1. *Continued.*

Continued on next page.

e

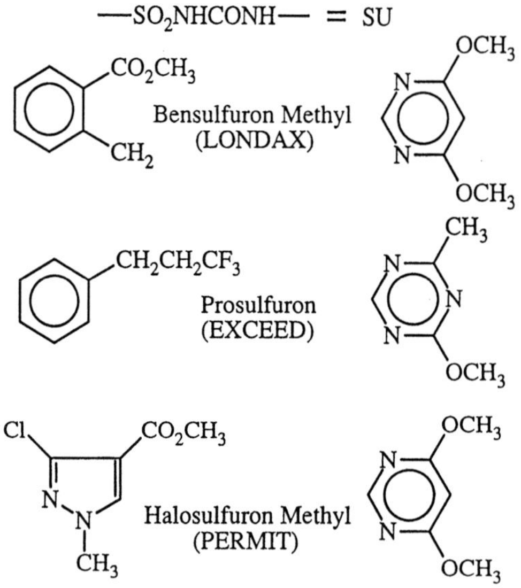
—SO2NHCONH— = SU
CO2CH3
CH2
Bensulfuron Methyl
(LONDAX)
OCH3
N
N
OCH3
CH2CH2CF3
Prosulfuron
(EXCEED)
CH3
N
N
N
OCH3
Cl
CO2CH3
N
N
CH3
Halosulfuron Methyl
(PERMIT)
OCH3
N
N
OCH3

Figure 1. *Continued.*

Table 1. Chemical Properties and Important Environmental Characteristics of the Sixteen Target Analytes[1].

Compound	MW	pK_a	Water Solubility, mg/l	Recommended Rate, kg/ha, a.i.[2]	Soil Half-Life
Imazapyr	261.3	2.8-3.1, 3.8-3.9, 11.4	11,272	0.50-1.50	6-24 months
Flumetsulam	325.3	~3 and ~5	49 (pH 2.5)	0.0784	1-2 months
Imazethapyr	289.3	same as Imazapyr	1,400	0.0706	1-3 months
Nicosulfuron	410.4	4.3	18,000 (pH 7.2)	0.0347	3-8 weeks
Imazaquin	311.3	same as Imazapyr	60-120	0.0706	1-3 months
Thifensulfuron methyl	387.4	4.0	2,400 (pH 6)	0.0045	6-12 days
Metsulfuron methyl	381.4	3.3	9,500 (pH 6.1)	0.0045	1-4 weeks
Sulfometuron methyl	364.4	5.3	70 (pH 7)	0.84	4 weeks
Chlorsulfuron	357.8	3.8	27,900 (pH 7)	0.026	4-6 weeks
Triasulfuron	401.8	4.64	815 (pH 7)	0.029	69-139 days
Bensulfuron methyl	411.4	5.2	120 (pH 7)	----	4-20 weeks
Prosulfuron	419.4	3.76	4,000 (pH 6.8)	0.0202	70-150 days
Halosulfuron methyl	434.8	----	349 (pH 6)	0.0706	25-30 days
Chlorimuron ethyl	414.8	4.2	1,200 (pH 7)	0.0134	1-9 months
Triflusulfuron methyl	492.4	4.4	110 (pH 7)	0.0175	2-4 days
Primisulfuron methyl	468.3	5.1	243 (pH 7)	0.0403	30-63 days

(1) *1998 Cultural & Chemical Weed Control in Field Crops*, BU-3157-S, University of Minnesota, Extension Service, St. Paul, MN; *1998 Herbicide Manual for Agricultural Professionals*, WC-92, Revised 11/97, Iowa State University, Extension Service, Ames, IA; *Herbicides: Chemistry, Degradation, and Mode of Action*, Volume 3, Marcel-Dekker, Inc., New York, NY, 1988; *The Imidazolinone Herbicides*, CRC Press, Boca Raton, FL, 1991; *The Agrochemicals Handbook*, 3rd Edition, 5th Update, Royal Society of Chemistry, Cambridge, England, U.K., 1994; Personal Communication with Novartis Crop Protection, Monsanto Agricultural Group, American Cyanamid Company, and DuPont Agricultural Products Company.

(2) active ingredient.

---- not available

agricultural commodities, large amounts of pesticides, predominantly herbicides, are applied to cropland. Gradually, midwestern producers are replacing some of their older choices with these newer chemicals. By 1994, both Accent[R] (nicosulfuron) and Pursuit[R] (imazethapyr) had captured a significant share of the herbicide market for grass and broadleaf control in corn and soybean, respectively.

Reconnaissance Survey

In general, the overall goal of the project was to enhance the understanding of the movement of these new herbicides into nontarget water resources. Presently, little information is published describing the occurrence, fate, or movement of SU, IMI, or SA herbicides from production agriculture landscapes into either surface water or groundwater across the Nation. The purpose of this study was to begin the process of development of that understanding. To achieve that goal, the reconnaissance survey established two objectives:

Objective 1
Analyze water samples using a new method of analysis developed by DuPont employing electrospray mass spectrometry (ESMS).

Objective 2
Begin development of a database of herbicide residue occurrences for these new herbicide classes in water resources by analyzing surface water and ground water from representative drainage areas across the midwest.

Study Plan and Sampling Sites. Samples were collected from small streams, larger rivers, and wells. The majority of the surface-water samples were collected from locations within the Eastern Iowa Basin, Upper Mississippi River Basin, and the Lower Illinois River Basin as defined by the U. S. Geological Survey (USGS) and inclusive in the study units within its National Water Quality Assessment (NAWQA) program (*13*). Samples were also taken at selected National Stream-Quality Accounting Network (NASQAN) sites defined by USGS along with other sites in the midwestern United States. Figure 2 shows the sampling site locations in Illinois, Indiana, Iowa, Louisiana, Minnesota, Mississippi, and Nebraska, and the proximity of each to Mississippi River River drainage. Site names, site-identifications, estimated drainage basin drainage areas, and dates of collection for all surface water samples are listed in Table 2. Surface water sites were selected to represent a range of basin areas and river sizes, and to cover the large geographic regions within which the use of these chemicals is growing. Information on recent use patterns of sulfonamide, sulfonylurea or imidazolinone herbicides was used to target specific areas for sampling activities. All of the rivers that were sampled drain largely agricultural watersheds. When possible, samples were collected in conjunction with NASQAN or NAWQA activities, both to reduce the cost of sample collection and to insure compliance with other NAWQA data-quality asessment protocols.

The majority of groundwater samples were collected from a network of wells in Iowa that are part of the Iowa Groundwater Monitoring (IGWM) program. IGWM is a joint effort of the University of Iowa Hygienic Laboratory (UHL), Iowa Dept. Of Natural Resources (IDNR), Iowa Geological Survey Bureau (IGSB), and USGS. Wells from this network have been sampled systematically for more than 15 years (*14,15*). Groundwater samples were also taken from selected wells within the Lower Illinois River Basin NAWQA study. The wells that were sampled in Illinois were newly installed monitoring wells, while the wells sampled in Iowa were generally public supply or municipal wells. Table 3 provides details for the well water samples.

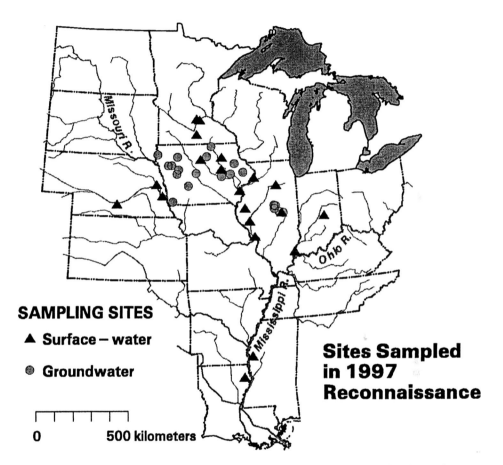

Figure 2. 1997 Midwest Reconnaissance Sampling Site Locations Showing the Drainage Relationship to the Mississippi River.

Table 2. Surface-Water Sampling: Station ID, Name, Location, Date and Estimated Drainage Area.

Station Identification	Name & Location	Drainage Area (sq.mi.)	First Sample	Second Sample
03377500	Wabash River at New Harmony, IN	29,234	6/11/97	6/19/97
05553500	Illinois River at Ottawa, IL	11,000	6/18/97	---------
05586100	Illinois River at Valley City, IL	27,000	6/17/97	---------
05572000	Sangamon River at Monticello, IL	550	8/18/97	---------
05584500	LaMoine River at Colmar, IL	655	6/13/97	[1] 8/19/97
05288705	Shingle Creek at Minneapolis, MN	28	6/15/97	6/24/97
05320270	Little Cobb River at Beauford, MN	130	6/23/97	7/16/97
05330000	Minnesota River near Jordan, MN	16,200	[1] 6/24/97	7/01/97
05331570	Mississippi River at Hastings, MN	37,000	6/26/97	9/02/97
05420500	Mississippi River at Clinton, IA	85,600	6/17/97	6/27/97
05420680	Wapsipinicon River near Tripoli, IA	343	6/22/97	7/09/97
05449500	Iowa River near Rowan, IA	429	6/23/97	7/15/97
05464220	Wolf Creek near Dysart, IA	299	6/16/97	6/24/97
05465500	Iowa River at Wapello, IA [2]	12,499	6/19/97	6/25/97
06770195	North Dry Creek near Kearney, NE	~40	7/17/97	8/11/97
06800000	Maple Creek near Nickerson, NE	450	7/23/97	8/06/97
06805500	Platte River at Louisville, NE [3]	6,800	6/23/97	7/10/97
07288650	Bogue Phalia near Leland, MS	484	6/18/97	7/09/97
07369500	Tensas River at Tendal, LA	309	6/11/97	7/01/97

[1] replicate sample also collected

[2] additional duplicates collected on 7/31/97

[3] additional sample collected on 8/11/97

--------- not collected

Table 3. Well Water Sampling: Station ID, Name, Depth and Date.

Station Identification	Name, State	Depth (ft)	Sample Date
402304089032601	LUS2-6, IL	20.50	8/20/97
402059088571701	LUS2-7, IL	28.08	9/17/97
401327089025201	LUS2-9, IL	17.5	(1) 9/18/97
400345088444901	LUS2-20, IL	22.92	10/22/97
400033088300301	LUS2-21, IL	16.33	9/09/97
415417092180101	Belle Plaine 4, IA	42.	6/13/97
420005091431201	Cedar Rapids S6, IA	65.	6/13/97
420451093561301	Boone 20, IA	63.7	7/22/97
422915095323504	Holstein 3, IA	54.	6/30/97
425344095090401	Sioux Rapids 2, IA	54.	7/02/97
426020091273701	Manchester 7, IA	270.	(1) 8/20/97
425341093132501	Sheffield 2, IA	27.	7/23/97
433224092550802	Saint Ansgar 2, IA	240.	8/26/97
421617095051001	Wall Lake 3, IA	43.	6/30/97
412852094275101	Menlo 3, IA	30.	8/28/97
430017096285301	Hawarden 2, IA	36.	7/03/97
420352092552401	Marshalltown 14, IA	160.	(2) 6/24/97
422831095465102	Correctionville 1W, IA	26.	7/01/97
420336095115601	Vail 1, IA	32.	6/30/97
413040093290501	Carlisle 5, IA	30.	8/26/97
411727094374001	Fontanelle 5, IA	39	8/28/97
404327095284801	Farragut, 79-2 (North), IA	65	8/26/97

(1) replicate sample also collected

(2) replicate sample collected as Myersville 2 @1300 on 6/24/97

ft., feet below land surface

Sampling Schedule, Sampling Procedure, and Field Quality Assurance.
Sixty-two water samples (38 surface water and 24 groundwater) were collected between June and October for the 1997 reconnaissance. Each sample was sent to the National Soil Tilth Laboratory, Ames, Iowa, (NSTL). In most cases, two samples were collected at each surface-water site (one on each of two site visits) and one sample was collected at each groundwater site. Surface-water samples were collected during runoff events after most of the post-emergent application of SU's, IMI's, and SA's had occurred in 1997. The timing of sample collection was dependent on streamflow conditions and weather. In most cases, samples were collected when streamflow was at or above the 50th percentile on the flow duration curve. A second sample was collected from many of the sites within 1-4 weeks of the first. Groundwater samples were collected later in the summer and early fall.

NAWQA protocols have been developed for both surface water (*16*) and groundwater (*17*) as well as for the general field measurements which accompany sampling (*18*). The equal-width-increment sampling method (*19*) was used on most streams, but equal-discharge-increment sampling was used on some larger rivers. All equipment was precleaned with a Liquinox/tap-water solution, rinsed with tap water, then rinsed with deionized water, followed by methanol, and air dried. Sample volume was 1-liter. All samples were filtered through a 0.7 μ baked glass-filter using a metering pump and teflon tubing. Filtrate was collected in precleaned 1-liter amber glass bottles. Samples were immediately chilled and shipped on ice to the NSTL. Analysis of each for all 16 compounds was then carried out using electrospray mass spectrometry (ESMS). Concurrent samples were collected and forwarded directly to the USGS National Water Quality Laboratory (NWQL) for pesticide analysis by gas chromatography-mass spectrometry (GC-MS) with selected-ion monitoring (SIM) techniques to determine both triazine and chloroacetanilide residues (*20*). Field quality assurrance for these samples consisted of both concurrent replicates and field blanks.

Analytical Method

Because of the diversity of structure and chemical property encompassing IMI's, SU's and SA's (see Figure 1 and Table 1), the development of a single pesticide residue analytical method to determine sixteen target analytes simultaneously is a daunting challenge. Furthermore, development of any method for surface water generally requires additional preparatory clean-up steps due to the presence of dissolved salts, dissolved organic carbon, and suspended fine particulate materials that are not normally required for groundwater samples. A method for quantifying nine SU's in soil and water has been developed using high-performance liquid chromatography (HPLC) with ultraviolet (UV) detection. With a limit of quantitation (LOQ) of 0.1 μg/l for water it is intended for screening purposes and requires confirmation by liquid chromatography-mass spectrometry (LC-MS) techniques (*21*). The target analyte list does not include nicosulfuron. Similar approaches have been applied to the IMI's (*22*).

A multi-compound method for surface water developed by DuPont in conjunction with a consortium of herbicide manufacturer's together with the USEPA was used (*23*). This method is intended for use on surface water receiving drainage from agricultural landscapes, either overland flow or spring seepage. This multianalyte method utilizes liquid chromatography interfaced with mass spectrometry (LC/MS) using electrospray ionization (ESI) on a single quadrupole instrument. Solid-phase extraction (SPE) is used extensively throughout the sample preparation steps. Samples are acidified, the analytes extracted on an RP-102 cartridge, and the eluate further cleaned-up on a strong anion exchange cartridge stacked directly above an alumina cartridge. Dry residue from the cleanup cartridge eluates is dissolved in LC initial conditions mobile phase and separated by gradient elution on an n-octylsilica analytical column. For confirmation of identity for each analyte, three criteria are imposed: (1) detection of three characteristic ions, (2) ion ratios relative to the molecular ion (MH+) must be ±20% of the ion ratios monitored for each corresponding standard, (3) LC/MS retention time for each analyte must be ±0.3% of the retention time for each of the corresponding standards.

The mass spectrometer is operated in the positive ion mode using programmed selected ion monitoring (SIM). In order to generate fragment ions in addition to the hydrogen-atom adduct of the molecular ion [MH]+ for each compound, the capillary exit voltage (CapEx) is changed throughout each chromatographic run. Changing this voltage is necessary in order to optimize collision-induced dissociation (CID) resulting in the fragment ions. The voltage difference between the capillary exit and the first skimmer determines the amount of fragmentation; the greater the voltage difference the greater the fragmentation. This voltage difference generates sufficient energy to rupture chemical bonds by accelerating ions into collisions with molecules of the drying gas. Since the skimmer voltage is fixed during tuning, the amount of fragmentation can be controlled by changing the capillary exit voltage. In addition, in order to capture each of the three ions chosen to represent characteristic fragementation for each analyte, the selected-ion monitoring function of the mass spectrometer must also be programmed to capture the appropriate ions within the designated retention window for each analyte.

Equipment. The mass spectrometer was a Hewlett-Packard 5989B MS Engine Quadrupole unit equipped with an HP 59987A API-Electrospray MS Interface. The liquid chromatograph is a Hewlett-Packard 1090 L Series II unit with a 79847A Autosampler, a 79846A Autoinjector, and a 79835A (DR5 SDS) solvent delivery system with a static mixer. For all LC separations, the analytical column was Zorbax RX-C8, 250 X 4.6 mm., 5 μ spherical particles, used with a 12.5 X 4.6 mm. guard column containing the same packing.

Standards, Instrument Tuning, Calibration. The preparation of both stock and working standards, the tuning of the mass spectrometer, the gradients used both for the analytical separation and the programming of the capillary exit voltage, and the water sample treatment and extraction process itself, were carried out exactly as described in the method report (23). With only minor modifications the sample clean-up procedure is the same as the capillary electrophoresis (CE) method developed and published by the USEPA-OPP&TS (24). Capillary electrophoresis was the primary technique and LC/MS was the confirmatory method. Quantitative data obtained by CE agreed very well with data obtained by LC/MS.

Figure 3 illustrates the analytical separation which is achieved for all sixteen of the target analytes using the C8 column. Under a flow rate of 1.0 ml/min, with the mobile phase split so that 90% is directed to waste and 10% to the API-interface, all 16 analytes are distinctly separated from each other within a 5-45 minute time window during which the mobile-phase gradient proceeds from 20% MeCN in 0.15% acetic acid-amended water to 50% MeCN in acetic acid-amended water. Furthermore, peak shape is undistorted and symmetrical with no evidence of fronting or tailing. Width at half-height is <0.5 minute for the last two peaks in the chromatogram. This suggests no mixed-mode mechanism of separation is operative. Dissolved constituents occurring as chemical artifacts and behaving as interferrents, such as those often encountered in surface water, can behave unpredictably in unintentional mixed-mode separations. Three pair of analytes nearly coelute; chlorsulfuron-triasulfuron, prosulfuron-halosulfuron methyl, and triflusulfuron methyl-primisulfuron methyl. However, this appears to be of little consequence for the prosulfuron-halosulfuron methyl pair and for the triflusulfuron methyl-primisulfuron methyl pair because these analytes do not generate the same ions. Therefore, their near-coelution will not affect detection or quantitation. In contrast, chlorsulfuron and triasulfuron both generate the same

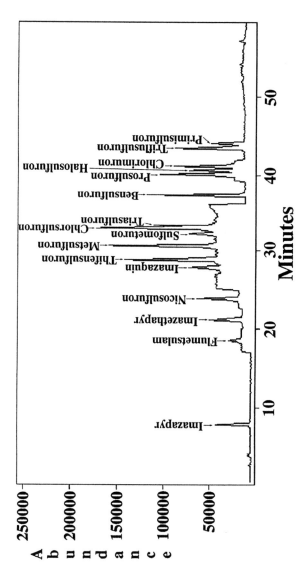

Figure 3. HPLC Separation For All Sixteen Target Analytes.

confirming ions under the protocol and differ only in the mass of their respective hydrogen-atom adduct molecular ions.

Results

Method Validation. NSTL's recovery on laboratory spikes of surface water and ground water from agricultural drainage yielded a mean of 89% for all sixteen analytes. Generally, the IMI's furnished about 20% lower recovery that the SU's and flumetsulam. Subsequent experiments have shown that low recoveries for the IMI's can be improved somewhat by adjusting the sample acidity to pH 3 just prior to extraction on RP-102 cartridges. Experience has also shown that the IMI's tend to elute later from the alumina cartridge than do the SU's or flumetsulam. Increasing the elution volume for the 0.5% glacial acetic acid in dichloromethane results in improved recoveries for the IMI's. It has also been observed that when moisture is introduced into the alumina cartridge, recoveries are higher for the IMI's. In contrast to the IMI results, recoveries for the SU's approached theoretical spike values with the exception of halosulfuron methyl.

By summer 1997, the Soil Tilth laboratory was one of the first facilities outside of the development laboratories to use this protocol, others being formally involved in multi-laboratory, multi-operator precision and accuracy collaborations. Based upon that experience, we suggest two alternative practices which may enhance the method performance. First, the use of isotopically enriched analogs of the target analytes as recovery surrogates or for quantitation either by the internal standard technique or the isotopic dilution method. Surrogates would serve to alert the analyst to gross sample processing errors which may affect the analyte recovery. Internal standard quantitation, either directly or through isotopic dilution, may increase the accuracy of the determination or at least reduce the variability associated with quantitation routines used on the MS. Second, quantitative sensitivity could be improved without changing the sample preparation portion of the method by adoption of a "two-pass" approach to quantitation. In the first-pass, data for quantitation are acquired for all analytes at a capillary exit voltage fixed at low potential. This would result in a high abundance of the hydrogen-atom adduct of the parent molecule [MH]+ with little or no accompanying fragmentation. This, in turn, would yield highest sensitivity. In the second-pass, following concentration of the extract, the capillary exit voltage is programmed according to method protocol, using the daughter ion ratios for confirmation of identity. This approach does require additional instrument time for each sample, but would allow for lowering the LOQ below 0.10 ppb without significant procedural changes.

Reconnaissance Study. Sixty-two surface and groundwater samples collected from the midwest during June-September 1997 had been analyzed by the end of October. Instrumental analysis involves operating the electrospray mass spectrometer with a variable capillary exit voltage in order to generate daughter ions for confirmation of identity. Residues of the sixteen target analytes were identified and confirmed in 16% of the river water samples and in 8% of the well water samples. Target herbicides were detected in small tributary streams, larger river basins, and two Iowa municipal wells pumping alluvial aquifers. The only analyte not detected in any of the samples was sulfometuron methyl, a herbicide used on noncropped pasture or rangeland. A representative chromatographic profile typical of all river water extracts is shown in Figure 4, the Iowa River near Rowan in north central Iowa, sampled on June 23, 1997. This chromatogram shows two confirmed "hits", imazethapyr and nicosulfuron, with [MH]+ ions seen at m/z 290 and 411, respectively. The box plots indicate the presence of the fragment ions based upon comparison to standards. Confirmation of identity is achieved when the ion abundance falls within the box plot distribution for each ion. In the case of prosulfuron and primisulfuron, both daughter ion ratios do not fall within

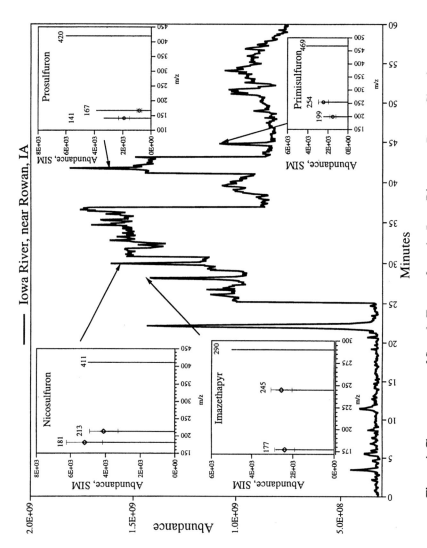

Figure 4. Chromatogram of Sample Extract from the Iowa River near Rowan Showing Imazethapyr and Nicosulfuron Detects Based Upon Identifying and Confirming Ions.

the box plot range; therefore, presence of these compounds is not confirmed, even though the presence of the [MH]+ ion would suggest otherwise. The use of confirming ions in this manner may be subject to error when interpreting chromatograms containing coeluting peaks. This results from alteration of fragmentation patterns, which, in turn, may effect ratio distributions. The baseline offset seen in these and other chromatograms is in response to changes in the capillary exit voltage during the 45-minute run.

The remainder of the original extracts were stored in a freezer for 11 months. In order to reassess our original results and for the purposes of further confirmation of our findings, each of the original extracts with confirmed hits was rerun in August 1998. One year of experience working with this new method had suggested that meeting the confirming ion criteria may have been responsible for missing some of the low level detects. Accordingly, the post acquisition data processing was modified. Quantitation was done using a 5-point calibration on only the [MH]+ ion acquired at constant capillary exit voltage for each analyte throughout the LC run. As shown in Figure 5, the chromatographic profile has changed, signal-to-noise ratios are improved for most analytes, and several small peaks are now discernable.

Advantages of using the single-ion criterion mode of mass spectral analysis for detection of very low concentrations of target analytes is illustrated in Figure 5. Displayed together are the [MH]+ ion chromatogram for the sixteen analytes of the standard mixture and for the extract from the sample collected on June 23, 1997 at the Iowa River near Rowan. Within the 5-20 minute retention time window, one target analyte is identified by both its mass and retention time characteristics. Data acquisition within the retention window reveals a peak at m/z 262, both qualifiers corresponding to imazapyr. Within the 20-35 minute retention time window, flumetsulam at m/z 326, imazethapyr at m/z 290, nicosulfuron at m/z 411, and imazaquin at m/z 312, each are identified at their respective retention times. Within the 35-50 minute retention time window, prosulfuron at m/z 420 and primisulfuron at m/z 469 are detected. Several of these very low-level detects, imazapyr, imazaquin, flumetsulam, prosulfuron, and primisulfuron, would have probably been missed by using the (MH)+ ion plus the two confirming ion fragments approach

A comparison of the two SIM approaches to identification is illustrated in Figure 6. The sample is an extract of well water from a municipal well near Correctionville in northwestern Iowa. This well, pumping an alluvial aquifer of the Little Sioux River, was sampled on July 1, 1997. The chromatogram compares the three-ion SIM from the well water and the [MH]-ion SIM of the well water to the sixteen compound standard mixture for the [MH] ion only. The [MH]-ion chromatogram is easier to interpret because it is much less complex. Overlay of the [MH]+ SIM standard with the [MH]+ sample extract clearly identifies both nicosulfuron and imazethapyr both as prominent peaks in the chromatogram. While this example confirms the presence of these two post-emergence chemicals in a shallow groundwater resource, this finding should not be interpreted as generally representative of shallow alluvial groundwater in Iowa. Shallow water tables together with rather unique hydrologic characteristics observed in alluvial aquifers elsewhere along the Iowa River have resulted in dramatic artesian responses to drilling (25). Although, most alluvial aquifers are not artesian. Notwithstanding our detection in two shallow municipal wells in different areas of the state, a small scale laboratory column study evaluating the leaching potential of nicosulfuron through six different Iowa loamy and clayey soils has concluded that chemisorption on smectites greatly diminishes and probably prevents the contamination of well water by this herbicide (26). However, this study did not take into account

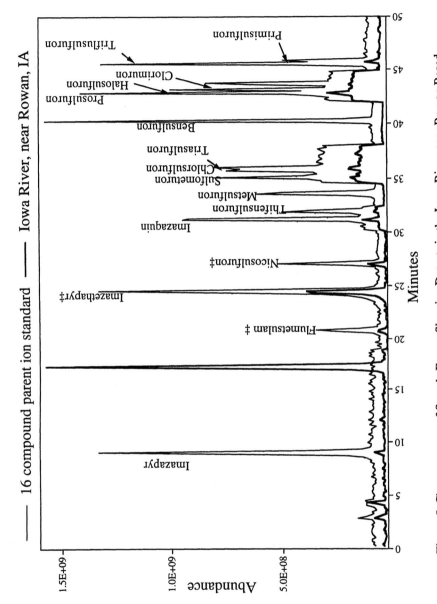

Figure 5. Chromatogram of Sample Extract Showing Detects in the Iowa River near Rowan Based Upon the (MH)+ Ion Only.

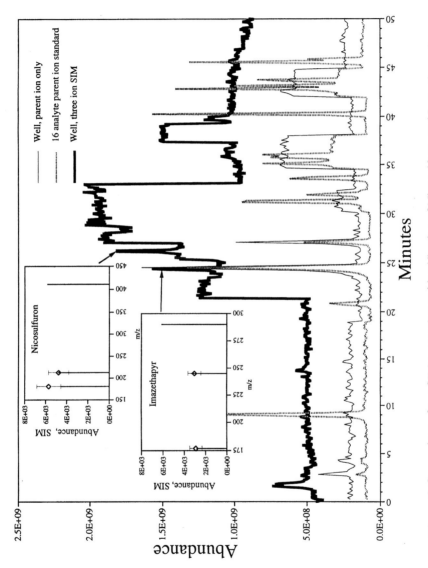

Figure 6. Comparison of two SIM approaches to identification of Nicosulfuron and Imazethapyr in Extract of Corectionville, IA., Well Water.

other factors such as macropore flow, fractured clays, organic carbon influences, and other variables which affect transport.

The chromatograms shown in Figures 4, 5 and 6 confirm that this method is capable of detecting and quantifying IMI and SU families of low-use rate herbicides at or below the stated method LOQ. Furthermore, the rigorous qualitative criteria can be applied to authentic water resources receiving agricultural drainage. The results from the NSTL for each of the sixteen target analytes are presented in Table 4. The chemicals detected most frequently and in highest concentration are fully consistent with best estimates of actual producer usage patterns based upon recent surveys. 1997 surveys suggest the greatest usage areas for flumetsulam, imazethapyr and nicosulfuron to be acreage encompassing the southern half of Minnesota, the northern half of Iowa, and the southern half of Illinois (Peter, C. J., DuPont Agricultural Products Company, personal communication, 1997). This is an area dominated by corn/soybean rowcropping on landscapes which discharge agricultural drainage into waterways and tributary streams which ultimately drain into the central and/or lower Mississippi River basin. Also, it is important to note that confirmed detects in the early June sampling, often within 30-45 days following a post-emergence broadcast application, were not repeated in a later sampling, often as soon as 10-30 days later. Although data from one sampling year cannot support long-term multi-year conclusions, it appears that occurrence of these herbicides in surface water is a seasonally transient phenomenon. Therefore, the data is not necessarily suggestive of long-term cumulative water quality problems resulting from the introduction of these relatively new "low-use-rate" chemicals.

Throughout the field sampling phase, replicate samples were sent to the USGS laboratory to be analyzed for older families of herbicides, emphasizing triazines and chloroacetanilides. Those results from the same sites are presented in Table 5. In general, both frequency of occurrence and measured concentration for atrazine, metolachlor, and acetochlor are greater than for the IMI's and SU's. While most of the IMI's and SU's have only been in use some 15 years or less, this is a reflection of current herbicide use patterns. However, comparisons drawn between the two classes of herbicides must take into account the limits of detection, reporting limits, and limits of quantitation associated with two different methods of analysis. Furthermore, application rates per acre differ markedly between the two chemical classes. It is also likely that a portion originates from release of bound and largely undegraded pesticide residue stored in the soil volume for these older herbicides introduced in the 1950's, 60's and 70's (27). A second reconnaissance study of somewhat larger scale conducted by the USGS in 1998 across the same area and sampling many of the same sites one year later resulted in similar findings corroborating the 1997 study.

An ideal quantitative method for pesticide residue determination in water should possess certain attributes and advantages. Among these are (a) rapid sample throughput with minimum cost, manpower, and instrument requirement, (b) high accuracy and precision, (c) freedom from matrix and reagents borne interference and contamination, (d) accounting for target analyte recovery, (e) accounting for analyte loss during sample preparation, and (f) simple routine operation by trained technicians (28). For any chromatographic method, analysts may use any technique which demonstrates that the peak is pure and homogeneous and originates only from the correct analyte. Typically, this proof is provided by mass spectrometry. For the pesticides coming into usage today as well as those under development in the future, the technique of LC/MS and liquid chromatography-mass spectrometry-mass spectrometry (LC/MS/MS) is rapidly

Table 4. Reconnaissance Results for SU's, IMI's, and SA.

μg/L[†]	NSTL		Surface Water						Groundwater	
	MDL	LOQ	Tensas River near Tendal, LA	Iowa River near Rowan, IA	Mississippi River near Clinton, IA	Wabash River near New Harmony, IN	Illinois River, site 1	Illinois River, site 2	Belle Plaine, IA	Correctionville, IA
Imazapyr	0.02	0.08	nd	<0.02*	nd	nd	nd	0.03*	nd	nd
Flumetsulam	0.04	0.12	0.5‡	0.05‡	nd	nd	nd	0.04*	nd	nd
Imazethapyr	0.02	0.08	nd	0.21‡	0.05‡	0.03*	nd	0.11‡	<0.02*	0.09‡
Nicosulfuron	0.03	0.08	0.2*	0.18‡	0.05‡	0.65	.05‡	0.04*	0.03*	0.12‡
Imazaquin	0.02	0.08	0.7*	0.04*	0.02	0.17	nd	0.02*	nd	nd
Thifensulfuron methyl	0.03	0.12	nd	0.06*	nd	nd	nd	nd	nd	nd
Metsulfuron methyl	0.03	0.08	nd	0.04*	0.05*	nd	nd	0.04*	nd	nd
Sulfometuron methyl	0.02	0.12	nd	nd	nd	nd	nd	nd	nd	nd
Chlorsulfuron	0.02	0.12	nd	0.05*	nd	nd	nd	nd	nd	nd
Triasulfuron	0.03	0.12	0.05	0.04*	nd	nd	nd	0.05*	nd	nd
Bensulfuron methyl	0.01	0.08	nd	0.04*	nd	nd	nd	nd	nd	nd
Prosulfuron	0.02	0.08	nd	0.04*	0.02*	nd	nd	0.04*	nd	nd
Halosulfuron methyl	0.03	0.08	nd	<0.03*	nd	nd	nd	nd	nd	nd
Chlorimuron ethyl	0.03	0.08	0.4*	0.03*	0.03*	0.06*	nd	nd	nd	nd
Triflusulfuron methyl	0.02	0.08	nd	<0.04*	nd	nd	nd	nd	nd	nd
Primisulfuron methyl	0.04	0.12	nd	<0.04*	nd	nd	nd	0.05*	nd	nd

[†] all values quantitated by the external-standard method on (MH)+ using a five-point calibration.
MDL - method detection limit estimated as 3X signal-to-noise ratio for flumetsulam.
LOQ - limit of quantitation at which confirming ion ratios consistently fall within 20% windows.
* confirming ions criteria not met.
Illinois River site 1 is located near Valley City, IL.
Illinois River site 2 is located near Ottawa, IL.
‡ confirmed by ion ratios on initial analysis within 4 days of sample receipt.
nd - not detected.

Table 5. Results on Reconnaissance Duplicates for Triazines and Chloroacetanilides.

µg/L	USGS[1] LOQ	Surface Water						Ground Water	
		Tensas River near Tendal, LA	Iowa River near Rowan, IA	Mississippi River near Clinton, IA	Wabash River near New Harmony, IN	Illinois River, site 1	Illinois River, site 2	Belle Plaine, IA	Correctionville, IA
Atrazine	0.005	39	6.7	0.4	19.8	7.96	5.65	0.16	0.11
Cyanazine	0.005	0.3	0.14	0.1	3.3	1.68	1.14	nd	nd
Simazine	0.005	0.1	0.05	0.05	1.3	0.06	0.04	nd	nd
Acetochlor	0.005	nd	0.25	0.18	2.9	0.74	0.78	nd	nd
Metolachlor	0.005	8.1	5.1	0.2	10.3	1.75	1.15	nd	nd
Metribuzin	0.005	3.7	0.01	nd	0.2	0.02	0.01	nd	nd

[1] for details of methodology, see reference (20).
LOQ - limit of quantitation.
Illinois River site 1 is located near Valley City, IL.
Illinois River site 2 is located near Ottawa, IL.
nd - not detected

becoming the standard approach, as many of these new chemicals are not detected very well by gas chromatography without derivatization. This is especially true as the limit-of-detection for environmental monitoring purposes continues its progression to quantify lower and lower values; now approaching 5-25 part-per-trillion (ppt). For external-standard quantitation, a 5-point calibration line should bracket expected concentrations for samples, and should be completely characterized as to line equation, y intercept, and linear range. For internal-standard quantitation, an isotopic analog of the analyte which coelutes with the analyte but is distinguishable by mass spectrometry is recommended.

Summary

While only fourteen more surface water samples were collected than groundwater samples, herbicides were detected twice as frequently in surface water. One or more of the sixteen target analytes were detected in 16% of the surface water samples and in 8% of the well water samples taken from across the upper midwest between June and October, 1997. Most of the detects occurred in surface water early in the growing season and none were observed above 1.0 µg/L. Flumetsulam, nicosulfuron, and imazethapyr were detected most frequently.

The DuPont-multianalyte method can be used for monitoring water resources for residues of sixteen ALS-inhibiting herbicides. Although developed for surface water, our experience suggests it can be used on groundwater without modification. While the several sample-preparation steps are somewhat tedious, experienced analysts can expect consistent results with recoveries ranging between 75-125% of theoretical at the LOQ and at ten times the LOQ for all sixteen analytes. It should become a widely used standard technique because it is a monitoring method which is well suited to address the in-stream "total maximum daily load" requirement for determination of pesticide runoff under the "impaired waterways list" as mandated by Section 303(d) of the Clean Water Act (*29*). Another value to analysts of this method lies in its adaptability. It can easily be simplified to accommodate a smaller number of analytes required by more focused studies carried out at scales other than large-scale monitoring efforts. As the USEPA moves toward adoption of "performance-based measurement systems" (PBMS) for environmental monitoring, this method is easily modified to meet data-quality objectives by demonstrating its equivalency to the different techniques specified by regulatory agencies (*30*).

Acknowledgments

We wish to thank Dr. Dayan B. Goodnough for his many helpful suggestions and especially want to note his willingness to visit NSTL and provide guidance on using the unpublished method. The authors also wish to acknowledge the field sampling efforts of the staff from the NAWQA study units in the USGS District Offices in IL, IN, IA, LA, MN, MS, and NE.

Disclaimer

Mention of specific products, suppliers, or vendors is for identification purposes only and does not constitute an endorsement by the U.S. Dept. Of Agriculture to the exclusion of others.

Literature Cited

1. *National Crop Residue Management Survey*, prepared by the Conservation Technology Information Center, West Lafayette, IN., **1995**, 75 pp.
2. *National Crop Residue Management Survey,* prepared by the Conservation Technology Information Center, West Lafayette, IN., **1996**, 78 pp.

3. Harrison, S. K.; Loux, M. M. In *Handbook of Weed Management Systems*; Smith, A. E., Ed.; Marcel Dekker: New York, NY. **1995**; Chapter 5.
4. Obrigawitch, T. T.; Cook, G. and Wetherington, J., *Pestic. Sci.*, **1998**, 52, 199-217.
5. Beyer, E. M.; Duffy, M. J.; Hay, J. V.; Schlueter, D.D. In *Herbicides: Chemistry, Degradation, and Mode of Action*, Kearney, P. C. and Kaufman, D. D., Eds.; Marcel Dekker: New York, NY. **1988**; Volume 3.
6. Blair, A. M.; Martin, T. D., *Pestic. Sci.*, **1988**, 22, 195-219.
7. Shaner, D. L.; O'Connor, S. L. In *The Imidazolinone Herbicides*, CRC Press: Boca Raton, FL., **1991**, 290 pp.
8. *Agricultural Chemical Usage, 1996 Field Crops Summary*, prepared by the National Agricultural Statistics Service, Washington, D.C., September **1997**, 92 pp.
9. *Agricultural Chemical Usage, 1995 Field Crops Summary*, prepared by the National Agricultural Statistics Service, Washington, D.C., March **1996**, 100 pp.
10. *Agricultural Chemical Usage, 1994 Field Crops Summary*, prepared by the National Agricultural Statistics Service, Washington, D.C., March **1995**, 106 pp.
11. *Agricultural Chemical Usage, 1993 Field Crops Summary*, prepared by the National Agricultural Statistics Service, Washington, D.C., March **1994**, 114 pp.
12. Agricultural Chemical Usage, 1992 Field Crops Summary, prepared by the National Agricultural Statistics Service, Washington, D.C., March **1993**, 118 pp.
13. Gilliom, R. J.; Alley, W. M.; Gurtz, M. E., *Design of the National Water-Quality Assessment Program: Occurrence and Distribution of Water-Quality Conditions*, **1995**, U. S. Geological Survey Circular 1112, 33 pp.
14. Detroy, M.G.; Hunt, P.K.B.; Holub, M.A., *Ground-Water Quality-Monitoring Program in Iowa: Nitrate and Pesticides in Shallow Aquifers,* **1988**, U.S. Geological Survey Open-File Report 88-4123, 32 pp.
15. Kolpin, D.W.; Kalkhoff, S.J.; Goolsby, D.A.; Sneck-Fahrer, D.A.; Thurman, E.M., *Ground Water*, **1997**, 35, 679-688.
16. Shelton, L. R., *Field Guide for Collection and Processing Stream-Water Samples for the National Water-Quality Assessment Program,* **1994**, U.S. Geological Survey Open-File Report 94-455, 42 pp.
17. Koterba, M. T.; Wilde, F. D.; Lapham, W. W., *Ground-Water Data-Collection Protocols and Procedures for the National Water-Quality Assessment Program: Collection and Documentation of Water-Quality Samples and Related Data,* **1995**, U. S. Geological Survey Open-File Report 95-399, 113 pp.
18. Wilde, F. D.; Radtke, D. B., *National Field Manual for the Collection of Water-Quality Data, Chapter A6, Field Measurements,* **1998**, Techniques of Water Resource Investigations, Book A9, U. S. Geological Survey, 20 pp.
19. Edwards, T. K.; Glysson, D. G., Field Methods for Measurement of Fluvial Sediment, **1988**, U. S. Geological Survey Open-File Report 86-531, 118 pp.
20. Zaugg, S.D; Sandstrom, M.W.; Smith, S.G.; Fehlberg, K.M., *Methods of Analysis by the U.S. Geological Survey National Water Quality Laboratory--Determination of Pesticides in Water by C-18 Solid-Phase Extraction and Capillary Column Gas Chromatography/Mass Spectrometry with Selected-Ion Monitoring.* **1995**, U. S. Geological Survey Open-File Report 95-181, 49 pp.
21. Powley, C. A. and de Bernard, P. A., *J. Agric. Food. Chem.*, **1998**, 46, 514-519.
22. Stout, S. J.; daCunha, A. R.: Picard, G. L., and Safarpour, M. M., *J. Agric. Food Chem.* **1996**, 44, 2182-2186.
23. Rodriguez, M.; Orescan, D.B., *Anal. Chem.*, **1998**, 70, #13, 2710-2717.
24. Krynitsky, A. J., *J. Assoc. Official Anal. Chem.*, **1997**, 80, #2, 392-400.
25. Libra, R. D., in *Iowa Geology*, Iowa Dept. Of Natural Resources, 1995.
26. Gonzalez, J. and Ukrainczyk, L., *J. Environ. Qual.*, **1999**, 28, 101-107.
27. Steinheimer, T.R.; Scoggin, K.D., *J. Environ. Monit.* **1999**, *in press.*
28. Krull, I. and Swartz, M., *LC-GC*, **1998**, 16, #12, 1084-1090.
29. *Pesticide and Toxic Chemical News*, CRC Press LLC, Washington, D.C., May 21, 1998.

21. Powley, C. A. and de Bernard, P. A., *J. Agric. Food. Chem.*, **1998**, 46, 514-519.
22. Stout, S. J.; daCunha, A. R.: Picard, G. L., and Safarpour, M. M., *J. Agric. Food Chem.* **1996**, 44, 2182-2186.
23. Rodriguez, M.; Orescan, D.B., *Anal. Chem.*, **1998**, 70, #13, 2710-2717.
24. Krynitsky, A. J., *J. Assoc. Official Anal. Chem.*, **1997**, 80, #2, 392-400.
25. Libra, R. D., in *Iowa Geology*, Iowa Dept. Of Natural Resources, 1995.
26. Gonzalez, J. and Ukrainczyk, L., *J. Environ. Qual.*, **1999**, 28, 101-107.
27. Steinheimer, T.R.; Scoggin, K.D., *J. Environ. Monit.* **1999**, *in press.*
28. Krull, I. and Swartz, M., *LC-GC*, **1998**, 16, #12, 1084-1090.
29. *Pesticide and Toxic Chemical News*, CRC Press LLC, Washington, D.C., May 21, 1998.
30. Keith, L., *Update on PBMS*, in *EnvirofACS*, Division of Environmental Chemistry, American Chemical Society, December, 1998 ; Hansen, G. A., *An Update on U.S. EPA Implementation Efforts Regarding Performance-Based Measurement Systems as Reported at WTQA '98, Amer. Lab.*, December, 1998.

Chapter 18

The Potential of Vegetated Filter Strips to Reduce Pesticide Transport

J. L. Baker, S. K. Mickelson, K. Arora, and A. K. Misra

Department of Agricultural and Biosystems Engineering,
Iowa State University, Ames, IA 50011

Reduction of pesticide transport to surface water resources to avoid water quality problems can involve both in-field and off-site approaches. One promising practice that is receiving increased emphasis, and that bridges both approaches, is the use of vegetated filter strips, either within the field or between the field border and a water resource of concern. Two main factors that affect the efficiency of pesticide transport reduction are the cropping/filter strip area ratio and the properties, and therefore major transport mechanism, of the pesticide. Area ratios as low as 1:1, for example with strip cropping, where one of the two equal strips is cropped without use of the pesticide of concern, can be very effective and much more so than say for a situation where the ratio is > 50. Because vegetated filter strips are generally much more effective at reducing sediment transport than runoff volume, losses of pesticides that are more strongly adsorbed are more readily controlled.

Pesticides transported from treated fields with sediment and water to surface water resources can cause concerns for human health with respect to drinking water, and for the viability of acquatic life that must exist in waters receiving agricultural drainage. Because of at least some adsorption of pesticides by most soils (the degree depending primarily on pesticide and soil properties), the major pesticide transport mechanism is through overland flow as opposed to transport with subsurface drainage water reaching the stream as either artificial tile flow or natural base flow.

To reduce the nonpoint source pollution represented by pesticide transport to surface water resources, various in-field and off-site best management practices

(BMPs) have been proposed. Because pesticide transport or loss is a product of pesticide concentrations and the masses of the water and sediment carriers, practices that reduce concentrations and/or carriers reduce losses. Two significant in-field practices that can reduce concentrations involve rate and method of application.

Studies (*1,2*) have shown that pesticide losses are roughly proportional to the application rate used, so that practices such as herbicide banding in rowcrops, where the area treated and thus the field rate of application may be only 25 to 50% of a broadcast application, are good BMPs. Other studies (*3,4*) have shown that mechanical incorporation, where the pesticide is physically mixed in the top 5 to 10 cm of soil, decreases runoff losses significantly, particularly for runoff events that take place within a short period of time after pesticide application with no intervening rainfall. Incorporation reduces the amount of pesticide in the thin (about 1 cm deep) surface soil "mixing zone" that interacts with rainfall and runoff, and thus causes a reduction in concentration. The concept of a mixing zone at the soil surface is used to understand and model the process by which pesticide is transferred from surface and near-surface soil to water. In this mixing zone, turbulence, caused by raindrop splash and flowing water, works in conjunction with dispersion and diffusion in establishing an equilibrium between pesticide adsorbed to soil and that in solution. The amount of pesticide in the mixing zone at the time of a runoff event can be significantly reduced when an extended period of time between application and a runoff event allows for more diffusion (and dissipation by various means) of pesticide from the surface soil. Similarly, rainfall events with insufficient intensity to cause runoff can also dissolve and move pesticide down out of the mixing zone.

A significant practice that can reduce the mass of water and particularly sediment carriers is conservation tillage (*5-7*). For strongly adsorbed pesticides, such as glyphosate, trifluralin, and chorpyrifos, the erosion control benefits of conservation tillage make it a very good BMP to control pesticide runoff losses. For weakly and moderately adsorbed pesticides, such as 2,4-D, alachlor, and atrazine, the reduction in runoff volume/mass usually associated with conservation tillage, at least on an annual basis, can also make it a BMP for those pesticides. However, there are specific cases, such as washoff of pesticides residing on surface crop residue, when pesticide concentrations are higher for conservation tillage (as well as the case where the desire to leave crop residue on the soil surface prohibits mechanical pesticide incorporation). Furthermore, there are specific cases when the water carrier volume is greater for conservation tillage, such as when runoff for a single event is greater for untilled or lesser-tilled soil versus soil that has been recently tilled and is rough and temporarily has a much higher porosity and infiltration rate.

Vegetated Filter Strip Function

When in-field practices, such as those cited above or described by Baker and Mickelson (*8*), are not feasible, or when their effects as BMPs are not sufficient to solve the pesticide loss problem, off-site practices may need to be considered. One off-site BMP for reducing field-to-stream pesticide transport, that is receiving renewed attention, is the use of buffer or vegetated filter strips (VFS). In actuality, VFS bridge both in-field and off-site approaches because they can be either within a field or between the field border and a water resource of concern.

In addition to being untreated areas, the presence of dense, close-grown permanent vegetation in VFS is critical to their effectiveness. VFS have the potential to reduce pesticide transport by reducing both concentrations and masses of carriers. As runoff (from a treated field) containing pesticide dissolved in water and adsorbed to sediment passes through a VFS, concentrations can be reduced through pesticide adsorption to living and dead vegetation and to non-treated in-place soil within the VFS. The water carrier mass can be decreased through infiltration into the soil in the VFS, as it is quite likely that the infiltration rate in the VFS is higher than in the field from which the runoff is originating (because of better structure and lower antecedent moisture content of the VFS soil; sod in the case of grass filter strips). There is some question of the final fate of the infiltrating water, and the pesticide dissolved in it, but water quality evidence from studies in which the pesticide content of subsurface drainage is measured would indicate considerable attenuation before that water reaches the surface again somewhere down gradient. The sediment carrier mass will generally be decreased as the runoff water flows across the rough soil and among the stems of the close-grown vegetation. The capacity of flowing water to keep sediment in suspension is proportional to the sixth power of the velocity. Therefore, if the VFS decreases the flow velocity by half (doubling the depth of flow), the sediment carrying capacity would be decreased by a factor of 64.

Since the contact time or time of passage of runoff flowing through a VFS will determine how much concentrations are reduced or how much water and sediment is retained, there are several physical factors that are important in determining the effectiveness of a VFS as a BMP. Some of those are the soil type, slope, length, and vegetation type/vegetation density of the VFS; and the soil type, area (relative to the VFS), and point/uniformity of runoff entry into the VFS from the source field.

Results of Pesticide Transport Reduction/VFS Studies

Although there have been several studies on the effects of VFS on reducing sediment (9) and (10) nutrients transport, the number of studies on pesticide transport reduction are more limited. Some of the most relevant ones are discussed below.

Grassed Waterway/Simulated Rainfall/2, 4-D Asmussen et al. (*11*) conducted a rainfall simulation study on the reduction of 2,4-D in runoff as it passed through a grassed waterway. The waterway was 4.6 x 24.4 m long, with vegetation consisting of bermudagrass and bahiagrass. The soil was a Cowarts loamy sand with 0.5% organic matter. 2,4-D was applied at a rate of 0.56 kg/ha to corn plots with an area of 30.2 m^2. Therefore the field drainage area to VFS area ratio was 1:3.7. Rainfall was simulated at a rate of 25.4 mm/h for 30 min on the corn plots, and was applied under dry and wet antecedent soil moisture conditions for both the corn plots and the waterway. Table I shows the percent reductions of the runoff (water), sediment, 2,4-D in both the water phase and sediment phase, and total 2,4-D. The grassed waterway was shown to be effective in reducing 2,4-D in the plot runoff, by approximately 70%, for the area ratio studied. As is evident in the data, the reduction in transport came as a result of reductions in both concentrations and carriers.

Table I. Reduction in water, sediment, and 2,4-D in a grassed waterway

	% Reduction	
	Dry condition	Wet condition
Runoff (water)	50	2
Sediment	98	94
2,4-D in water phase	71	69
2,4-D sediment phase	>99	>99
Total 2,4-D	72	69

SOURCE: Adapted with permission from reference 11. Copyright 1977.

Grassed Waterway/Simulated Rainfall/Trifluralin Rhode et al. (*12*) conducted a rainfall simulation study on the reduction of the herbicide trifluralin in runoff as it was directed through a 4.6 x 24.4-m long waterway. As in the study of Asmussen et al. (*11*), the soil was a Cowarts loamy sand with 0.5% organic matter, and the vegetation consisted of bermudagrass and bahiagrass. The day before rainfall, 1.12 kg/ha was applied to 28 m^2 plots; therefore the source area to VFS area ratio was 1:4.0. Rainfall was applied at 191 mm/h for about 30 min; two tests were run, one with a wet waterway and one with it dry. Table II shows the percent reductions of runoff (water and sediment mixture); of trifluralin in the mixed water/sediment runoff, as attributed to infiltration and to adsorption; and of the total trifluralin. Again for the area ratio studied, the grassed waterway was shown to be effective in reducing trifluralin in runoff.

Table II. Reduction in runoff and trifluralin in a grassed waterway

	% Reduction	
	Dry condition	Wet condition
Runoff (water plus sediment)	73	44
Trifluralin (due to infiltration)	43	29
Trifluralin (due to adsorption)	53	57
Total trifluralin	96	86

SOURCE: Adapted with permission from reference 12. Copyright 1980.

VFS/Simulated Rainfall/Atrazine Mickelson and Baker (*13*) conducted a rainfall simulation on the reduction of atrazine as it passed through a VFS consisting of 59% smooth brome, 35% bluegrass, and 6% tall fescue. The VFS plots were either 1.5 x 4.6 m or 1.5 x 9.1 m long, with a constant amount of simulated runoff added to the tops of the VFS (while simulated rainfall was being applied to them) to represent source area to buffer strip area ratios of 10:1 and 5:1, respectively. The simulated runoff had a nominal atrazine concentration in water of 1 mg/L, and either 0 or ~ 10,000 mg/L suspended sediment to simulate runoff for no-till and conventional tillage systems, respectively. Simulated rainfall was added to the VFS plots at an intensity of 66 mm/h for 1 h. Simulated runoff was added to the top of the plots after 10 min of wetting rain. This inflow was added to represent runoff from a 69 m^2 area at 2.5 cm/h for approximately 50 min. The percent reduction of atrazine and sediment in overland flow by the VFS is shown in Table III. Both the 5:1 and 10:1 area ratio plots were effective in reducing the sediment from the conventional tillage simulated runoff, with over 70% of the sediment retained in the plots. The 5:1 ratio plots were able to reduce the atrazine losses to a greater degree than for the 10:1 plots. There was no significant difference between reductions of atrazine with the no-tillage runoff versus the conventional tillage runoff. This can probably be attributed to the fact that atrazine is moderately adsorbed to the sediment, and is lost mainly dissolved in runoff water.

Table III. Percent reduction of atrazine and sediment with 5:1 and 10:1 area ratio

Simulated Tillage[a]	Area Ratio	% Reduction Atrazine	Sediment
NT	10:1	35.0	---
NT	5:1	59.0	---
CT	10:1	28.3	72.2
CT	5:1	51.3	75.7

[a]NT indicates no-tillage; CT indicates conventional tillage.
SOURCE: Adapted with permission from reference 13. Copyright American Society of Agricultural Engineers 1993.

VFS/Simulated Rainfall/Atrazine, Metolachlor, and Cyanazine Misra et al. (14) conducted a rainfall simulation study on the reductions in transport of atrazine, metolachlor, and cyanazine in runoff as it passed through a VFS consisting of 100% brome grass on a Storden loam soil (5.8% organic matter). The VFS plots were 1.5 x 12.2 m long, with the amount of simulated runoff added to the tops of the VFS plots (while simulated rainfall was being applied to them) adjusted to represent source area to buffer strip area ratios of 15:1 and 30:1. The second treatment variable in this 2 x 2 factorial experiment was herbicide concentrations in inflow, which were nominally either 1.0 or 0.1 mg/L for each herbicide. Simulated rainfall, with an intensity of 63.5 mm/h, was applied to the plots for 15 min before inflow runoff was added, and then for an additional 45 min after this inflow began. The VFS removal efficiencies in herbicide removal for the different area ratios and inflow concentrations are given in Table IV. Infiltration was one of the main factors for reduction of herbicide transport within a treatment; the greater the infiltration, the greater the herbicide removal. Although there was no significant difference in herbicide removal between the 15:1 and the 30:1 area ratios, there was a significant difference found between having high inflow concentration (1.0 mg/L) as compared to low inflow concentration (0.1 mg/L), with greater retention with the higher concentration (although unexplained differences in infiltration/runoff retention seemed to result in the concentration effect).

Table IV. Buffer strip efficiency in herbicide removal with 15:1 and 30:1 area ratios and 1.0 and 0.1 inflow concentrations

Area Ratio	Inflow Conc.	% Removal			
		Runoff Water	Atrazine	Metolachlor	Cyanazine
15:1	0.1 mg/L	35.6	31.2	32.0	26.2
15:1	1.0 mg/L	47.8	49.8	46.8	46.8
30:1	0.1 mg/L	25.4	26.4	27.4	25.6
30:1	1.0 mg/L	40.1	47.8	41.7	42.4

SOURCE: Adapted with permission from reference 14. Copyright American Society of Agricultural Engineers 1996.

In a separate but companion study in the same area (15), the effects of buffer reductions in transport of the same three herbicides were determined. Nominal herbicide concentrations were each 1 mg/L, and for runoff inflow with sediment, the sediment concentration was ~ 10,000 mg/L. The results for runoff, sediment, and herbicide reductions are shown in Table V. Runoff from the end of the buffer strips began the quickest and was the greatest for the bare plots with runoff inflow containing sediment, a combination which would lead to more "surface-sealing" and a lower infiltration rate. Runoff reduction and herbicide retention was greatest for the grassed VFS with runoff inflow containing no sediment. As for the other study, infiltration was one of the main factors for reduction of herbicide transport within and strips with and without vegetation, and runoff with and without sediment, on among treatments.

Table V. Buffer strip efficiency in herbicide retention with strips with and without vegetation and for runoff inflow with and without sediment

Surface	Sediment	% Reduction				
		Runoff	Sediment	Atrazine	Metolachlor	Cyanazine
Bare	None	49.1	-	50.7	45.2	50.1
Bare	With	29.8	32.2	33.5	27.5	35.6
Grass	None	83.1	-	85.2	82.6	84.1
Grass	With	54.3	77.8	53.6	53.3	57.5

SOURCE: Adapted with permission from reference 15. Copyright 1995.

Turf Strip/Simulated Rainfall/Dicamba, 2,4-D, Mecoprop, and Chlorpyrifos
Cole et al. (16) conducted two rainfall simulation studies (dry and wet antecedent moisture conditions) on the reduction of four pesticides, dicamba, 2,4-D, mecoprop, and chlorpyrifos, in runoff water as it passed through strips of bermudagrass turf. The plots were on a Kirkland silt loam soil with an average slope of 6%. The treated field/source area was 1.8 x 4.9 m long, and the VFS below the source area were either 2.4 or 4.9 m long, with the grass clipped to either a 1.3 or 3.8 cm height (in a 2x2 factorial experiment; an aerification treatment was also performed on an additional set of plots, 4.9 m long with both grass heights, but the data were not significantly different, and therefore, were

averaged with the non-aerification treatment). The source area to VFS area ratios were then 2:1 and 1:1. Rainfall was simulated at a rate of 51 mm/h for the dry run and 64 mm/h for the wet run, with rainfall being applied to both the source and VFS areas. For the dry run, an average of 92% of the rainfall on plots with and without buffer strips infiltrated; for the wet run, the average was much less at 42%. Table VI shows the percent reductions of dicamba, 2,4-D, mecoprop, and chlorpyrifos transported in runoff water. The most obvious effect is that of more infiltration and thus greater reductions in pesticide transport under the dryer antecedent moisture conditions. A second observation is that for the pesticide chlorpyrifos, with a K_{oc} value hundreds of times higher than those for the other three pesticides, there was significant reduction even for the wet run when over half the rainfall applied ran off. Differences between VFS lengths and between grass heights were generally not statistically significant.

Table VI. Reduction in dicamba, 2,4-D, mecoprop, and chlorpyrifos transport as affected by buffer length, grass height, and antecedent moisture conditions

Length m	Height cm	Antecedent m.c.	% Reduction			
			Dicamba	2,4-D	Mecoprop	Chlorpyrifos
2.4	1.3	dry	43	45	40	64
4.9[a]	1.3	dry	99	94	93	64
2.4	3.8	dry	78	81	81	100
4.9	3.8	dry	95	91	92	85
2.4	1.3	wet	0[b]	0	0	73
4.9	1.3	wet	27	35	33	100
2.4	3.8	wet	11	26	18	100
4.9	3.8	wet	0	0	0	80

[a]Values for 4.9-m length include plots with and without aerification.
[b]For some treatments, losses were > those without VFS (shown as 0%).
SOURCE: Adapted with permission from reference 16. Copyright 1997.

VFS/Natural-Simulated Rainfall/Metolachlor and Metribuzin Webster and Shaw (*17*) conducted a three-year natural-simulated rainfall study on the reduction of metolachlor and metribuzin in runoff water as it passed through a VFS. The VFS planted to tall fescue were 4 x 2 m long downslope of herbicide treated soybean plots (under three different tillage systems) that were 4 x 20 m long; thus the source area to VFS area ratio was 10:1. The plots were on a Brooksville silty clay soil (3.2% organic matter) with a 3% slope. To supplement natural rainfall to obtain a minimum of 50 mm rainfall every 2 wk, a rainfall simulator was used with a 76 mm/h rate. Table VII shows the annual percent reductions of runoff and metolachlor and metribuzin transport in runoff water as affected by the tillage/cropping system used on the source area. On average, the absolute losses of metolachlor and metribuzin from the source areas were generally lower for the conventional till/soybeans plots than for either the no-till mono- or double-crop system (data not shown), but the percent reduction in herbicide transport by the VFS was not related to tillage. The percent reduction in herbicide transport was more closely related to the percent reduction in runoff volume for any tillage/year combination, with statistical analyses indicating no difference in concentration of metolachlor and metribuzin in runoff with or without a VFS for any tillage system.

Table VII. Annual reduction in runoff, metolachlor and metribuzin by a VFS as affected by tillage/cropping in the source area

Year	Tillage	% Reduction		
		Runoff	Metolachlor	Metribuzin
1991	No-till/soybeans	65	58	57
1991	Conv.-till/soybeans	52	50	56
1991	No-till/soybeans-wheat	0	0	20
1992	No-till/soybeans	33	75	78
1992	Conv.-till/soybeans	14	33	45
1992	No-till/soybeans-wheat	4	29	22
1993	No-till/soybeans	42	58	57
1993	Conv.-till/soybeans	30	50	22
1993	No-till/soybeans-wheat	42	88	90

SOURCE: Adapted with permission from reference 17. Copyright 1996.

Oats Strip/Natural Rainfall/Atrazine Hall et al. (*4*) reported the percent of atrazine lost in runoff water and sediment from runoff plots, under natural rainfall, from with a 14% slope on Hagerstown silty clay loam soil. Atrazine was applied at two rates, 2.2 and 4.5 kg/ha, either as a preemergence (PRE) or pre-plant-incorporation (PPI) application on corn plots. The plots measured 1.8 x 22 m long, with five plots corn only and four plots of corn with a 6 m oat strip within the plots, located at the bottom of the plot. For these plots, the source area to VFS area ratio was then 2.7:1. The results are given in Table VIII for three months in the summer for the plots with the oats strips at the bottom. As shown, the reduction of atrazine was found to be significant when the runoff passed through the oat strips. They concluded that "maximum reduction of water, soil, and atrazine losses from this hillside was achieved with a conventional-tillage management system that combined pre-plant-incorporation of atrazine residue with strip cropping on the plot tiers. This system reduced atrazine losses in runoff water and sediment by 87-97%, compared with the preemergence-sprayed, non-stripped plot tiers."

Table VIII. Percent reduction in atrazine transport using oat strips

	% Reduction		
	Water	Soil	Total
PRE/2.2 kg/ha	85.3	100.0	90.6
PRE/4.5 kg/ha	59.5	83.8	65.5
PPI/2.2 kg/ha	64.8	66.7	64.9
PPI/4.5 kg/ha	82.3	83.3	82.4

SOURCE: Adapted with permission from reference 4. Copyright 1983.

VFS/Natural Rainfall/Atrazine, Metolachlor, and Cyanazine Arora et al. (*18*) conducted a natural rainfall study on the reductions in transport of herbicides atrazine, metalochlor, and cyanazine in runoff as it passed through a 20.1 m (66') long VFS consisting of 87% brome, 11% blue grass, and 2% other grasses on a Nicollet silt loam soil. The 20.1 m length meets the requirement of a "66-ft buffer strip" between a treated field and a direct point of entry of runoff into a stream on the labels for atrazine and cyanazine. The source area and flow-dividing equipment were such that runoff from a 0.41-ha herbicide-treated corn field would flow onto six 1.5 x 20.1 m long VFS with 15:1 and 30:1 area ratios (three replications of each area ratio). The results are given in Table IX for six runoff events in 1993 and 1994. The values are averaged for the two area ratios because the differences were not statistically different; however, by doubling the source area to VFS area ratio the runoff water, sediment, and herbicide retention percentages were generally reduced. The results again show that for these moderately adsorbed herbicides, the reduction in transport is closely related to the reduction in runoff volume due to infiltration in the VFS. It is also evident from the data that infiltration, and the reduction in herbicide transport, is a more dominant process where less than 20% of the rainfall runs off from the source area for any given event.

Table IX. Reduction of runoff, sediment, and herbicides for runoff events in 1993 (2) and 1994 (4)

		% Reduction				
Event	Runoff[a]/Rainfall	Runoff	Sed	Atra	Met	Cyan
# 1/93	6.5/15.3	9	44	13	22	15
# 2/93	2.6/10.2	34	57	44	33	37
# 1/94	2.5/15.3	97	100	100	100	100
# 2/94	19.9/45.9	44	65	54	51	49
# 3/94	0.9/20.3	98	98	98	99	98
# 4/94	2.3/23.1	69	86	58	73	69

[a]In mm for runoff from the source area (and for rainfall).
SOURCE: Adapted with permission from reference 18. Copyright American Society of Agricultural Engineers 1996.

Riparian Forest Buffer System/Natural Rainfall/Atrazine and Alachlor Lowrance et al. (*19*) conducted a natural rainfall study on the reduction of atrazine and alachlor transport by a riparian forest buffer system (RFBS) that included an 8 m wide Bermudagrass and Bahia grass strip adjacent to the field, a 40 to 55 m wide pine forest strip, and then a 10 m wide hardwood forest containing a stream channel. The soils were Alapaha and Tifton loamy sands. Measurements of herbicide concentrations (in runoff water/sediment mixtures) and runoff volumes (and therefore, by calculation, loads) were measured at the boundaries between the fields and grass strips, the grass strip and the pine forest strips, and the pine forest strip and the hardwood strip. Table X shows the data for average atrazine and alachlor concentrations and percent reductions in transport for the three-year study.

Both herbicide concentrations and loads were reduced significantly during transit through the RFBS. Calculations were performed on the atrazine concentration data to estimate what part of the change in concentration was due to dilution and what part was due to other factors. Overall, roughly 50% of the concentration changes for individual events would be ascribed to dilution meaning that other processes such as adsorption to vegetation and in-place soil must also be operative. Atrazine and alachlor were detected in shallow groundwater samples from under the pine forest, but generally at less than 1 µg/L values; whereas in comparison, surface runoff concentrations from the treated corn field were much higher, averaging 34 and 9 µg/L for atrazine and alachlor, respectively, during the after application periods.

Table X. Concentrations and reduction in atrazine and alachlor transport as runoff passes through grass and forest buffer strips (1992-1994)

Position	Avg. Concentration µg/L		% Reduction[a]	
	Atrazine	Alachlor	Atrazine	Alachlor
Field edge	5.5	1.4	-	-
Downslope edge of grass	3.8	0.69	70	81
Middle of pine forest	0.99	0.48	76	73
Downslope edge of pine forest	0.26	0.25	97	91

[a]Total reduction in load relative to what is leaving the field.
SOURCE: Adapted with permission from reference 19. Copyright American Society of Agricultural Engineers 1997.

Buffer Strips-Tile Outlet Terrace/Natural Rainfall/Atrazine, Metolachlor, and Cyanazine Mickelson et al. (*20*) conducted a natural rainfall study relative to the label requirement that for atrazine and cyanazine uses with a tile-outlet terrace system, a set-back or buffer area is required around the outlet stand-pipe. In this case, the setback or buffer area is a 20.1 x 40.2 m rectangle of untreated crop area as generally the standpipe is next to the terrace bank. The objectives of the study were to determine if alternative BMP's to the setback could reduce herbicide runoff losses as much or more, without the perceived increased economic costs and hardship of the setback. Three management practices were put into place on six isolated subwatershed areas draining to six individual intakes in two terraces. The treatments included: 1) herbicides preemergence surface broadcast applied without incorporation to the whole subwatershed area (no setback); 2) herbicides preemergence surface broadcast applied without incorporation to the watershed area, excluding a 20.1 x 40.2 m setback adjacent to the standpipe (setback); and 3) herbicides pre-plant surface broadcast applied with incorporation to the whole watershed area (incorporation-no setback). The herbicides were atrazine, metolachlor, and cyanazine. All the plots received tillage with a single pass of a tandem disk. For the non-incorporation treatment, this occurred before herbicide application, whereas for the incorporation treatment, disking followed herbicide application. Four events occurred over a 53-day period of time following herbicide application, with the first event occurring 31 days after application. These events ranged from 22 mm to 172 mm of rainfall. The total amount of rainfall that occurred before the first runoff event was 118 mm. The total percent herbicide loss from the terraces for the four events is given in Table XI. Due to intervening rain and the long period of time before the first runoff event after application, the benefits of herbicide incorporation were minimized when compared the other treatments (due to the movement of the herbicides out of the mixing zone with rainfall/infiltration/diffusion). The setback treatment on the average did not reduce herbicide outflow losses versus the no setback treatment for the four events beyond what would be expected from the reduction of area treated.

Table XI. Percent herbicide loss from the terraces after four runoff events

No Setback			Setback			Incorporation-No Setback		
Atra	Meto	Cyan	Atra	Meto	Cyan	Atra	Meto	Cyan
3.2%	1.5%	1.5%	3.5%	1.8%	1.6%	3.7%	1.5%	1.5%

SOURCE: Adapted with permission from reference 20. Copyright 1998.

Summary

As can be observed from the studies presented, several different methods can be used to estimate the effects of VFS on pesticide transport, including use of both natural and simulated rainfall. One of the primary concerns in design of such experiments is how closely can true field conditions be represented. For simulated rainfall or runoff, important considerations are antecedent moisture content in the VFS and whether rainfall is applied to the VFS during the study, both of which significantly affect the hydrology and the reality of the conditions. If rainfall is added during passage of runoff through VFS, pesticide dilution from the rainfall needs to be considered. As discussed below, the source area-VFS area ratio is another major factor. Runoff sampling and analysis and measurement of flow volumes needs to be accurate as possible because removal amounts will generally be calculated as differences between inflow and outflow. However, the monitoring and redistribution of inflow will require special attention to avoid influencing the results by the methods used. A decision will also have to be made on whether and how to handle sediment separate from runoff waters. In terms of sampling, the first runoff event, and flow at the beginning of each event, should be given more attention because pesticide concentrations are generally higher and more variable at these times (there is no preset standard methodology; however, experience and flexibility can be very useful).

From the results of the studies presented, it is evident that buffer strips can be effective in reducing pesticide transport in runoff from treated fields, particularly if covered with close-grown vegetation (a VFS). These can take the form of grassed waterways, contour buffer strips, vegetative barriers, and terrace tile inlet buffers within fields; or as field-borders, filter strips, set-backs, and riparian forest buffers at the field edge or offsite. A major factor in determining their effectiveness is the field runoff source area to buffer strip area ratio. As this area ratio increases, the effectiveness of the buffer strips in retaining pesticides decreases. Except for strip-cropping and some other special cases, area ratios of 1:1 or 2:1 are generally economically not feasible. If 5% of a field is in a VFS, this would give an area ratio of 20:1; however, if a 32-ha (80 ac) field drains to one point protected by a 20.1 x 40.2 m (66 x 132') rectangular VFS, the area ratio would be high at 400:1. A corollary to this is that uniform inflow and passage of runoff through the VFS is necessary to get the full benefit of the VFS area (VFS maintenance to avoid "concentrated flows" is necessary). While no studies can be cited that show that concentrated flow reduces VFS effectiveness in reducing pesticide transport, other studies (*10, 21*) have shown that concentrated flow reduces infiltration and sediment deposition.

A second major factor is the pesticide sorption potential for soil and sediment. For weakly to moderately adsorbed pesticides, the major carrier is runoff water, and infiltration of runoff into the VFS is a major removal process. Reduced transport because of reduced concentrations due to pesticide adsorption by vegetation and in-place soil on passage through a VFS can also be a major removal process. However, results are variable as to the degree of this process (and more information is needed to explain the variability). Pesticides strongly adsorbed by soil have more potential to be adsorbed by vegetation and in-place soil in VFS, and probably even more important, because sediment retention by VFS is much greater than for runoff water, strongly adsorbed pesticides which are transported mostly on sediment are much better candidates for this method of control. Other factors related to VFS efficiency include: antecedent soil moisture content, soil texture, pesticide concentration levels in the runoff, plant population, and buffer strip slope as discussed by Fawcett, et al. (*22*). The effects of adsorption and reductions in masses of carriers and pesticide concentrations in water as affected by these factors are shown in Table XII which gives calculated percent pesticide transport reductions based on hypothetical reductions in runoff and sediment masses upon flow through a VFS, for pesticides with different partition coefficients (K= concentration in sediment/concentration in water). It was assumed that pesticide concentrations in sediment were not changed on flow through a VFS, although if concentrations in water decrease, desorption from sediment and reduction in concentration is possible. However, it is also known that the selective deposition process, where larger, more dense particles settle faster, can result in "enrichment" where pesticide concentrations in sediment may increase.

Based on the wide range of values shown in Table XII, buffer strips should not be considered an "answer-all" BMP, but instead should be considered a BMP that could be used in conjunction with in-field BMPs to reduce pesticide and sediment losses. These might include: contour farming, strip cropping, conservation tillage, pesticide banding, pesticide incorporation, terraces, and choice of product/formulation based on more favorable properties.

Table XII. VFS Pesticide Transport Reduction for Hypothetical Conditons[a]

Water conc.[b] reduction	K= 1	10	100	1000
------------------------------------%--------------------------------				
Large runoff event or large area ratio (10% runoff and 60% sediment reductions[c]):				
none	10	15	35	55
10%	19	23	40	56
25%	33	35	46	58
50%	55	56	58	60
Moderate runoff event or moderate area ratio (25% runoff and 80% sediment reductions[c]):				
None	26	30	52	75
10%	33	37	56	76
25%	44	47	62	77
50%	63	64	71	78
Small runoff event or small area ratio (50% and 95% sediment reductions[c]):				
None	50	54	72	91
10%	55	59	75	91
25%	63	65	79	92
50%	75	77	85	93

[a]All runoff events with mass of runoff 100 times mass of sediment entering VFS.
[b]Concentration reduction assumed (only for water) due to possible adsorption.
[c]Percent reductions in runoff and sediment masses roughly estimated for differences of a factor of 5 in the volumes of runoff entering the VFS going from small to moderate to large.

Literature Cited

1. Hall, J.K.; Pawlus M.; Higgins E.R. *J. Environ. Qual.* **1972**, 1:172-176.
2. Lemke, D.W.; Baker J.L.; Melvin S.W. *Agricultural Drainage Well Annual Report*; Iowa Dept. Agric. Land Steward, Des Moines, IA, **1993**, 66 pp.
3. Baker, J.L.; Laflen M.; *J. Environ. Qual.* **1979**, 8:602-607.
4. Hall, J.K.; Hartwig N.L.; Hoffman L.D. *J. Environ. Qual.* **1983**, 12:336-340.
5. Laflen, J.M.; Colvin T.S. *Trans. ASAE*, **1981**, 24:505-509.
6. Baker, J.L.; Johnson, H.P. *Trans. ASAE*, **1979**, 22:554-559.
7. Baker, J.L.; Laflen, J.M.; Johnson, H.P. *Trans. ASAE*, **1978**, 21:886-892.
8. Baker, J.L.; Mickelson, S.K. *Weed Tech.* **1994**, 8:862-869.
9. Meyer, L.D.; Dabney, S.M.; Harmon, W.C. *Trans. ASAE*, **1995**, 38:809-815.
10. Magette, W.L.; Brinsfield, R.B.; Palmer, R.E.; Wood, J.D. *Trans. ASAE*, **1989**, 32:663-667.
11. Asmussen, L.E.; White, A.W. Jr.; Hauser, E.W.; Sheridan, J.M. *J. Environ. Qual.* **1977**, 6:159-162.

12. Rhode, W.A.; Asmussen, L.E.; Hauser, E.W.; Wauchope, R.D.; Allison, H.D. *J. Environ. Qual.* **1980**, 9:37-42.
13. Mickelson, S.K.; Baker, J.L. Paper No. 932084. *American Society of Agricultural Engineers,* **1993**, St. Joseph, MI.
14. Misra, A.K.; Baker, J.L.; Mickelson, S.K.; Shang, H. *Trans. ASAE,* **1996**, 39:2105-2111.
15. Misra, Akhilesh Kumar. *M.S. Thesis,* **1995**, Iowa State Univ., Ames, IA.
16. Cole, J.T.; Baird, J.H.; Basta, N.T.; Huhnke, R.L.; Storm, D.E.; Johnson, G.V.; Payton, M.E.; Smolen, M.D.; Martin, D.L.; Cole, J.C. *J. Environ. Qual.* **1997**, 26:1589-1598.
17. Webster, E.P.; Shaw, D.R. *Weed Sci.* **1996**, 44:662-671.
18. Arora, K.; Mickelson, S.K.; Baker, J.L.; Tierney, D.P; Peter, C.J. *Trans. ASAE,* **1996**, 30:2155-2162.
19. Lowrance, R.; Vellidis, G.; Wauchope, R.D.; Gay P.; Bosch, D.D. *Trans. ASAE,* **1997**, 40:1047-1057.
20. Mickelson, S.K.; Baker, J.L.; Melvin, S.W.; Fawcett, R.S.; Tierney, D.P.; Peter, C.J. *J. Soil Water Conserv.* **1998**, 53:18-25.
21. Dillaha, T.A.; Reneau, R.B.; Mostaghimi, S.; Lee D. *Trans. ASA,* **1989**, 32:513-519.
22. Fawcett, R.S.; Tierney, D.P.; Peter, C.J.; Baker, J.L.; Mickelson, S.K.; Hatfield, J.L.; Hoffman, D.W.; Franti, T.G. *Proceedings of the National Agricultural Management,* New Orleans, LA, December 13-15, **1995**, Conservation Technology Information Center, West Lafyette, IN.

CALIFORNIA'S SAN JOAQUIN VALLEY

Chapter 19

Organophosphorous Insecticide Concentration Patterns in an Agriculturally Dominated Tributary of the San Joaquin River

N. N. Poletika[1], P. L. Havens[1], C. K. Robb[1], and R. D. Smith[2]

[1]Dow AgroSciences LLC, 9330 Zionsville Road, Indianapolis, IN 46268
[2]Reed D. Smith Associates, Inc., 11507 Silver Oak Road, Oakdale, CA 95361

Previous monitoring for organophosphorous (OP) insecticides in the San Joaquin River basin of California did not adequately characterize the sources and patterns of exposure in agriculturally dominated streams. In this study, intensive daily sampling was conducted for one year in Orestimba Creek, a tributary of the San Joaquin River. Three commonly used OP insecticides were monitored, along with weather data, stream discharge, and pesticide use in the watershed. Heavy rainfall during the dormant tree application season severely restricted the number of dormant treatments applied. Spray drift appeared to contribute to water contamination. Irrigation tail water may have also contributed to the chemical concentrations found in the creek. A worst-case exposure assessment using measured pesticide mass export values and discharge data suggests that, for this study year, maximum concentrations in the receiving water of the San Joaquin River would have been significantly lower than those in the creek water.

Several recent studies report residues of OP insecticides in surface waters of the San Joaquin River basin of California (1, 2, 3). Toxicity testing with the aquatic invertebrate *Ceriodaphnia dubia* indicates that some of the samples collected during different use seasons have the potential to exert adverse acute effects on sensitive nontarget aquatic organisms. Frequently, the samples shown to be acutely toxic to *Ceriodaphnia dubia* also contain OP insecticides at concentrations sufficient to account for much of the toxicity (1). In general, the detections of insecticides tend to be episodic in nature, and the peak observed concentrations, although potentially toxic, are not so great as to suggest that reasonable and practical best management practices (BMPs) cannot be implemented to reduce residues to nontoxic levels.

Sampling in the San Joaquin River basin has typically been conducted at a single point on a tributary near its confluence with the river at intervals of one to two weeks. Data generated with this sampling strategy can describe general trends of chemical transport within the local drainage basin. However, little information is available to characterize specific crops and fields contributing insecticide mass or the relative contributions of different transport processes such as surface runoff, spray drift, or irrigation tail water.

This study was designed to provide more detailed monitoring information for the Dow AgroSciences product, chlorpyrifos (CPF) ([O,O-diethyl-O-(3,5,6-trichloro-2-pyridinyl) phosphorothioate], CAS No. 2921-88-2), one of the insecticides detected in previous surface water monitoring studies. Additional monitoring was conducted for two other OP products commonly used in California agriculture and detected in surface water, diazinon (DZN) ([O,O-diethyl-O-(2-isopropyl-4-methyl-6-pyrimidinyl) phosphorothioate], CAS No. 333-41-5) and methidathion (MET) ([O,O-dimethyl phosphorodithioate, S-ester with 4-(mercaptomethyl)-2-methoxy Δ^2-1,3,4-thiadiazolin-5-one], CAS No. 950-37-8). The primary objective was to obtain a detailed data set that can be used to identify specific use patterns, agricultural practices, and transport processes contributing the bulk of off-site chemical movement into a primary agricultural drain that is tributary to the San Joaquin River. Although the study goals were specific to OP product use in the San Joaquin Valley, we expect that much of the information obtained will be generally applicable to insecticide use in western irrigated agriculture. In addition, the data should be helpful in identifying BMPs with potential to reduce off-site chemical movement.

Materials and Methods

Study Area. The lower reach of Orestimba Creek, an agriculturally dominated natural drainage in Stanislaus County, California, was chosen as the study location (Figures 1 and 2). Orestimba Creek originates in the Coast Range of mountains in western Stanislaus County, passes through irrigated farm land in the San Joaquin Valley, and terminates at its confluence with the San Joaquin River.

The most important crops in the study area that receive insecticide applications are alfalfa, walnuts, almonds, and dry beans (4). These crops commonly are grown on medium- to fine-textured soils that are classified as loams, clay loams, and clays (5). Slopes in the valley floor range from 0 to 2%. Irrigation methods used in the various crops include furrow, flood, sprinkler, and low-volume sprinkler. Fields located roughly between the Central California Irrigation District (CCID) Main

290

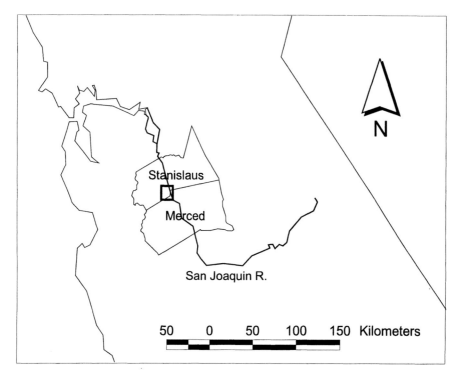

Figure 1. Map of central California showing the study area in southern Stanislaus County.

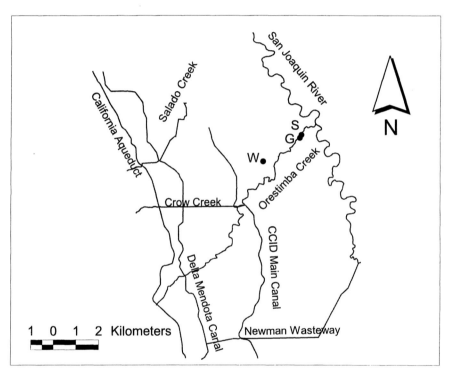

Figure 2. Water features and measurement stations in the study area. G = Gaging Station, S = Sampling Station, W = Weather Station.

Canal and the San Joaquin river are served by the CCID (Figure 2). Those west of the CCID Main Canal obtain water primarily from the Del Puerto Water District. Private wells may also provide supplemental irrigation.

In general, this site can be described as an intensively farmed area characterized by highly engineered water delivery and drainage systems required for large-scale irrigation of crops. The slopes of fields in the region typically fall from west to east, from the coastal mountains toward the San Joaquin River. However, precision land-leveling is commonly practiced, and many community drains exist, both surface and subsurface. Crow Creek is a major underground drain that empties into Orestimba Creek (Figure 2). The drainage pattern for individual fields is not predictable from elevation maps, and the actual drainage from fields into the creek is complex (Figure 3). The map in Figure 3 was generated from observations of individual fields by study personnel during irrigation events and from information obtained from interviews with growers. Geographical coverages used in all study area maps were constructed by digitizing satellite imagery and aerial photography obtained in the spring of 1996.

Instrumentation. Near the confluence with the San Joaquin River is the United States Geological Survey (USGS) gaging station 11274538 (Orestimba Creek At River Rd Nr Crows Landing California). This station is operated intermittently, depending on need for discharge data. Except during drought periods, flow is year-round. Winter rains provide most of the measured flow during the winter and spring months, while irrigation return water contributes to flow during the spring, summer, and fall (D. Sparks, USGS, personal communication, August 1997).

A weather station was installed in an open field about 0.7 km from the north bank of the creek (Figure 2) and equipped with electronics to measure rainfall, wind speed and direction, solar radiation, air temperature, and soil temperature at a depth of 2.5 cm.

Sampling was conducted using an ISCO Model 2700 autosampler (ISCO, Inc., Lincoln, Nebraska) installed at the location shown in Figure 2, about 20 m below USGS Station 11274538.

Water Sampling. Surface water samples were collected and analyzed for trace levels of commercially applied CPF, DZN, and MET. These applications were documented from pesticide use reports collected by the office of the county agricultural commissioner and geographically referenced to individual fields in the study area.

Daily and weekly samples were collected for a period of one year (5/1/96 through 4/30/97). Daily samples consisted of approximate 10 mL volumes collected hourly and composited over a 24-h period to produce a total sample volume of about 250 mL. Time-proportional sampling was used rather than a flow-proportional scheme because relatively large fluctuations in stream flow were anticipated during the unattended operation of the autosamplers, and there was no reliable method to obtain consistent, appropriate sample sizes under these conditions. Weekly grab samples of similar volume were also collected. Discrete sampling is more typical of historical surface water monitoring studies conducted in the area. This sampling strategy allowed comparing daily and weekly collection frequency and also time-weighted composite and grab sampling.

Sample Handling and Shipping. During periods where ambient temperatures exceeded 25 °C, composite sample containers were bathed in cubed ice or ice packs during the 24-h sampling period. Plastic-coated clear-glass jars with screw caps and Teflon® liners were used as sample containers. Samples were placed on ice in coolers during transport from the field to temporary storage. After removal from the field the samples were stored in refrigerators at a temperature near 4 °C. Samples were shipped to the Dow AgroSciences analytical laboratory in Indianapolis at weekly intervals. The samples were packed on wet ice packs in insulated coolers and shipped by overnight air express. Storage was in the dark at all locations at all times.

Chemical Analysis. Aliquots (100-mL) of gravity-settled samples were extracted from water by partitioning into hexane twice. The resulting extracts were combined, an internal standard was added (butathiophos), and the extract was concentrated under a stream of nitrogen to a known volume. Aliquots of the final solution were analyzed by capillary gas chromatography using flame photometric detection of the three analytes, CPF, DZN, and MET. Detections were confirmed using a second capillary column, or, for a select number of samples, by gas chromatography with mass selective detection. Method detection limits were 10, 10, and 24 ng L^{-1}, for CPF, DZN, and MET, respectively. Field spikes were used to assure good quality control during sample collection, shipping, storage, and analysis, and repeated analysis of field spike samples demonstrated storage stability for the short periods of chilled storage required to efficiently process the samples.

Results and Discussion

Chemical Applications. Only insecticide applications to fields that drain into the creek (Figure 3) or had the potential to contribute spray drift directly to the creek were recorded in the study. A buffer zone extending 305 m on either side of the mid-stream line was assumed to identify all fields capable of producing spray drift contamination. This buffer is based on Tier 1 predictions in the AgDRIFT model, version 1.02 (Spray Drift Task Force). Tables 1 through 3 list all of the applications relevant to the study and also provide a qualitative assessment of the likelihood of spray drift contamination in the creek. As shown in the tables, a few fields received two or three applications during the study period. Total masses of chemical applied to these fields were 1308, 669, and 315 kg of CPF, DZN, and MET, respectively. Because the field half-life of MET is approximately 7 to 10 d (6,7), applications of this chemical in the 1995-1996 dormant season probably did not provide much residue available for runoff by May. Therefore, a more realistic mass estimate for this compound is the 74 kg applied during the study period. Alfalfa applications made in March of 1996 were considered to be contributors to later runoff loadings.

All CPF treatments were made to alfalfa and walnut during the irrigation season. DZN was applied primarily to almond, both in-season and during the dormant period, and also to other tree and field crops. The few MET treatments were reported on tree crops as foliar and dormant applications. Alfalfa treatments were aerially applied, while air-blast speed sprayers were used in the tree crops.

Chemical Concentrations. Because none of the suspended solids contained in the original samples were extracted, some error in estimating total chemical mass present in the creek water was introduced by the analytical procedure. This is unimportant

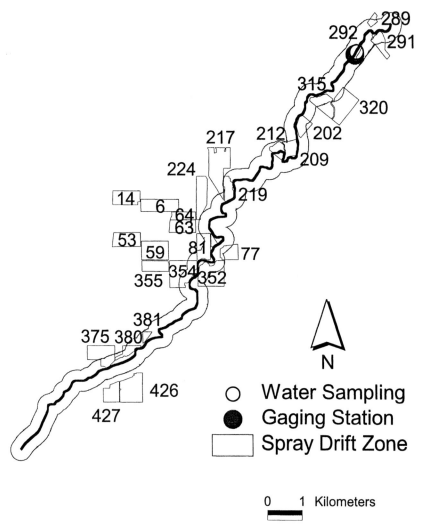

Figure 3. Agricultural fields receiving insecticide applications that drain into Orestimba Creek or fall within the indicated spray drift zone. Field identification numbers correspond to those given in Tables 1-3.

Table 1. First chemical applications to fields that drain into Orestimba Creek or fall within the spray drift zone

Field	Crop	Compound	Product	Date	Day	Method[a]	Drift Potential[b]	Mean Wind Speed (m s⁻¹)	Mean Direction[c] (°)	Drift Likelihood[d]
6	ALFALFA	Chlorpyrifos	LORSBAN 4E	3/6/96	-56	A	-			
63	ALFALFA	Chlorpyrifos	LORSBAN 4E I	3/6/96	-56	A	-			
64	ALFALFA	Chlorpyrifos	LORSBAN 4E I	3/6/96	-56	A	-			
224	ALFALFA	Chlorpyrifos	LORSBAN 4E I	3/6/96	-56	A	-			
355	ALFALFA	Chlorpyrifos	LORSBAN 4E I	3/6/96	-56	A	-			
202	ALFALFA	Chlorpyrifos	LOCK-ON INSE	3/8/96	-54	A	-			
291	ALFALFA	Chlorpyrifos	LORSBAN 4E I	3/8/96	-54	A	-			
320	ALFALFA	Chlorpyrifos	LORSBAN 4E I	3/11/96	-51	A	-			
354	ALFALFA	Chlorpyrifos	LORSBAN 4E I	3/16/96	-46	A	-			
217	WALNUTS	Chlorpyrifos	LORSBAN-4E	4/25/96	-6	G	-			
53	WALNUTS	Chlorpyrifos	LORSBAN 4E-H	4/26/96	-5	G	-			
77	WALNUT	Chlorpyrifos	LORSBAN 4E-H	4/30/96	-1	G	-			
380	WALNUT	Chlorpyrifos	LORSBAN 4E-H	4/30/96	-1	G	-			
381	WALNUT	Chlorpyrifos	LORSBAN 4E-H	4/30/96	-1	G	-			
14	WALNUT	Chlorpyrifos	LORSBAN-4E	5/3/96	2	G	-			
289	WALNUT	Chlorpyrifos	LORSBAN 4E-H	5/29/96	28	G	+?	1.8	03	-s -d -p
292	WALNUT	Chlorpyrifos	LORSBAN 4E-H	5/29/96	28	G	+	1.8	03	-s -d +p
219	WALNUT	Chlorpyrifos	LORSBAN 4E I	7/3/96	63	G	+	1.7	346	-s +d +p
209	ALFALFA	Diazinon	HELENA DIAZI	3/7/96	-55	A	-			
315	APRICOT	Diazinon	DIAZINON 50W	5/8/96	7	G	+	2.1	10	-s -d +p
212	WALNUT	Diazinon	D.Z.N DIAZIN	6/14/96	44	G	+	2.0	02	-s +d +p
59	MELON	Diazinon	HELENA DIAZI	7/27/96	87	A	-			
352	SPINACH	Diazinon	CLEAN CROP D	3/26/97	329	A	+	1.4	353	-s -d +p
375	ALMONDS	Methidathion	SUPRACIDE 25	1/31/96	-91	G	-			
426	ALMONDS	Methidathion	SUPRACIDE 25	2/1/96	-90	G	-			
427	ALMONDS	Methidathion	SUPRACIDE 25	2/1/96	-90	G	-			
81	WALNUTS	Methidathion	SUPRACIDE 25	5/3/96	2	G	+	3.7	342	+s +d +p

[a] A = Aerial, G = Ground.
[b] Field located within spray drift zone and application made during monitoring period.
[c] From true North.
[d] s = wind speed (+ = speed > 2.2 m s⁻¹), d = wind direction (+ = creek downwind), p = proximity (+ = field border on creek bank).

Table 2. Second chemical applications to fields that drain into Orestimba Creek or fall within the spray drift zone

Field	Crop	Compound	Product	Date	Day	Method[a]	Drift Potential[b]	Mean Wind Speed (m s^{-1})	Mean Direction[c] (°)	Drift Likelihood[d]
77	WALNUT	Chlorpyrifos	LORSBAN 4	5/19/96	18	G	+	3.6	308	+s -d -p
53	WALNUTS	Chlorpyrifos	LORSBAN-4	7/25/96	85	G	-			
81	WALNUTS	Chlorpyrifos	LORSBAN 5	7/26/96	86	G	+	1.3	343	-s +d +p
212	WALNUT	Chlorpyrifos	LORSBAN-4	8/3/96	94	G	+	2.0	343	-s +d +p
291	ALFALFA	Chlorpyrifos	LOCK-ON I	8/18/96	109	A	+?	1.4	349	-s -d -p
380	WALNUT	Chlorpyrifos	LORSBAN 5	9/5/96	127	G	+	1.9	354	-s +d +p
381	WALNUT	Chlorpyrifos	LORSBAN 5	9/5/96	127	G	+	1.9	354	-s +d +p
6	ALFALFA	Chlorpyrifos	LOCK-ON I	9/12/96	134	A	-			
224	ALFALFA	Chlorpyrifos	LOCK-ON I	9/12/96	134	A	+?	2.5	352	+s ?d -p
63	ALFALFA	Chlorpyrifos	LORSBAN 4	3/5/97	308	A	-			
64	ALFALFA	Chlorpyrifos	LORSBAN 4	3/5/97	308	A	-			
355	ALFALFA	Chlorpyrifos	LORSBAN 4	3/5/97	308	A	+	2.1	348	-s -d +p
202	ALFALFA	Chlorpyrifos	LOCK-ON I	3/11/97	314	A	+	3.0	329	+s +d +p
217	WALNUTS	Chlorpyrifos	LORSBAN-4	4/29/97	363	G	+	0.9	149	-s -d +p
375	ALMONDS	Diazinon	DIAZINON	1/27/97	271	G	-			
426	ALMONDS	Diazinon	DIAZINON	1/27/97	271	G	-			
427	ALMONDS	Diazinon	DIAZINON	1/27/97	271	G	-			
315	APRICOT	Methidathion	SUPRACIDE	1/1/97	245	G	+	5.6	163	+s +d +p

[a] A = Aerial, G = Ground.
[b] Field located within spray drift zone and application made during monitoring period.
[c] From true North.
[d] s = wind speed (+ = speed > 2.2 m s^{-1}), d = wind direction (+ = creek downwind), p = proximity (+ = field border on creek bank).

Table 3. Third chemical applications to fields that drain into Orestimba Creek or fall within the spray drift zone

Field	Crop	Compound	Product	Date	Day	Method[a]	Drift Potential[b]	Mean Wind Speed (m s⁻¹)	Mean Wind Direction[c] (°)	Drift Likelihood[d]
355	ALFALFA	Chlorpyrifos	LOCK-ON I	3/22/97	325	A	-			
212	WALNUT	Chlorpyrifos	LORSBAN-4	4/22/97	356	G	+	2.6	350	+s +d +p
315	APRICOT	Diazinon	D.Z.N DIA	4/30/97	364	G	+	--[e]	--[e]	--[e] +p

[a] A = Aerial, G = Ground.
[b] Field located within spray drift zone and application made during monitoring period.
[c] From true North.
[d] s = wind speed (+ = speed > 2.2 m s⁻¹), d = wind direction (+ = creek downwind), p = proximity (+ = field border on creek bank).
[e] No observation. Note: 0830 sampling time on 4/30/97 may have precluded capture of chemical mass resulting from drift.

relative to characterizing chemical toxicity to organisms in the water column, where it is generally assumed that sediment-sorbed chemical residues are not bioavailable. However, estimates of chemical loads exported from the creek would be biased downward, because the sediment-sorbed loads were not measured. The reported K_{oc} values for CPF, DZN, and MET, 5051, 666, and 289, respectively (8), indicate the relative importance of this bias in the mass loading estimation.

Figure 4 summarizes the chemical concentrations observed. The reported data are uncensored (all quantifiable chromatographic peaks reported, and unquantifiable peaks were assigned the value of zero). Any errors associated with interpreting this uncensored data is limited to the extremely small concentrations that are below levels of probable toxicological significance. Validation of the analytical method proceeded in stages for the three analytes, and this is reflected in the period of observations reported: days 1-364 for CPF, days 21-364 for DZN, and days 56-364 for MET.

Comparison of weekly grab samples with the corresponding weekly 24-h time-proportional composite samples in Figure 4 showed that the concentrations in the grab samples were more variable (n = 1 rather than n = 24). This indicates that the time-proportional composite samples provided a more accurate estimate of concentration, particularly with respect to defining exposures to aquatic organisms in ambient water samples over a specific period of time (24 h in this case). Obviously, the daily sampling regime characterized pulse duration, whereas the weekly regime afforded no information of this kind. Subset analysis should suggest the optimal sampling frequency for this watershed.

Relating the information on chemical applications and the potential for spray drift inputs into creek water (Tables 1-3) to the concentrations found in the daily composite samples (Figure 4) suggests the following interpretations regarding sources of chemical contamination. For CPF, the day 28 peak was linked to a possible spray drift event from the walnut application to fields 289 and 292. Drift from the walnut application to field 212 on day 94 may be implicated with the peak found on day 95. The only remaining spray event (day 356) associated with a peak (day 357) came from a walnut treatment, again in field 212. The other CPF peaks observed were definitely not linked to downstream spray drift events. These concentration peaks were hypothesized to result from upstream drift events or contributions from unconfined irrigation tail water leaving alfalfa fields or walnut orchards receiving additional CPF applications summarized in Tables 1 through 3. Detailed analysis of drift potential at upstream sampling locations and field-specific irrigation delivery records will be performed to test the hypothesis. Simulation modeling of chemical fate and transport is also in progress to assist with hypothesis testing. There were no records of dormant season use of CPF in tree crops, and no detections were associated with natural rainfall during this period (December 1996 through February 1997).

The only significant DZN application not included in the period of chemical sampling for spray drift was the day 7 apricot treatment in field 315 (Table 1). There were fewer total applications of DZN relative to CPF, and only one was associated with a possible spray drift event: apricot in field 315 on day 364 (Table 3). The largest concentration peaks not connected with spray drift appeared to be related to field crops rather than trees. For example, the day 88 peak followed a melon application on day 87 in field 59, and an early spring spinach treatment in field 352 on day 329 preceded the peak observed on day 331. Other applications clearly associated with concentration peaks included walnut in field 44 (day 46 peak) and almond in fields 426 and 427 (day 288 peak, possibly the day 305 peak). Some smaller peaks did not appear to be linked with specific events: days 82, 171, 269, and

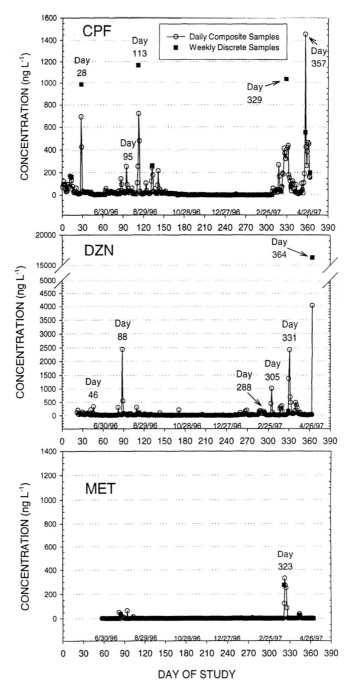

Figure 4. Chemical monitoring data.

320. Again, detailed analysis of upstream drift potential and irrigation records from individual fields is necessary to identify other events contributing to DZN mass found in creek water samples.

In most years, DZN is used primarily as a dormant season product in this area. Due to rainfall patterns in December 1996 and January and February of 1997 (Figure 5), very few intended dormant season treatments were applied (personal communications from several local pest control consultants). The three reported dormant applications in almond on day 271 were not followed by any significant rainfall events (Figure 5). Concentrations of DZN were therefore smaller during this period, relative to those reported from previous dormant seasons (1, 2, 3).

Although MET analysis did not begin until day 56, only one application was excluded in the spray drift monitoring, walnut on day 2 in field 81. Very few fields were treated with this product during the study period, and none were associated with spray drift events. Like DZN, MET is used principally in this watershed as a dormant spray. The single reported dormant treatment occurred on apricot in field 315 on day 245. A large rainfall event occurred on day 246 that generated large stream flow that same day (Figure 5). However, no MET peak was observed. Even though the analytical method was slightly less sensitive for this analyte, the lack of detection was probably related more to dilution of contaminated runoff from the single treated orchard. The source for the large MET peak found on day 323 is unknown.

Stream Hydrology and Chemical Loading into the San Joaquin River. Large stream flows resulted from rainfall events generated by area-wide storm systems. As a consequence, both the non-agricultural upper watershed and the agricultural area in the valley floor contributed runoff from natural rainfall (Figure 5, days 15-24 and 224-285). During the irrigation season most or all of the stream flow was from irrigation return water and seepage from irrigated fields (Figure 5, days 1-14, 30-180, 183-204, 206-222, and 319-364). For the greater part of the year, about 250 days, the stream channel was filled by wastewater coming from commercial agriculture, and thus the creek functions as a primary drain during this period. The ecological significance of water quality in this system should be considered if BMPs are implemented. For example, if irrigation efficiency is improved to minimize tailwater drainage, any benefits in reduction of chemical loading may be offset by decreased stream flow, which could more significantly impact habitat quality. Complicating this analysis is the fact that under natural conditions there is often no aquatic habitat in the channel. Arguably, however, the reach of the San Joaquin River to which Orestimba Creek is tributary is an important perennial aquatic ecosystem meriting protection during all seasons. Moreover, water quality in the San Joaquin River affects downstream surface water bodies such as the Delta region and the San Francisco Bay system.

The concentration data presented in Figure 4 can be combined with the discharge data from Figure 5 to compute the daily loading of dissolved mass for each chemical passing the sampling point. Total mass (dissolved + sorbed to suspended sediment) could not be estimated because we did not analyze the solid fraction of water samples for chemical residues, and there was no measurement of suspended sediment load in the stream daily discharge data. This site is located about 1.6 stream km from the confluence with the river. Although there are three treated fields that drain downstream of the sampling location (Figure 3), any error contributed by these fields to the mass export values based on sampling data probably is small, especially when summed over the entire study year. Figure 6 depicts the calculated

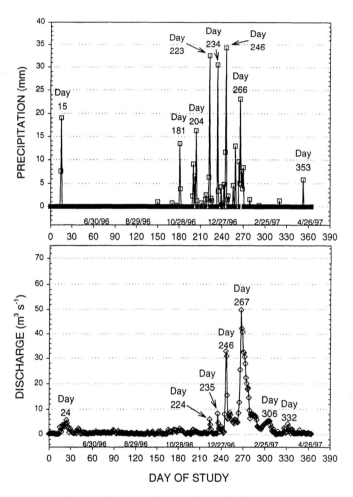

Figure 5. Rainfall data from weather station and discharge reported at USGS gaging station 11274538. Discharge data for the current water year is provisional and subject to revision.

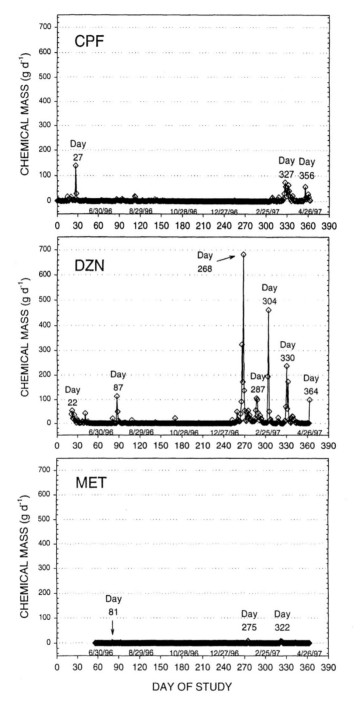

Figure 6. Chemical mass loading downstream from sampling location.

mass loadings assumed to have entered the San Joaquin River on a daily basis. Total mass for the year-long study period was 1.26, 4.70, and 0.06 kg, or 0.10, 0.70, and 0.08 % of applied mass for CPF, DZN, and MET, respectively. The percent figure for MET reflects the adjustment made to exclude the previous season (1995-1996) dormant applications. These results appear consistent with those from other studies reporting loss of chemical in surface runoff (6, 9, 10). Contributions from spray drift events may have compensated for the exclusion of sediment-sorbed mass in our estimates.

Comparison of Figures 4 and 6 shows a definite relationship between the season (rainfall vs. irrigation) and concentration/mass loading for the three products. CPF concentrations were greatest during the irrigation season, but due to low flows, the mass loading into the main-stem river was smallest at this time. DZN concentrations followed a similar temporal pattern. However, the presence of a few dormant applications, combined with presumably large runoff losses and stream flow during winter storms, produced the largest mass loading events. There was too little MET mass to generate a discernible pattern.

We performed a worst case exposure analysis for Orestimba Creek contributions to the San Joaquin River. The most recent historical mean daily discharge data from the closest USGS San Joaquin River gaging station downstream from Orestimba Creek was obtained and assumed to represent river volumes receiving discharge from the creek (11). The period of record for the river data was 10/1/1995 through 9/30/1996. Assuming no background levels of chemical were present in the river from upstream contamination, the maximum possible concentrations achievable by mass loading from the creek were calculated as follows. First, the maximum daily loading was found for each chemical from the present study. Second, the minimum 24-h flow volume passing the river gaging station during the period of record was computed. Third, the maximum loads were diluted by the 24-h minimum river flow volume to estimate a chemical concentration for this lowest flow day in the river. The estimated concentrations in the river were 92, 451, and 6 ng L^{-1}, respectively, for CPF, DZN, MET. These represent the largest possible concentrations for this study year and thus would lie in the extreme tail of a probability distribution. In contrast, the maximum observed concentrations in the creek during the study period were 1455, 4030, and 331 ng L^{-1}, respectively, for the compounds. This analysis of upper-bound river concentrations suggests that exposures to aquatic life in the main-stem river would tend to be lower than in the tributary due to dilution. Only if all upstream sources contributed the same amount of chemical mass equivalent to that from Orestimba Creek on a flow-proportional basis would the concentration in the river be the same as in the tributaries. Such a situation is unlikely to occur on any given river flow day.

Conclusion

Based on the use patterns practiced in this localized study area during the period of observation, we conclude that spray drift and natural runoff contributed to movement of OP insecticide residues into Orestimba Creek. Further analysis of field-specific observations may also implicate irrigation tailwater in transport of chemical mass. Changes in application practices and in field drainage management may therefore have value in controlling off-site chemical movement and reducing chemical concentrations in the water column. From a mass loading perspective, export of pesticide mass peaks into the San Joaquin River was greater during the winter rainy season, even in a year when dormant applications appeared to be fewer than usual. BMPs designed to reduce winter storm pesticide runoff from dormant treated orchards, such as cover crops (6), may help to decrease this loading into the San Joaquin River basin.

Uncertainties and Future Research Needs

The most significant uncertainties in this study were accuracy and completeness of records related to 1) chemical application, 2) details of the field drainage network under both irrigated and rainfall conditions, and 3) individual irrigation events in specific fields. We also were unable to quantify chemical inputs from CCID main canal spill water.

Analysis of irrigation records and visualization of event patterns in space and time utilizing concentration and discharge data from additional sampling locations on the creek remain to be completed in the present study. Other studies in progress include repeated dormant season monitoring, field evaluation of currently recommended general water quality irrigation BMPs for effectiveness in reducing CPF transport, and development of a watershed-scale simulation modeling system to gain a better understanding of the watershed and to evaluate watershed-scale effectiveness of management practices on water quality.

Acknowledgments

The authors thank Novartis Crop Protection, Inc. for financial support of this project and George Dial and Ed Olberding of Dow AgroSciences for chemical analysis and analyte confirmation.

References

1. Foe, C. 1995. *Insecticide Concentrations and Invertebrate Bioassay Mortality in Agricultural Return Water From the San Joaquin Basin.* Staff report to the Central Valley Regional Water Quality Control Board, Sacramento, CA, 1995.
2. Domagalski, J.L. Nonpoint Sources of Pesticides in the San JoaquinRiver,California: Input from Winter Storms, 1992-1993. U.S. Geological Survey Open File Report 95-165, 1995.
3. Ross, L.J., Stein, R., Hsu, J., White, J., Hefner, K. 1996. Distribution and Mass Loading of Insecticides in the San Joaquin River, California. Winter 1991-92 and 1992-93. California Environmental Protection Agency, Department of Pesticide Regulation Report EH 96-02.
4. Stanislaus County Agricultural Commissioner and Sealer of Weights and Measures. Pesticide use data reports for 1994 -1997. Modesto, CA.
5. Unpublished Soil Survey of Stanislaus County, California, Western Part (personal communication, Keith Azevedo, West Stanislaus Resource Conservation District), 1997.

6. Ross, L.J., Bennett, K.P., Kim, K.D., Hefner, K., Hernandez, J. Reducing Dormant Spray Runoff from Orchards. California Environmental Protection Agency, Department of Pesticide Regulation Report EH 97-03, 1997.
7. Wauchope, R.D., Buttler, T.M., Hornsby, A.G., Augustin Beckers, P.W.M., Burt, J.P. The SCS/ARS/CES Pesticide Properties Database for Environmental Decision-Making. Rev. Environ. Contam. Toxicol. 123:1-164, 1992.
8. Montgomery, J.H. 1997. Agrochemicals Desk Reference—2nd ed. Lewis Publishers, CRC Press, Boca Raton, FL, 656 pages.
9. Wauchope, R.D. 1978. The Pesticide Content of Surface Water Draining from Agricultural Fields – A Review. J. Environ. Qual. 7:459-472.
10. Spencer, W.F., Cliath, M.M., Blair, J.W., LeMert, R.A. 1985. Transport of Pesticides from Irrigated Fields in Surface Runoff and Tile Drain Waters. Conservation Research Report No. 31, USDA Agricultural Research Service.
11. United States Geological Survey. United States NWIS-W Data Retrieval for Station 11274550 (San Joaquin R Nr Crows Landing Ca). http://h20-nwisw.er.usgs.gov/nwis-w/US/. Accessed 3/7/98.

Chapter 20

Pesticide Transport in the San Joaquin River Basin

Neil M. Dubrovsky[1], Charles R. Kratzer[1], Sandra Y. Panshin[2], Jo Ann M. Gronberg[3], and Kathryn M. Kuivila[1]

[1]U.S. Geological Survey, Placer Hall, 6000 J Street, Sacramento, CA 95819–6129
[2]U.S. Geological Survey, 5957 Lakeside Boulevard, Indianapolis, IN 46278–1996
[3]U.S. Geological Survey, Menlo Park, CA

Pesticide occurrence and concentrations were evaluated in the San Joaquin River Basin to determine potential sources and mode of transport. Land use in the basin is mainly agricultural. Spatial variations in pesticide occurrence were evaluated in relation to pesticide application and cropping patterns in three contrasting subbasins and at the mouth of the basin. Temporal variability in pesticide occurrence was evaluated by fixed interval sampling and by sampling across the hydrograph during winter storms. Four herbicides (simazine, metolachlor, dacthal, and EPTC) and two insecticides (diazinon and chlorpyrifos) were detected in more than 50 percent of the samples. Temporal, and to a lesser extent spatial, variation in pesticide occurrence is usually consistent with pesticide application and cropping patterns. Diazinon concentrations changed rapidly during winter storms, and both eastern and western tributaries contributed diazinon to the San Joaquin River at concentrations toxic to the water flea *Ceriodaphnia dubia* at different times during the hydrograph. During these storms, toxic concentrations resulted from the transport of only a very small portion of the applied diazinon.

The San Joaquin Valley is one of the most important agricultural areas in the United States. Almost the entire valley floor is agricultural land, and its agricultural history dates back to the 1870s. The combination of abundant water and the long growing season results in an exceptionally productive agricultural economy. In 1987, gross sales from agricultural products from the San Joaquin Valley accounted for about 5 percent of the total value of agricultural production in the United States. In 1993, a total of 16.6 million pounds (lb) of pesticides (1,800 different compounds) were applied to agricultural land in the San Joaquin River Basin, with an additional 3 million lb of nonagricultural application (*1*).

Previous studies have detected pesticides in water samples from the San Joaquin River and its tributaries (*2–8*). Studies also have demonstrated that water in the San Joaquin River is sometimes toxic to the test organism *Ceriodaphnia dubia* (*2–4*). Foe (*3*) examined the seasonality of pesticide concentrations, *C. dubia* mortality, and pesticide applications to different crops. He identified seven insecticides — carbaryl, chlorpyrifos, diazinon, fonofos, malathion, methomyl, and parathion — as most likely responsible for the toxicity of the water at different times of the year and associated these pesticides with the crops to which they were applied. This link between agricultural pesticide use and toxicity to aquatic organisms underscores the importance of understanding the factors that cause pesticide transport to streams.

Purpose and Scope

The purpose of this report is to examine the spatial and temporal variability of dissolved pesticide occurrence and concentrations in surface water within the San Joaquin River Basin and, to the extent possible, report the sources and transport mechanisms responsible for their presence. Data were collected on the concentrations of 83 compounds (76 pesticides and 7 degradates) in surface-water samples from four sites within the San Joaquin River Basin (Figure 1) during January through December 1993. The San Joaquin River near Vernalis site is located at the mouth of the San Joaquin River and characterizes water quality in the basin as a whole. The other three sites were in subbasins selected to characterize one type of physiography, localized pesticide application, and specific land use— Orestimba Creek at River Road near Crows Landing (small western tributaries), Salt Slough at Highway 165 near Stevinson (wetlands areas receiving subsurface drainage in the southwest), and the Merced River at River Road near Newman (large eastern tributaries). Data also were collected from the three major east-side tributaries to the San Joaquin River — the Merced, Stanislaus, and Tuolumne rivers — to evaluate diazinon transport during winter storms in 1994. These studies are part of an integrated study of the water quality of the San Joaquin–Tulare Basins conducted as part of the U.S. Geological Survey's (USGS) National Water-Quality Assessment Program (*9*).

Methods

Most samples were flow weighted and cross-sectionally integrated by standard USGS methods (*10*). Some of the storm samples were grab samples collected with a 3-L Teflon bottle, or were single vertical depth-integrated samples collected near the center of flow. Samples were split into two 1-L aliquots, and the aliquots were filtered through a 0.7 micrometer glass-fiber filter. One liter was extracted by solid-phase extraction (SPE) cartridges containing porous silica coated with a C-18 phase, and analyzed by gas chromatography/mass spectrometry (GC/MS) for 47 compounds at the USGS National Water Quality Laboratory (NWQL) according to

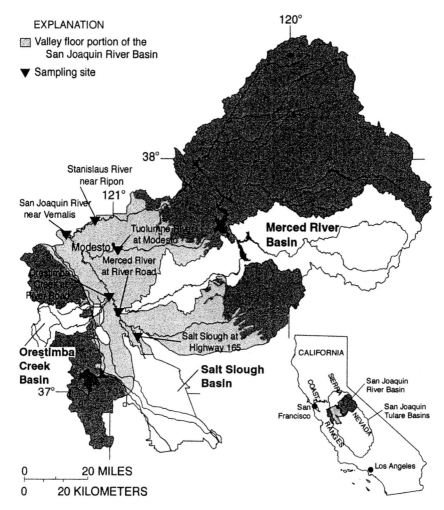

Figure 1. Study area map, including basin boundaries and sample site locations, San Joaquin River Watershed, California.

the method of Zaugg and others (*11*). One liter was extracted by passing the sample through a graphitized carbon solid-phase extraction cartridge, and analyzed by high-performance liquid chromatography (HPLC) for 36 compounds at the NWQL according to the method of Werner and others (*12*). Supplementary samples collected at the San Joaquin River at Vernalis during winter storms by the USGS Toxic Substances Hydrology Program were analyzed by SPE onto a C-8 bonded phase and analyzed by GC/MS at the USGS laboratory in Sacramento, California, according to the method of Crepeau and others (*13*). The 83 compounds analyzed (Table I) will be referred to as the target compounds in the following text.

Sampling frequency was varied throughout the year to target different types of pesticides during different seasons. Sample collection was most frequent during the winter (one to two times each week) because it was hypothesized that offsite movement of pesticides would be facilitated by rainfall. Sampling frequency decreased from weekly to once every 3 weeks from April through September. Sample collection was least frequent (about once a month) during the autumn when there is neither rainfall nor irrigation. In addition, samples were collected during winter storms to study transport of diazinon during two periods: January and February 1993, to evaluate transport from the western San Joaquin Valley; and January and February 1994, to evaluate transport from the eastern San Joaquin Valley.

The quality assurance of pesticide data collected in 1993, a summary of which follows, was evaluated in Panshin and others (*14*). Out of a possible 913 analyses of 22 field blank samples, there were only five detections. This very low rate of false positives indicates that no systematic contamination was caused by the sampling or equipment cleaning protocols. Mean percent recoveries ranged from 86 to 144 percent for most compounds in the 13 spike samples analyzed by the GC/MS method; and mean percent recoveries ranged from 31 to 112 percent for most compounds in the 5 spike samples analyzed by the HPLC method. The lower spike recoveries for the HPLC method indicate that this method has an increased chance of not detecting pesticides present in environmental samples at low, but initially detectable, concentrations. Finally, analysis of 13 replicate samples show that the results are reproducible. Pesticides not detected in the environmental sample were also not detected in the paired replicate sample in 97 percent of the cases; and pesticides detected in environmental samples also were detected in the paired replicate sample in 89 percent of the cases.

Overall Occurrence

Pesticides were detected in all but 1 of the 143 surface-water samples collected from the four sites during calendar year 1993 (*14*). Fifty percent of the samples contained seven or more pesticides. Forty-nine — 33 herbicides and 16 insecticides — of the 83 pesticides analyzed for were detected (Table I). Six pesticides were detected in more than 50 percent of the samples — the herbicides simazine, EPTC, dacthal, and metolachlor; and the insecticides diazinon and chlorpyrifos. Twenty-two compounds had a frequency of detection of at least 20 percent in one or more of the subbasins (Figure 2).

Table I. Summary of pesticides analyzed and results for the four sites — Orestimba Creek, the Merced River, Salt Slough, and the San Joaquin River — sampled in 1993: method, method detection limit (MDL), number of samples, frequency of detection, and maximum concentration

Pesticide	Method	MDL (µg/L)	Number of Samples	Frequency (%)	Maximum Concentration (µg/L)
2,4,5-T	HPLC	0.035	78	nd	ld
2,4-D	HPLC	0.15	78	14	1.2
2,4-DB	HPLC	0.24	78	nd	ld
2,6-Diethylaniline	GC/MS	0.003	142	7.0	0.007
Acetochlor	GC/MS	0.002		nd	ld
Acifluorfen	HPLC	0.035	78	nd	ld
Alachlor	GC/MS	0.002	142	15	0.31
Aldicarb	HPLC	0.55	76	1.3	e0.46
Aldicarb sulfone	HPLC	0.10	76	nd	ld
Aldicarb sulfoxide	HPLC	0.021	76	nd	ld
Atrazine	GC/MS	0.001	142	40	0.13
Atrazine, desethyl	GC/MS	0.002	142	9.2	e0.005
Azinphos-methyl	GC/MS	0.001	142	9.2	e.0.39
Benfluralin	GC/MS	0.002	142	0.70	0.007
Bentazon	HPLC	0.014	78	nd	ld
Bromacil	HPLC	0.035	76	nd	ld
Bromoxynil	HPLC	0.035	78	nd	ld
Butylate	GC/MS	0.002	142	7.8	0.010
Carbaryl	GC/MS	0.003	142	23	e5.2
Carbofuran	GC/MS	0.003	142	13	e0.097
Carbofuran, 3-hydroxy	HPLC	0.014	76	nd	ld
Chloramben	HPLC	0.42	76	nd	ld
Chlorothalonil	HPLC	0.48	76	nd	ld
Chlorpyrifos	GC/MS	0.004	142	64	0.26
Clopyralid	HPLC	0.23	78	nd	ld
Cyanazine	GC/MS	0.004	142	35	1.3
DDE, p,p´-	GC/MS	0.006	142	23	0.062
DNOC	HPLC	0.42	78	nd	ld
Dacthal	GC/MS	0.002	142	64	0.22
Dacthal, mono-acid	HPLC	0.017	78	nd	ld
Diazinon	GC/MS	0.002	142	76	3.8
Dicamba	HPLC	0.035	78	nd	ld
Dichlobenil	HPLC	1.2	76	nd	ld
Dichlorprop	HPLC	0.032	78	3.9	0.11
Dieldrin	GC/MS	0.001	142	9.9	0.021
Dinoseb	HPLC	0.035	78	nd	ld
Disulfoton	GC/MS	0.017	142	nd	ld
Diuron	HPLC	0.02	76	37	1.9
EPTC	GC/MS	0.002	142	54	2.2
Ethalfluralin	GC/MS	0.004	142	7.0	0.13
Ethoprop	GC/MS	0.003	142	0.70	0.003
Fenuron	HPLC	0.013	76	nd	ld
Fluometuron	HPLC	0.035	76	nd	ld

Table I. Summary of pesticides analyzed and results for the four sites — Orestimba Creek, the Merced River, Salt Slough, and the San Joaquin River — sampled in 1993: method, method detection limit (MDL), number of samples, frequency of detection, and maximum concentration—*Continued*

Pesticide	Method	MDL (µg/L)	Number of Samples	Frequency (%)	Maximum Concentration (µg/L)
Fonofos	GC/MS	0.003	142	13	0.26
HCH, alpha-	GC/MS	0.002	142	0.70	0.002
HCH, gamma-	GC/MS	0.004	142	0.70	0.005
Linuron	GC/MS	0.002	142	0.70	0.29
MCPA	HPLC	0.17	78	1.3	e0.12
MCPB	HPLC	0.14	78	nd	ld
Malathion	GC/MS	0.005	142	8.5	0.39
Methiocarb	HPLC	0.026	76	nd	ld
Methomyl	HPLC	0.017	76	9.2	0.67
Methyl parathion	GC/MS	0.006	142	nd	ld
Metolachlor	GC/MS	0.002	142	71	1.6
Metribuzin	GC/MS	0.004	142	8.5	0.047
Molinate	GC/MS	0.004	142	12	4.0
Napropamide	GC/MS	0.003	142	20	0.14
Neburon	HPLC	0.015	76	nd	ld
Norflurazon	HPLC	0.024	76	1.3	0.44
Oryzalin	HPLC	0.31	76	nd	ld
Oxamyl	HPLC	0.018	76	nd	ld
Parathion	GC/MS	0.004	142	nd	ld
Pebulate	GC/MS	0.004	142	20	0.24
Pendimethalin	GC/MS	0.004	142	4.2	0.054
Permethrin, *cis*-	GC/MS	0.005	142	0.70	0.013
Phorate	GC/MS	0.002	142	nd	ld
Picloram	HPLC	0.05	78	nd	ld
Prometon	GC/MS	0.018	142	28	0.021
Pronamide	GC/MS	0.003	142	11	0.022
Propachlor	GC/MS	0.007	142	0.70	e0.002
Propanil	GC/MS	0.004	142	0.70	0.004
Propargite	GC/MS	0.013	142	22	e20
Propham	HPLC	0.035	76	nd	ld
Propoxur	HPLC	0.035	76	nd	ld
Silvex	HPLC	0.021	78	nd	ld
Simazine	GC/MS	0.005	142	94	1.4
Tebuthiuron	GC/MS	0.01	142	2.8	e0.008
Terbacil	GC/MS	0.007	142	0.70	e0.008
Terbufos	GC/MS	0.013	142	nd	ld
Thiobencarb	GC/MS	0.002	142	7.0	0.51
Triallate	GC/MS	0.001	142	0.70	0.003
Triclopyr	HPLC	0.25	78	1.3	e0.010
Trifluralin	GC/MS	0.002	142	44.4	0.11

[e, estimated; nd, not detected; ld, less than the MDL]

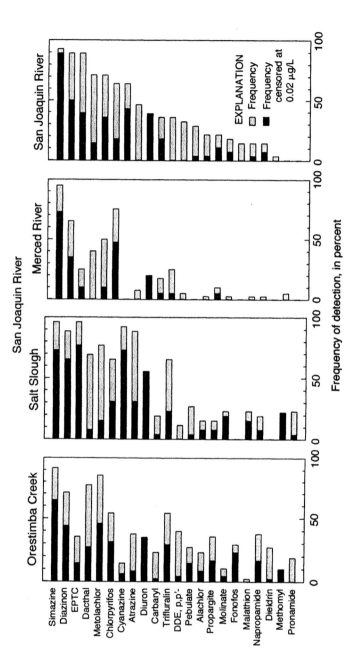

Figure 2. Frequency of detection for each subbasin site and the San Joaquin River site for each pesticide with a frequency of detection of at least 20 percent at a site. Both the total frequency of detection and the frequency of detection when all data are censored to a common reporting level of 0.02 µg/L are shown.

Four of the six pesticides that were detected in more than 50 percent of the samples — simazine, EPTC, diazinon, and chlorpyrifos — are also among the 10 most heavily applied target pesticides in the San Joaquin River Basin. More than 75,000 lb of active ingredient of each of these pesticides were applied in the basin in 1993 (*1*). Conversely, 3 of the 10 most heavily applied target pesticides — azinphos-methyl, malathion, and oryzalin — were detected in 10 percent or less of the samples. Overall, 38 of the 54 target pesticides with known application (70 percent) were detected during this study, indicating that, in most cases, application of a pesticide resulted in its detection in surface water in the basin.

Concentrations of the detected pesticides usually were low (less than 0.1 µg/L), but highly variable — median concentrations of the six most frequently detected pesticides ranged from 0.004 µg/L for dacthal to 0.050 µg/L for simazine, and 10 pesticides had maximum concentrations greater than 1 µg/L (2,4-D, carbaryl, cyanazine, diazinon, diuron, EPTC, metolachlor, molinate, propargite, and simazine) (Table I). In general, the compounds that were detected most frequently had the highest concentrations.

The concentrations of pesticides were compared to established water-quality criteria to indicate where a pesticide might cause adverse effects. In the following discussion, the term criteria is used in a general sense to refer to standards or guidelines established by national or international agencies or by organizations in North America that have regulatory responsibilities or expertise in water quality. Although U.S. Environmental Protection Agency drinking-water standards were not exceeded, concentrations of seven pesticides exceeded criteria for aquatic life: the herbicides diuron and trifluralin; and the insecticides azinphos-methyl, carbaryl, chlorpyrifos, diazinon, and malathion (*14*). A criteria for the protection of freshwater aquatic life were exceeded in 37 percent of the 143 stream samples; 40 percent of these exceedances — 15 percent of all samples — are attributed solely to diazinon concentrations exceeding 0.08 µg/L. Over half of the pesticides detected have no established aquatic life criteria, and the potential for these compounds to induce toxicity, endocrine disruption, impaired immune response, or other adverse biological effects is not well known.

Spatial Variation in Pesticide Occurrence and Concentrations

The spatial distribution of detections of pesticides in surface waters are related to where pesticides are applied, as well as the crop type and hydrologic characteristics of the basin. The California Department of Pesticide Regulation maintains detailed information on pesticide application, including type of compound, location, date, amount applied, and target crop for each pesticide application. This information was used, along with detailed information on crop distribution and hydrologic characteristics, to describe the relation between crop type, pesticide application, and pesticide occurrence in the three subbasins studied in 1993.

Orestimba Creek is typical of the small western tributaries to the San Joaquin River where streamflow is almost exclusively agricultural runoff during the summer, but may also include large amounts of runoff from the nonagricultural

Coast Ranges during the winter. The greatest variety of pesticides were detected here (28 herbicides and 12 insecticides) compared with the other sites. During the winter precipitation season, the high variability in the number of detections in samples from Orestimba Creek is attributed to rapid changes in the source of streamflow during a storm (*15*)(Figure 3). Samples with many pesticides are attributed to the first flush of pesticides off the fields, a phenomenon observed in other agricultural settings (*16*). On the basis of hydrologic and chemical evidence, samples with few compounds are believed to represent streamflow primarily derived from the nonagricultural Coast Ranges part of the basin (*15*). During the irrigation season, samples usually contained detectable quantities of 15 to 22 pesticides (Figure 3). Analysis of samples collected in 1992 also detected a consistently high number of compounds during the irrigation season (*7*). Pesticides detected more frequently in Orestimba Creek than at the other sites include azinphos-methyl, DDE, dieldrin, ethalfluralin, fonofos, napropamide, and propargite (Figure 2). Data on pesticide application and land use indicate that the presence of most of these pesticides primarily is due to application on dry beans and truck crops (*14*). In contrast, dieldrin and the parent compound of DDE (DDT) are no longer applied, and their detection is due to persistence following historical application.

Salt Slough drains a low-lying part of the San Joaquin Valley, which includes large areas of wetlands and cotton. There is no significant upland area within this subbasin, and streamflow is dominated by agricultural runoff, agricultural subsurface drainage, and wetlands drainage. Twenty-five herbicides and eight insecticides were detected at this site. The numbers of pesticides detected in samples from this site are consistently high; seldom are fewer than eight different pesticides detected (Figure 3). Pesticides detected more frequently at Salt Slough than at other sites were atrazine, cyanazine, diuron, EPTC, malathion, and molinate (Figure 2). The presence of all of these pesticides, except atrazine, can be attributed to application primarily on cotton, rice, alfalfa, and truck crops.

The Merced River is one of three major eastern tributaries that carry runoff from the Sierra Nevada year round, often as reservoir release, and agricultural runoff during the summer. Although 26 pesticides were detected in this river, the frequency of detection and concentrations were usually much lower than corresponding levels in the other two basins (Figure 2). Similarly, samples from this site usually contain fewer than eight detectable pesticides (Figure 3).This relatively low occurrence is due to a combination of two factors; the generally coarse-grained soils of the eastern San Joaquin Valley result in little surface runoff during rainfall or irrigation, and pesticides that do reach the Merced River are diluted by the release of the relatively pesticide-free water from a reservoir in the Sierra Nevada foothills (*14*).

Because of the contrasts in the physical characteristics of the three subbasins, there is a clear spatial correspondence between pesticide application and pesticide occurrence among the subbasins only for those compounds applied to crops grown exclusively, or dominantly, in one of the subbasins. Pesticides found at the San Joaquin River near Vernalis, the site selected to integrate the impact of land use in the basin, provide a good indication of pesticide occurrence, as well as the

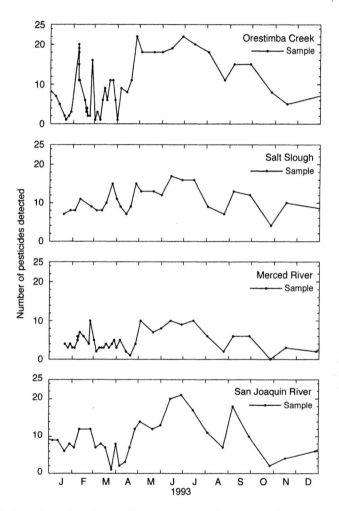

Figure 3. Number of pesticides detected per sample at each subbasin site in 1993.

frequency of detection of the most commonly occurring pesticides in the basin (Figure 2) (*14*).

Temporal Variation In Pesticide Occurrence and Concentrations

The number of pesticides present in each sample can vary widely during the year and depends in part on the source of water to the stream, as well as the timing of application. Runoff from precipitation on nonagricultural land, which occurs in the upper part of the Orestimba Creek and the Merced River subbasins during winter storms, results in a more variable number of pesticide detections during the winter than during the summer (Figure 3). The number of pesticide detections is consistently high in Orestimba Creek and Salt Slough during May, June, and July when these streams receive irrigation return flow.

Many pesticides show a clear correspondence between the time of application and occurrence. This is particularly true for pesticides applied during the summer. Fourteen pesticides, including EPTC (Figure 4*A*), had high application rates and correspondingly high concentrations during the summer irrigation season (*14*). The occurrence or highest concentrations of compounds applied before or during the winter precipitation season, for example diazinon (Figure 4*B*), also generally matches the period of application. Conversely, several pesticides exhibited maximum concentrations during winter storms, even though maximum application occurred at some other time of year. Chlorpyrifos attains its maximum concentration in the San Joaquin River in January and February (Figure 4*C*), rather than at the time of maximum application during the spring and summer. Similarly, cyanazine concentrations are much higher in January and February than during peak application in June and November (Figure 4*D*). The data indicate that both winter rainfall and irrigation tailwater may transport pesticides from the site of application to the receiving river or stream. In addition, during the autumn there is neither rainfall nor irrigation, resulting in relatively few detections (Figure 3) and low concentrations, despite high application of some compounds (Figure 4*D*).

Both occurrence and temporal distribution of pesticides are influenced by the physical and chemical properties of the pesticide. Although three specific properties — solubility, half-life, and K_{oc} (the organic–carbon–normalized adsorption coefficient) — are weakly correlated with transport of pesticide loads (*14*), more investigation is needed before these relations can be used to predict transport of pesticides accurately.

Transport of Diazinon by Storm Runoff

The main factors involved in the transport of the organophosphate insecticide diazinon in the San Joaquin River are the timing of diazinon applications and the occurrence of storms during January and February (*17*). Transport of diazinon was evaluated by first determining the temporal variability of concentrations during individual storm runoff events by collecting multiple samples across the hydrograph at selected sites, and then assessing that data on a basin-wide context using data from dye-tracer studies (*18*).

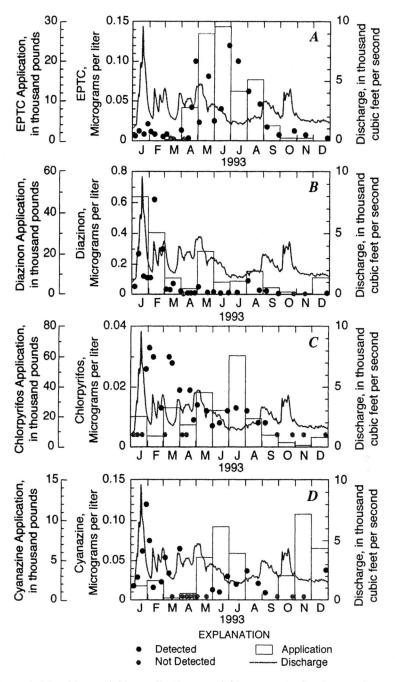

Figure 4. Monthly pesticide application, pesticide concentration in samples, and discharge for the San Joaquin River near Vernalis, California, in 1993 for (A) pebulate, (B) EPTC, (C) chlorpyrifos, and (D) cyanazine.

On February 7 and 8, 1993, 1.9 inches of rainfall followed a dry period in which insecticides were heavily applied to dormant orchards in the San Joaquin Valley. In response to the rainfall, the streamflow in Orestimba Creek rapidly increased from less than 100 cubic feet per second (ft^3/s), to greater than 1,000 ft^3/s (Figure 5A). Analysis of samples collected across the hydrograph showed that diazinon concentrations increased from 0.54 µg/L at the outset of the storm, to a maximum of 3.8 µg/L early on the rising limb of the hydrograph (Figure 5A). Concentrations then decreased to 0.17 µg/L prior to a steep rise in discharge. Samples collected at three other west-side drainages near peak discharge also showed very high concentrations of diazinon (15). As described above, the high concentrations are attributed to runoff from the agricultural portion of the valley, and the low concentrations at peak discharge are due to the contribution of pesticide-free runoff from the large, nonagricultural portion of the drainage basin in the Coast Ranges.

Diazinon contributed from small west-side watersheds such as Orestimba Creek reach the mainstem of the San Joaquin River early in the storm hydrograph. Data for the San Joaquin River site showed concentrations reaching a peak of 0.77 µg/L on February 8 (Figure 5B). This first peak is attributed to runoff from small west-side creeks (15). The second peak of 1.07 µg/L on February 11 may be due to runoff from the larger eastside tributaries. A lagrangian study of insecticide mass loading during the same period indicated that the Newman Wasteway, a west-side drain, also contributed a significant load of diazinon to the San Joaquin River during the latter part of this storm (6).

In 1994, about half of the annual diazinon application in agricultural areas of the San Joaquin River Basin occurred during two dry periods preceding sampled storms during January and February (17). Samples were collected from the three major eastside tributaries — the Merced, Tuolumne, and Stanislaus rivers — to evaluate the variation in diazinon concentrations across the storm hydrographs for both the January and February storms. In both storms, the Tuolumne River had the highest concentrations and transported the highest load to the San Joaquin River of the three eastside tributaries (17). For example, in response to the rainfall on February 8, the diazinon concentrations in the Tuolumne River reached a peak of 0.9 µg/L on the rising limb of the hydrograph (Figure 5C). The diazinon transported from this site was largely responsible for a peak concentration of 0.35 µg/L that was measured at the San Joaquin River near Vernalis during this storm (Figure 5D).

On the basis of storm sampling during 1993–94 and estimated traveltimes determined from dye-tracer tests (18), ephemeral west-side creeks and drains probably are the main source of diazinon to the San Joaquin River early during winter storms, whereas the Tuolumne and Merced rivers and other east-side sources that drain directly to the San Joaquin River were the main sources later (15, 17).

The overall amount of diazinon transported in the San Joaquin River during the January and February 1994 storms was only about 0.05 percent of the amount applied during the preceding dry periods. During 1991–93, 74 percent of diazinon

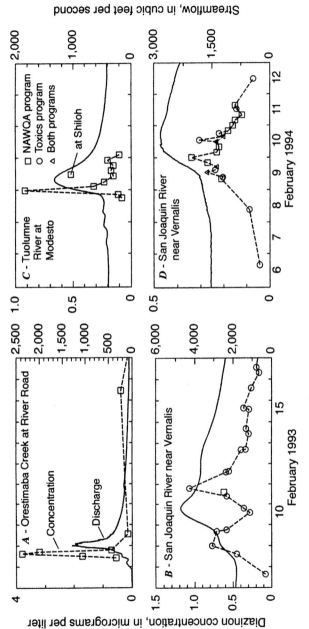

Figure 5. Concentration of diazinon in samples and streamflow showing response to winter storm runoff in a western tributary (*A*, Orestimba Creek) and the San Joaquin River (*B*) in February, 1993; and in an eastern tributary (*C*, Tuolumne River) and the San Joaquin River (*D*) in February, 1994.

transport in the San Joaquin River occurred during January and February. On the basis of daily samples from the San Joaquin River during 1991–94 (5), diazinon concentrations only exceeded 0.35 µg/L, a concentration shown to be acutely toxic to the water flea *Ceriodaphnia dubia*, during January and February storm runoff (4, 15, 17). *C. dubia* is a standard test organism (19, 20), and the effect of these concentrations on other aquatic organisms is largely unknown. These observations show that offsite movement of only a very small fraction of the applied pesticide can produce concentrations of concern under specific conditions. The fact that such a small proportion of the applied amount can result in toxicity makes the task of identifying the sources, and designing remedial strategies, very difficult.

Summary and Conclusions

Several factors that affect the spatial and temporal occurrence of pesticides in surface water were examined during this study. These factors include the location and timing of pesticide application in the different subbasins, and the hydrology of these subbasins. Because of the contrasts in the physical characteristics of the three subbasins, there is a clear spatial correspondence between pesticide application and pesticide occurrence among the subbasins only for those compounds applied to crops grown exclusively or dominantly in one of the subbasins.

The number of pesticides present in each sample also varies widely during the year and depends on the source of water to the stream. Specifically, individual pesticides often show a clear correspondence between the time of application and occurrence. For example, 14 pesticides had high application rates and correspondingly high concentrations during the summer irrigation season. Conversely, several pesticides exhibited maximum concentrations during winter storms, even though maximum application occurred at some other time of year. These data indicate that both winter precipitation and irrigation tailwater may transport pesticides from the site of application to a receiving river or stream.

Concentrations of diazinon at the mouth of the San Joaquin River only exceeded concentrations shown to be acutely toxic to *Ceriodaphnia dubia* during January and February storm runoff. Evaluation of the temporal variation of diazinon concentrations along with travel-time data from dye-tracer tests showed that different parts of the valley contribute high concentrations of diazinon to the mainstem of the San Joaquin River at different times during a winter storm. The data also show that concentrations of diazinon sufficient to be toxic to *C. dubia* can result from the transport of only a very small part of the total amount of pesticide applied, making identification of sources, and hence design of remedial strategies, difficult.

Although this study examined several of the links between pesticide occurrence and some causal factors, other potentially important factors were not examined. These factors include the method of pesticide application and the method of crop irrigation. Both of these factors could be important influences on the occurrence of pesticides in surface water. The San Joaquin River Basin is a complicated hydrologic system with extremely heterogeneous agricultural land uses, and the other causative factors need to be examined to understand the transport processes of pesticides to streams and to achieve the ultimate goal of minimal transport from the fields to surface water.

Literature Cited

(1) California Department of Pesticide Regulation. Pesticide use data for 1993 [digital data]. California Department of Pesticide Regulation, Sacramento, CA., unpublished data, 1994.

(2) Foe, C.G. *Insecticide concentrations and invertebrate bioassay mortality in agricultural return water from the San Joaquin Basin*; Central Valley Regional Water Quality Control Board, Sacramento, CA.,1995.

(3) Foe, C.G.; Connor, V. *San Joaquin watershed bioassay results, 1988–90*; Central Valley Regional Water Quality Control Board, Sacramento, CA. Staff Memorandum, 1991.

(4) Kuivila, K.M.; Foe, C.G. *Environ. Toxicol. and Chem.*, **1995**, 14, pp. 1141–1150.

(5) MacCoy, D.; Crepeau, K.L.; Kuivila, K.M. *Dissolved pesticide data for the San Joaquin River at Vernalis and the Sacramento River at Sacramento, California, 1991–94*; U.S. Geological Survey Open-File Report 95-110, 1995.

(6) Ross, L.J., Stein, R., Hsu, J., White, J., Hefner, K. *Distribution and mass loading of insecticides in the San Joaquin River, California: Winter 1991-92 and 1992-93*; Report EH96-02; Sacramento, CA., California Department of Pesticide Regulation, Environmental Monitoring and Pest Management Branch, Environmental Hazards Assessment Program, 1996.

(7) Domagalski, J.L. *J. of Hydrol.*, **1997a**, v. 192, pp. 33–50.

(8) Domagalski, J.L. *Pesticides in surface and ground water of the San Joaquin-Tulare Basins, California: Analysis of available data, 1966 through 1992*; Water-Supply Paper 2468, U.S. Geological Survey, 1997b.

(9) Dubrovsky, N. M.; Kratzer, C.R.; Brown, L.R.; Gronberg, J.M.; Burow, K.R. *Water Quality in the San Joaquin-Tulare Basins, California, 1992–95*; Circular 1159, U.S. Geological Survey, 1998.

(10) Shelton, L.R. *Field guide for collecting and processing stream-water samples for the National Water-Quality Assessment Program*; U.S. Geological Survey Open-File Report 94-455, 1994.

(11) Zaugg, S.D.; Sandstrom, M.W.; Smith, S.G.; Fehlberg, K.M. *Methods of analysis by the U.S. Geological Survey National Water Quality Laboratory: Determination of pesticides in water by C-18 solid-phase extraction and capillary-column gas chromatography/mass spectrometry with selected-ion monitoring*; U.S. Geological Survey Open-File Report 95-181, 1995.

(12) Werner, S.L.; Burkhardt, M.R.; and DeRusseau, S.N. *Methods of analysis by the U.S. Geological Survey National Water Quality Laboratory-- Determination of pesticides in water by Carbopak-B solid-phase extraction and high-performance liquid chromatography*; U.S. Geological Survey Open-File Report 96-216, 1996.

(13) Crepeau, K.L.; Domagalski, J.L.; Kuivila, K.M. *Methods of Analysis and Quality-Assurance Practices of the U.S. Geological Survey Organic Laboratory, Sacramento, California - Determination of Pesticides in Water by Solid-Phase Extraction and Capillary-Column Gas Chromatography/Mass Spectrometry*; U.S. Geological Survey Open-File Report 94-362, 1994.

(14) Panshin, S.Y.; Dubrovsky, N.M.; Gronberg, J.M.; Domagalski, J.L. *Occurrence and distribution of dissolved pesticides in the San Joaquin River Basin, California*; U.S. Geological Survey Water-Resources Investigation Report 98-4032, 1998.

(15) Domagalski, J.L.; Dubrovsky, N.M.; Kratzer, C.R. J. Environ. Qual., **1997**, 26, 2, pp. 454–465.

(16) Leonard, R.A. In *Movement of pesticides into surface water*; Cheng, H.H., Ed.; Soil Science Society of America Book Ser. 2., *Soil Sci. Soc of America*, Madison, WI., 1990, pp. 303–348.

(17) Kratzer, C.R. *Transport of diazinon in the San Joaquin River Basin, California*; U.S. Geological Survey Open-File Report 97-411, 1997.

(18) Kratzer, C.R.; Biagtan, R.N. *Determination of travel times in the lower San Joaquin River Basin, California, from dye-tracer studies during 1994–1995.* U.S. Geological Survey Water-Resources Investigation Report 97-4018, 1997.

(19) *Short–term methods for estimating the toxicity of effluents and receiving water to freshwater organisms*; EPA-600/4-85-014, U.S. Environmental Protection Agency, Office of Research and Development, Environmental Monitoring and Support Laboratory, U.S. Government Printing Office: Washington, D.C., 1985.

(20) *Short–term methods for estimating the toxicity of effluents and receiving water to freshwater organisms* (2nd ed.); EPA-600/4-89-001, U.S. Environmental Protection Agency, Office of Research and Development, Environmental Monitoring and Support Laboratory, U.S. Government Printing Office: Washington, D.C., 1989.

Chapter 21

Transport and Fate of Pesticides in Fog in California's Central Valley

James N. Seiber and James E. Woodrow

Center for Environmental Sciences and Engineering, University of Nevada at Reno, Reno, NV 89557

Wet deposition, which includes the scavenging of particle bound pesticides and pesticide vapors into atmospheric moisture (cloud and fogwater, rain and snow), is a potentially major sink for airborne pesticides. The pervasive wintertime tule fogs in California's Central Valley, studied extensively in the past 12 years, accumulate organophosphorus, triazine, and other pesticide groups. Concentrations of some pesticides in fogwater can significantly exceed those expected based upon vapor-water distribution coefficients. Fogwater deposition has been implicated as a source of inadvertent residues to non-target foliage, and of high-risk exposures for raptors residing in and around treated areas. The pesticide residue content of fogwater, and its significance in terms of transport, fate, and exposure are reviewed.

Airborne pesticides may exist as vapors or associated with liquid or solid aerosols (*1*). Vapor-aerosol distribution and partitioning of airborne pesticides will occur, as it does with all volatile and semivolatile air contaminants (*2*). Liquid phase vapor pressure is the controlling physical property. In general, chemicals of vapor pressures less than about 10^{-10} atmospheres favor the particulate phase and those with vapor pressures greater than about 10^{-9} atmospheres favor the vapor phase. Pesticides cover a broad range of vapor pressure, extending from gaseous chemicals under ambient conditions (methyl bromide, sulfuryl fluoride, and phosphine) to essentially non-volatile salts such as paraquat. But the majority of pesticides are semivolatile organic compounds (SVOCs) with vapor pressure falling in the range of, approximately, 10^{-4} to 10^{-10} atm.

Airborne pesticides may be removed from the air by principally three processes, namely, degradation, wet deposition, and dry deposition. Degradation may follow oxidative, photooxidative and/or hydrolytic pathways. Only a few pesticide chemicals have been studied in detail, and they display the same types of reactions seen for other classes of organics in the air (*3,4*). In only a very few cases, such as the merphos→DEF, OP thion→oxon, and trifluralin N-dealkylation, are atmospheric vapor transformations rapid enough to be of significance in the time scale of minutes or hours. At the other end of the spectrum are chemicals such as DDE and methyl bromide. Methyl bromide has a tropospheric lifetime of approximately one year, which allows this chemical to diffuse, unreacted, to the stratosphere (5).

Dry deposition involves the settling of particles, which is strongly influenced by particle size and the nature of the meteorology and terrain, and by direct vapor-surface exchange. Wet deposition includes the scavenging of particle-bound pesticides and pesticide vapors into atmospheric moisture (cloud and fog water, rain and snow), followed by rainfall, snowfall, or fog coalescence on surfaces. This is potentially a major sink for airborne pesticides, a source of exposure to pesticides for vegetation, aquatic organisms, and watershed ecosystems, and a means of degrading hydrolytically labile airborne pesticides. The content of pesticides in rain, cloud, fog, and snow has been studied extensively only in the past 10-12 years (6). The accumulating information is quite compelling. Pesticides are measurably present in air and rainfall sampled throughout the U.S. (7,8). Pesticides are also found in snow and ice, including in remote regions of the earth (9). And pesticides used (emitted) in the southeast or southern U.S. ride the storm fronts presenting major deposition inputs to the Chesapeake Bay (10), Great Lakes (11), and other water bodies. Pesticides and other anthropogenic trace organics are found in cloud and fog water where they may achieve concentrations even greater than those expected based upon vapor-water distribution calculations (12). Fog water residues in particular have been implicated as sources of inadvertent residues to non-target crops (13) and of high-risk exposures for hawks residing around treated orchards (14). Fogwater is also an indicator of long-range transport of pesticides to remote regions of the earth (15). The present chapter will delve into the data and underlying methodology associated with these findings.

Pesticide Use in California

The Central Valley of California receives extensive year round use of a variety of pesticides and frequent occurrence of wintertime fogs. The Central Valley comprises a land mass of nearly 50,000 km^2 residing between California's coastal mountain range on the west, the Sierra Nevada mountain range foothills on the east, and extending roughly from the cities of Redding (north) to south of Bakersfield. The northern portion, the Sacramento Valley, is dominated by rice and wheat as major field crops, and almond and other fruit and nut orchards. The southern portion, the San Joaquin Valley, produces cotton, wheat, corn, and a variety of vegetables (carrots, broccoli, cabbage, melons) as major field crops while grape, citrus, almond, walnut, nectarine, peach, plum, and apricot are among the major vineyard-orchard crops.

Pesticide use in California is extensive. Table I provides a breakdown of the top twenty-five pesticide chemicals used in a single year in California (1993) (16). Soil fumigants (methyl bromide, metam sodium, chloropicrin), organophosphate and chlorinated hydrocarbon insecticides-fungicides, triazine and thiocarbamate herbicides, cotton defoliants such as sodium chlorate and Def (S,S,S-tributyl phosphorotrithioate), and many others are included on the list.

Air Residues

Fog episodes in the Central Valley occur predominately in the November-February period. This period coincides with the dormant spraying of fruit and nut orchards. These sprays

Table I. Top Twenty-five Pesticides Used in California, 1993 (lbs active ingredient). Does not include petroleum-based solvents and weed oils, or surfactants

Chemical	Pounds Applied
Sulfur	73,528,071
Methyl bromide	14,768,033
Metam sodium	8,589,016
Sodium chlorate	4,339,262
Glyphosate (2 salts)	3,918,198
Copper hydroxide	3,608,993
Cryolite	2,798,871
Copper sulfate (pentahydrate)	2,532,735
Chlorpyrifos	2,378,207
Chloropicrin	2,124,422
Calcium hydroxide	1,847,484
Ziram	1,771,370
Propargite	1,691,678
Molinate	1,533,103
Sulfuryl fluoride	1,502,091
Diazinon	1,491,709
Trifluralin	1,404,087
Simazine	1,129,946
Maneb	1,122,084
Chlorothalonil	1,095,651
Diuron	1,090,683
S,S,S-tributyl phorophorotrithioate	979,018
Ethephon	897,657
Carbaryl	786,395
Oryzalin	725,521

Compiled from: *(16)*

contain a mixture of petroleum oil, an organophosphate (OP) insecticide, and often a fungicidal copper salt, which together control a host of pest organisms. The OPs employed in 1985-1995 included primarily chlorpyrifos, diazinon and methidathion and, prior to its banning and stock depletion in the early 1990's, ethyl parathion. The principal application areas include the extensive almond, peach, nectarine, and other fruit orchards throughout the Central Valley. Other major winter use chemicals in the valley include trifluralin and other herbicides, chlorothalonil and other fungicides, and soil fumigants/sterilants.

Orchard dormant sprays are susceptible to entering the air. They are usually sprayed by orchard 'speed sprayers' which propel the spray up and out from the nozzles into the tree canopy. Significant proportions of the sprayed material can bypass limbs and twigs and become airborne. As with all pesticide spray aerosols, the very small particles (VMD < 100 μm) can remain airborne, and be carried by wind out of the sprayed area. Perhaps of greater importance is the post-application volatilization of residue from the surfaces of a treated orchard. Glotfelty et al. (17) determined the spray distribution, drift, and post-application volatilization of diazinon applied with an air-blast speed sprayer to a dormant peach orchard in the San Joaquin Valley. Airborne losses were calculated by the Integrated Horizontal Flux (IHF) method from measurements of windspeed and downwind pesticide concentration profiles. There was more diazinon in the soil of the orchard floor than in the tree canopies. The total measured airborne loss from the orchard by drift during application (60 g) was exceeded by the totals lost by volatilization in the two sampling periods after application ended on the same day (76 g) and the day after application (93 g for just the 2 hr measurement period). Volatilization continued for at least 4 days after application ended. These volatilization losses were primarily from tree canopy surfaces, but a slower volatilization occurred from the soil surface as well. Thus, at least for diazinon, most of the Central Valley atmospheric residue during the spray season results from post-application volatilization rather than from drift.

By collecting 24 hr air samples for 17 days in January, 1989, at an "ambient" station in a region of the San Joaquin Valley heavily planted to deciduous orchards, Seiber et al. (18) determined the daily average air concentrations of chlorpyrifos, diazinon, methidathion, and parathion during intense OP dormant spraying (Figure 1). Parathion gave the highest OP air concentration during the first third of the January sampling days, chlorpyrifos dominated in the second third, and diazinon in the third. The samplers were located about 1 km from the nearest orchards, and probably several km from actively sprayed orchards. The trends in residues reflected active or fresh spraying operations within a 1-50 km distance upwind from the sampling station.

Table II shows the 12 hr averages for the thion and oxon forms of the same four OPs. Generally the residue concentrations were higher at night than during the day, perhaps reflecting calmer conditions at night and less dilution by incoming wind. The oxon forms, on the other hand, were higher in the day than at night. Oxon/thion ratios averaged 0.52 for diazinon in the day and only 0.10 at night. This may reflect, at least in part, the higher oxidant levels, and sunlight during the day which are believed to play roles in thion-oxon conversion.

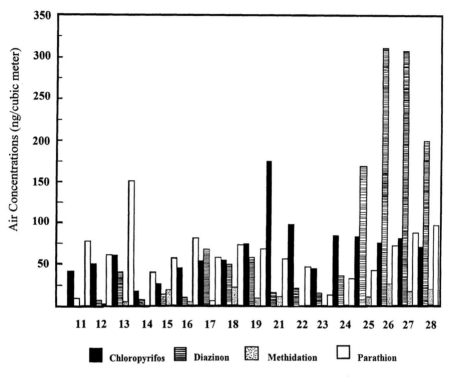

Figure 1. Diagram of residue concentrations of four OPs in air samplers operated for 24 h, Jan 11-28, 1989, at the Kearney Agricultural Center near Parlier, CA. (Reproduced with permission from reference 18. Copyright American Chemical Society 1993.)

Table II. Residue Averages of Four Organophosphates and Their Oxons in Day and Night XAD-4 Resin Air Samples, January, 1989, Kearney Agricultural Center *(18)*. Data are in ng/m^3

	Chlorpyrifos		Diazinon		Methidathion		Parathion	
	Day	Night	Day	Night	Day	Night	Day	Night
Avg thion	47.5	75.3	13.4	52.3	14.9	15.1	32.0	119.6
Std Dev	19.4	49.0	10.6	37.0	7.0	10.1	26.2	30.7
Avg oxon	12.2	5.3	7.2	4.9	7.4	3.5	12.3	6.0
Std Dev	7.3	2.3	3.8	2.8	6.5	0.5	10.5	2.7

Fogwater Concentrations

Methodology. Three types of fogwater collectors are depicted in Figure 2. The Caltech Rotating Arm Collector (RAC) (*19*) used a 1.5 HP motor to drive a 63 cm long stainless steel rod at 1700 rpm through the air. Each end of the rod has a slot milled into its leading edge to collect impacting fogwater droplets. High density polyethylene bottles mounted on each end of the rod catch the fogwater samples as they are accelerated from each slot. Collection rates up to 2 ml/min were obtained in the field (*20*).

The Caltech Active Strand Cloudwater Collector (CASCC) (*21*) employed a fan to draw air across six angled banks of 508 μm Teflon strands at a velocity of 20 cm/sec. Cloud- or fogwater droplets in the air are collected by the strands by inertial impaction. The collected droplets run down the strands, aided by gravity and aerodynamic drag, through a Teflon sample trough into a collection bottle. This instrument has a theoretical lower size-cut of 3.5 μm, based on droplet diameter, and has collected cloudwater at rates of up to 8.5 ml/min in the field (*20*).

A third design of high volume collector developed at USDA-ARS laboratory in Beltsville, MD (*22*) is a rotating screen device, 50 cm diameter, in which four layers of stainless steel screen are rotated around a central axis at 720 rpm. Fogwater obtained from droplets impacting on the screen is centrifuged to the periphery, collected in a slotted aluminum tube, and drained into a collection vessel. A large fan pulls air through the device at a sampling rate of ~ 160 m^3/min. The sampling rate typically allows collections of approximately 1 L/hr of fogwater depending on the suspended water content of the fog. This device is employed mounted on a pick-up truck, which moves slowly (5-15 km/hr) through the foggy area to be sampled. Vehicle and gas-powered generator fumes are thus continually swept toward the rear and away from the fog sampler intake. A useful adjunct to any fog- and cloudwater collection is the availability of a method for determining the presence and density of these atmospheric water suspensions. Mallant and Kos (*23*) described a low-cost optical fog detector which had many advantages over other detection devices described in the literature.

Fog sampling for pesticide analysis in the Central Valley was generally accompanied by simultaneous collection of interstitial air sampled by means of a high-volume dichotomous sampler (Figure 3) (*22*). This provided a sample of the interstitial vapor- and particle-phase pesticides by eliminating fog droplets of > 8 μm through the large particle orifice of this device. Fog droplets of < 8 μm, unactivated (dry) aerosol particles and vapor passed first through a precombusted (400° C overnight) glass fiber filter (GFF), which removed particulate matter, and then through a 7.5 cm diameter x 7.5 cm deep bed of porous polyurethane foam (PUF) which trapped the pesticide vapors. The PUF plugs were precleaned by the method of Bidleman and Olney (*24*). Alternatively, a Chromosorb 102 trap (*24*) was used to collect vapor samples at ca 1 L/min, for vapor analysis.

Extracts of fogwater, particle filters, and vapor traps were analyzed by gas chromatography directly, or following fractionation on a silica HPLC column using a hexane-to-methyl t-butylether (MTBE) gradient (*26,27*). Major components were identified by GC retention time, HPLC retention behavior, and GC-MS, in comparison with authentic samples.

Results. Using the USDA rotating screen fogwater collector and dichotomous air sampler mounted on a pickup truck, Glotfelty et al. (*12,22*) presented the landmark studies of pesticides in fog. They discovered that a variety of pesticides and their toxic alteration products are present in fog, and that they can reach high concentrations relative to rainwater and other atmospheric water samples. Also, in measuring the air-water distribution of pesticides between the suspended liquid phase in fogwater and the interstitial air, they found that some chemicals are enriched several thousand fold in the suspended liquid when compared with distributions expected based upon Henry's law coefficients for pure air-water distribution systems.

These fog samples were collected at the USDA Beltsville Agricultural Research Center in Beltsville, MD, and at several locations in California's San Joaquin Valley. Beltsville samples were often dark in color, with large amounts of black particles (largely carbonaceous) which settled out when left to stand, leaving an off-yellow supernatant. Several common pesticides such as malathion were present in measurable levels. Industrial organophosphorus chemicals, polynuclear aromatic hydrocarbons, and phthalate esters were present in these samples as well. Although the USDA Beltsville Center is agricultural, it is surrounded by a densely populated, partially industrial area with heavy vehicular traffic.

Table III lists the pesticides and alteration products identified in fog in California. Organophosphorus insecticides (diazinon, parathion, chlorpyrifos, methidathion, malathion, methyl parathion) and their oxygen analogues were most frequently found. But several types of herbicides, including the triazines (atrazine, simazine), dinitroaniline (pendimethalin), and chloroacetanilides (alachlor, metalochlor) were measurably present in some samples. The fog samples from California yielded more variety and higher concentrations of pesticides than the Beltsville samples, a result of the greater diversity and heavier volume of pesticide use in California's Central Valley. The OP insecticides frequently exceeded 10 μg/L in fog, which is 2-3 orders of magnitude greater than for these and similar compounds in rain from other locations, and in the Central Valley itself (*18*). Because these California samples were collected during wintertime dormant spraying in the Central Valley, the elevated levels of diazinon, parathion, methidathion, and chlorpyrifos were ascribed to this usage.

The California samples contained a number of oxygen analogues (OAs) of OP pesticides, with paraoxon being the most abundant. The source of these compounds is unknown. OAs form in the gas phase, by reaction of parent thions with atmospheric oxidants (*3*), and also on surfaces (*28,29*). OAs are potent inhibitors of cholinesterase and are generally responsible for most of the toxic effects of OPs (*30*). The fog collected in Lodi, which had the highest measured concentration of paraoxon (184 μg/L), yielded a lab-measured cholinesterase inhibition which matched that of equivalent concentrations of pure paraoxon standard.

The Henry's law constant, which generally describes the distribution of low solubility organics between the vapor and aqueous phase in equilibrium (*31*), was not a good quantitative indicator of the distribution of pesticides between the vapor and liquid phase in these fog samples. The measured air-water distribution coefficient (D) was frequently

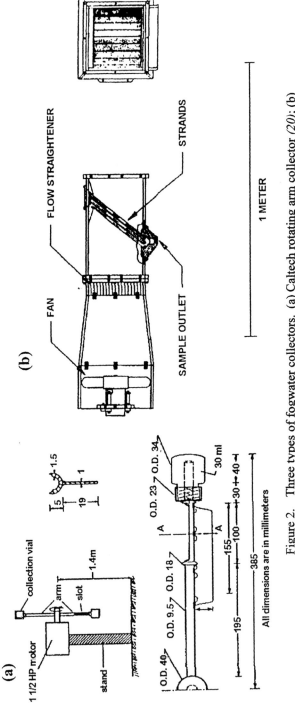

Figure 2. Three types of fogwater collectors. (a) Caltech rotating arm collector *(20)*; (b) Caltech active strand cloudwater collector *(20)*; (c) USDA high volume rotating screen atmospheric fog extractor *(22)*. (a and b reproduced with permission from reference *(20)*, Copyright 1990, Elsevier Publishing Company).

331

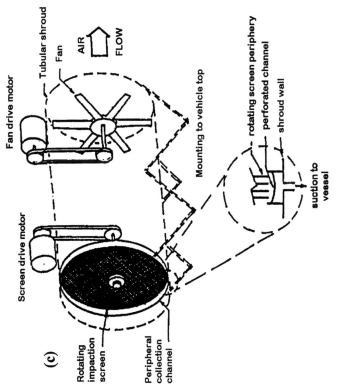

Figure 2. *Continued.*

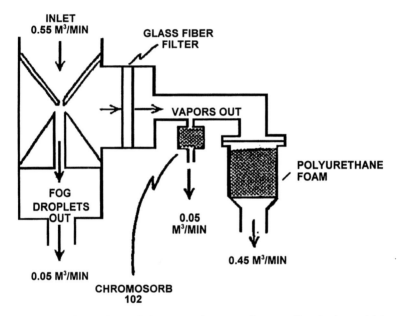

Figure 3. High-volume dichotomous impactor for sampling the interstitial pesticide vapors in fog *(22)*

Table III. Distribution of Pesticides and their Alteration Products Between Fogwater and Interstitial Air in California *(12)*

			Chemical concentration		Distribution ratio (x10⁶) air-water		Enrichment factor EF)
			fog water (ng/l)	Air (ng/m³)	measured (D)	literature (H)	
Organophosphorus Insecticides							
Diazinon		P	16,000	2.2	0.12	60	500
		C	11,800	4.2	0.36		170
		L	22,000	5.3	0.24		249
Parathion		P	12,400	3.2	0.25	9.5	38
		C	5,800	0.88	0.15		63
		L	51,400	7.9	0.15		63
Chlorpyrifos		P	1,020	3.3	3.2	500	156
		L	320	0.6	1.9		260
		C	6,500	14.7	2.3		217
Methidathion		P	840	<0.03	<0.04	0.07	>1.8
		L	570	<0.01	<0.02		>3.5
		C	15,000	<0.6	<0.04		>1.8
Malathion		P	70	<0.03	<0.4	2.4	>6
		L	110	0.02	0.18		13
		C	350				
Oxygen Analogues							
Parathion		P	9,000	0.21	0.02	0.25	11
OA		C	950	0.66	0.07		3.6
		L	184,000	0.25	0.00		250
Methidathion		P	120	0.03	<0.25	—	—
OA		L	8,200				
Chlorpyrifos		P	170	<0.03	<0.18	—	—
OA		L	800				
Diazinon		P	190	<0.03	<0.16	—	—
Herbicides							
DEF		P	250	<0.03	<0.7	0.2	—
		C	800	0.14	<0.4		
Atrazine		P	270	<0.2	<0.7	0.2	—
		C	320	<0.16	<0.4		—
		L	700	—	—		—
Simazine		P	390	<0.1	<0.3	0.025	—
		C	110	<0.06	<0.5		—
		L	1,200	—	—		—
Pendimethalin		P	1,370	0.64	0.47	1,500	3,200
		C	3,620	3.6	1.0		1,500

EF (= ratio of literature to measured distribution ratio) reflects aqueous-phase enrichment.
P = Parlier, CA; C = Corcoran, CA; L = Lodi, CA

very much less than H (Table III), especially for the California samples. D < H implies an aqueous-phase enrichment; more pesticide is dissolved in the aqueous phase than would be expected in an ideal solution at equilibrium. The extents of the aqueous phase enrichment were given as Enrichment Factors (EF = H/D), which were quite large, even several thousandfold, for chemicals such as pendimethalin and diazinon. The possible underlying reasons for enrichment, and its consequences, are discussed below.

The distribution of pesticides in foggy Central Valley atmospheres was studied in more depth (32). This study was conducted at one Central Valley sampling site (Kearney Agricultural Research Center, University of California, Parlier, CA) and included one foothill site about 50 km east of Parlier, at approximately 500 m elevation. The distribution of four OP insecticides (diazinon, parathion, chlorpyrifos, and methidathion) and their oxons between the droplet and air phases was studied during six fog events. Up to 50 μg/L was found for the total of the four OPs and up to 75 μg/L for the total of their oxons in the fogwater. Nearly all of the compounds exhibited aqueous phase enrichment, ranging from a mean of 1.4 (methidathion) to 58 (diazinon) (Table IV). The oxon to thion ratios in the fogwater was generally less than 1, with two exceptional sampling sites and dates. Atmospheric oxidation, especially during daylight hours, followed by uptake in the fogwater was implicated.

Even though there were high concentrations and high EF for the water phase, the very low volume of suspended water in fog leads to the highest proportion of all of the compounds, in all of the events, in the interstitial air phase, either as vapor or adsorbed to aerosol particles. Figure 4 shows this distribution for parathion for a single sample. The total concentration of parathion in the atmosphere was 9.4 ng/m^3. Of this, 78% (7.3 ng/m^3) was in the vapor phase. Only 10% (0.9 ng/m^3) was dissolved in the fog droplet, even though the concentration in the droplet was high -- 30 μg/L -- resulting from the high enrichment factor (18) and low concentration (<0.1 mg/m^3) of suspended water droplets in foggy atmospheres. Up to 11% (1.0 ng/m^3) appeared to be attached to aerosol particles. And only 0.6% was attached to solids filterable from the fogwater. It was not determined whether the particulate-bound pesticides were present adsorbed to particles before the fog formed, or were sorbed out from dissolved pesticides present in the droplets after the fog formed.

In order to determine whether the uptake of pesticides in fog was unique to the wintertime ground fog, or tule fog, of the inland Central Valley, Schomburg et al. (33) used a scaled-up teflon strand fog collector (18) to analyze air and fogwater in several spring advective oceanic fogs collected near Monterey, CA, and the heavily agricultural Salinas Valley and associated coastal plains. The pesticide content and distribution for several pesticides common to both areas (chlorpyrifos, diazinon, etc) were remarkably similar between the coastal and Central Valley samples. The conversion of thion to oxon, and the aqueous phase enrichments, were also very similar. The data also provided support for the hypothesis that non-filterable, strongly sorptive particles and colloids in the fogwater cause the enhancement of pesticides in water in the foggy atmosphere.

Fogwater entrainment and concentration of chemicals are not unique to pesticides. Sagebiel and Seiber (34) analyzed wintertime fog and interstitial air from a community in

Table IV. Aqueous-Phase Enrichment Factors (K_{AW}/D) for the Distribution of Pesticides between Fogwater and Air *(32)*

	date					
compound	1/8-9	1/9	1/11[a]	1/12	1/12-13	mean
diazinon	6	59	14	50	160	58
parathion	4	12	5	18	29	14
chlorpyrifos	7	55	3.5	40	74	42
methidathion	0.06	3	0.02	1.4	2.3	1.4
paraoxon	2.1	14	10	48	>69	>19

[a] Hills Valley Road site.

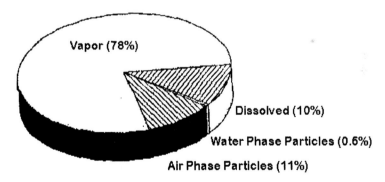

Figure 4. Distribution of parathion in the foggy atmosphere near Parlier, CA, on January 12, 1986 (Reproduced with permission from reference 32. Copyright American Chemical Society 1990.)

the Central Valley during a time when residential wood burning occurred. Guiacol, 4-methyl guaiacol, and syringol were the most commonly found among the 16 methoxylated phenolic lignin combustion products confirmed by GC/MS in fog samples. The distribution of methoxylated phenols generally followed Henry's law, that is, did not show the dramatic enrichments observed for less polar pesticides. This suggested that enrichment is a function of analyte hydrophobicity rather than any special structural features. Concentrations of methoxylated phenols in fogwater ranged to 1408 μg/L for syringol, and generally were in the 1-100 μg/L range when detected.

Seiber et al. (18) provided more detail on the distribution of pesticides in air, fogwater, and plant surfaces in a series of experiments carried out at the Kearney Agricultural Research Center at Parlier, CA. Fogwater contained residues of the four OP dormant spray insecticides (parathion, chlorpyrifos, diazinon, and methidathion) and their oxons (Table V). Concentrations ranged to 91 μg/L for parathion and 76 μg/L for diazinon, and were significantly lower for oxons (to 19 μg/L for paraoxon, 6.2 μg/L for methidathion oxon, 3.4 μg/L for chlorpyrifos oxon, and 3.0 μg/L for diazinon oxon), in fogwater sampled with the Teflon strand collector. When fogwater was collected by simply placing a collector beneath the drip lines of tree canopies, similar water concentrations were observed, but the ratio of oxon to thion increased dramatically, to 0.7 (diazinon) and 1.35 (methidathion) for the four insecticides. This indicated either that conversion of thion to oxon occurred in fogwater as it collected on, and passed through the foliage, or that conversion of thion to oxon occurred in the tree parts (leaves, needles, limbs) after the water evaporated as the fog lifted, with the surface-formed oxon being removed during the next episode of fog. The apparent tree surface catalysis of thion to oxon was noted previously (28).

Enrichment

The phenomenon of enrichment of chemical solutes in the suspended aqueous phase of foggy atmosphere has been the subject of much discussion since it was observed for several pesticides and related chemicals by Glotfelty et al. (12,22). It is generally assumed that the distribution of low solubility organic solutes between air and water is described by Henry's constant (= air/water distribution coefficient) and that this distribution coefficient, calculated from the ratio of the vapor density (a direct function of vapor pressure) to the water solubility of the pure organic chemical (11,31,35,36), would prevail in fog. Henry's law holds for vapor in equilibrium with water phase. This equilibrium is achieved rapidly when there is pronounced surface contact between the two, such as apparently occurs in atmospheric moisture in clouds and raindrops (37,8).

Because enrichment was more pronounced for hydrophobic than hydrophilic pesticides, it was hypothesized that fog droplets contained solutes, such as dissolved or colloidal organic matter, which increased the solubility of hydrophobic chemicals over pure water. Surface active material had previously been reported in fog (38). Common atmospheric organics, such as α-pinene, n-hexanol, eugenol, and anethole, produced films with surfactant-like properties. Such films may exist in cloud droplets and snowflakes, in addition to fog droplets. These surface active organics present at the air-water interface act to enhance the uptake of low-solubility organics into the aqueous phase (39).

Table V. Oxon/thion Ratios in Fogwater Collected Using Active Strand Fogwater Sampler, and as Tree Drip Beneath 3 Types of Trees, and in Non-fog, Clear Afternoon Air *(18)*

	Chlorpyrifos	Diazinon	Parathion	Methidathion
Clear afternoon air	0.45	0.52	0.62	0.58
Fogwater collected by Active Strand Sampling	0.21	0.07	0.60	1.8
Pine tree drip water	1.13	0.86	1.11	1.43
Deciduous tree drip water	1.08	0.64	0.71	1.15
Evergreen tree drip water	1.27	0.59	0.83	1.46

Although enrichment was reproducible in fog sampling conducted after that reported by Glotfelty et al. (*12,22*), it was generally less marked in subsequent studies. Aqueous phase enrichments for five chemicals ranged from 58 (diazinon), 42 (chlorpyrifos), 14 (parathion), and 1.4 (methidathion) in fog sampled in the San Joaquin Valley in 1986 (*17,32*) (Table IV). There was large variability in EF for the same chemical measured in different fog events which had no obvious explanation. The liquid water content, which ranged from 0.024 to 0.080 g/m^3, total organic carbon content (38 to 55 mg/l) and pH (5.4 to 7) did not vary that much between samples, and did not show correlational trends. Temperature effects were hypothesized to have played a small role. H generally increases with increasing temperature, and most of the calculated H constants were from data at 20 or 25° C. Since the fog collections were made in atmospheres of 1.5-8.5°C, measured distributions (D) should be lower than those calculated from H, but only by factors of 2 to 4.

Seiber et al. (*18*) found enrichment factors of only 5.7 (chlorpyrifos), 3.3 (diazinon), 2.7 (parathion), and 0.01 (methidathion). These magnitudes were considerably less than those of Glotfelty et al. (*12,22*), and almost within the range predicted for temperature effects on distribution. Seiber et al. (*18*) suggested the operation of variables, or sources or error, not yet considered or explained. It should be noted that Seiber et al. (*18*) used a Teflon strand collector (*32*) while the Glotfelty et al. (*12,22*) collections were made with the rotating stainless steel screen device (*22*).

Two primary hypotheses have emerged to explain the enrichment of organics in the aqueous phase of fogwater.

(1) Organic solutes or non-filterable colloids in the water phase enhance water solubility above that in pure water, or sorb organics so that more are bound in the fog water phase than in pure water.

(2) The air-water interface acts as a third phase, or compartment, in what had been assumed to be a simple 2-phase, or 2-compartment (viz, air and water) process.

Capel et al. (*40*) provided arguments in favor of the first hypothesis, essentially supporting the conclusions of Glotfelty et al. (*12,22*). They argued that, on a mass basis, the "dissolved" organic fraction in fogwater is 10-100 times typical values reported for rain, lakes, and rivers. They also showed that the surface tension of fog is less than that of pure water, reflecting the presence of surface active material. They assumed, then, that a portion of the fog droplet/air interface is covered with organic chemicals, so that airborne chemicals confront an organic/air interface rather than a water/air interface. This would provide a mechanism for the concentration or enrichment of hydrophobic organic contaminants in the fog droplet.

Sagebiel and Seiber (*34*) provided new evidence for the involvement of an interface in the enrichment process. They measured the air-water distribution of a number of phenolic products of incomplete combustion of wood in fogs collected in a winter-time residential setting in the Central Valley of California. The three most significant phenols, guaiacol,

4-methylguaiacol, and syringol, showed little to no enrichment. In plotting enrichment factor vs either octanol-water partition coefficient (Figure 5a) or water solubility (Figure 5b), it was clear that these wood smoke marker chemicals extended the range of solubilities represented by the pesticides studied by Glotfelty et al. (*12*) to much higher water solubility (Figure 5b) or much lower octanol-water partition coefficient (Figure 5a), demonstrating that enrichment was limited to non-polar organics in a regular fashion. This argued for a third phase in the air/fog system which might be made up of fine particulate matter, colloids, dissolved organic matter, or a surface film (first hypothesis). A theoretical treatment showed that this unknown sorptive phase could be described with the existing data for pesticides and wood smoke marker chemicals. The authors pointed out that they made no supposition regarding the composition of this third phase, but the concept of surface-active organics (first hypothesis) was one which was considered.

Goss (*41*) summarized arguments in favor of the second hypothesis. He argued that the existence of an inseparable organic phase in which non-polar chemicals accumulate provides only a partial explanation for the observations. Perona (*42*), Valsaraj (*43*), and Valsaraj et al. (*44*) pointed to the importance of adsorption at the air-water interface as the missing process which had been overlooked. The very large specific area of small fog droplets provides an adsorptive phase which contributes to enrichment. Adsorption at this interface can be described by a partition coefficient relating the amount adsorbed to the equilibrium concentration in the air (K_{ia}) or to the equilibrium concentration in the water (K_{iw}).

Valsaraj et al. (*44*) found a correlation existed between K_{iw} and K_{ow} (octanol-water partition coefficient) for selected hydrophobic organics:

$$K_{iw} = 3 \times 10^{-7} \, K_{ow}^{0.68}$$

They used this equation to predict the adsorption of pesticides at the air-water interface and predict the resulting enrichment in fog droplets. Goss (*41*) developed a more complex equation, relating K_{ia} and vapor pressure, which could be used in a similar way to that of Valsaraj et al. (*44*). Applying this equation, and some temperature corrections to adjust the vapor pressure and fog water collection, produced the results in Table VI. The calculated adsorption on the water surface led to a significant enrichment of chemicals in fogwater of the same magnitude as observed experimentally.

Hoff et al. (*45*) analyzed much of the same experimental data, but included new data on K_{ia} and K_{wa} determined by gas chromatographic retention times using water as a stationary phase. Their results supported that partitioning at the air-water interface can be appreciable and must be taken into account. Small droplets of water in air (fogs or clouds), small bubbles of air in water, and relatively dry low organic soils are examples. Correlations were derived for interface-air and interface-water coefficients for 44 polar and nonpolar organic chemicals.

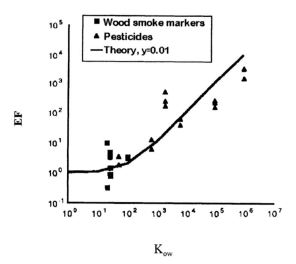
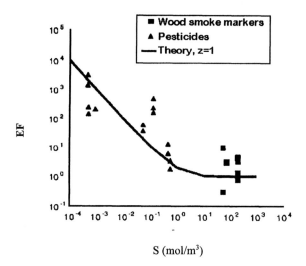

Figure 5. Enrichment Factor (EF) of pesticides and wood smoke markers plotted against octanol-water partition coefficient (top) and water solubility (bottom) (Reproduced with permission from reference *(34)*. Copyright 1993, SETAC)

Table VI. Comparison of the Calculated (8 μm Droplet Diameter) and Measured Enrichment Factors (41)

Substance	$EF_{calc.}$	EF_{lit}
Chlorpyrifos	142	7-74
		19-25
Parathion	39	4-29
Methyl parathion	11.7	0.7
Malathion	17.5	6
Paraoxon	1.4	2.1
Diazinon	273	6-160
		30-50
Atrazine	1.8	0.05
Alachlor	1.76	>4
Pendimethalin	2476	1500-3200
Fonofos	30	3-5
Guaiacol	1.2	4.2
	1.2	3.3
4-Methylguaiacol	1.4	3.0
	1.4	3.2
Syringol	10.0	9.6
	7.5	0.32

Significance

The significance of pesticide residue occurrence in foggy atmospheres and fogwater warrants discussion. From a human health viewpoint, residues of individual chemicals in fogwater probably do not pose a significant risk (*18*). The higher concentrations recorded, 100 ng/m^3 in air and 100 μg/l in fogwater, may be expressed in human respiratory exposures as:

Interstitial Air:
$$0.1\ \mu g/m^3 \times 15\ m^3/day \times \frac{1}{70\ kg} = \sim 0.02\ \mu g/kg/day$$

Suspended Water:
$$100\mu g/L \times 0.1 \times 10^{-3}\ L/m^3 \times 15\ m^3/day \times \frac{1}{70\ kg} = \sim 0.002\ \mu g/kg/day$$

That is, for a 70 kg person breathing at 15 m^3/day, the exposure is 0.02 μg/kg/day from breathing interstitial fog air, and 0.002 μg/kg/day for inhaling suspended water in the foggy air. For comparison, the acceptable daily intake (ADI) for parathion, perhaps the most toxic of the pesticides observed to date in fogwater, is 5 μg/kg/day, established by FAO/WHO based on an oral no-observed-effect-level (NOEL) of 0.05 mg/kg/day for red blood cell acetylcholinesterase inhibition (*46*). The exposure from breathing foggy air is thus less than 1/1000th of ADI under these high exposure conditions. People do not breathe foggy air 24 hours/day, and fogs tend to be transient by their very nature. Thus it is hard to envisage conditions under which single chemical risks from breathing foggy air become significant. But the data on pesticides in fogwater show that mixtures of several chemicals, including several OPs and their oxons, may be present simultaneously, so that it is not appropriate to dismiss fogwater altogether from the viewpoint of human health implications. More studies are needed, particularly in regards to mixtures and long-term exposures for people, including children and other sensitive subpopulations living in unusually fog-prone areas where pesticides are used.

Fogwater deposition to non-target food crops represents an indirect exposure for humans. Turner et al. (*13*) found that fogwater deposition was a source of inadvertent residues to non-target crops in California's Central Valley. But the residue contribution from this source is small, on the order of 0.01 - 0.1 ppm or less under worst-case conditions. Thus the concern is not on human health impacts but rather on the legal question of what to do with food crops which receive low-level inadvertent residues of chemicals for which no tolerance has been established on crops where the contamination is found.

For wildlife dwelling in or very near orchards, the risks may be significantly higher than those for humans. Wildlife, such as birds, are exposed constantly to residues in the air, at much higher levels for birds located within the canopy of treated trees. Also, deposition of airborne residue to the feathers or fur may be significant because deposited residue may be ingested during preening. Birds may also contact residue through their feet or talons when roosting on a treated or exposed branch. Finally, all wildlife could ingest residue in their food. For red-tail hawks, which frequent deciduous orchards in midwinter, inhalation, oral intake *via* preening, oral intake *via* food and water, and dermal intake through talons may all contribute to a residue exposure sufficiently high to produce biochemical and clinical signs of organophosphorus intoxication (*14,47*).

If one considers fog as simply one type of cloud, then the significance of fog- or cloudwater accumulation and transport of pesticides and other contaminants may assume a regional or global significance. The potential for regional or longer-range transport and deposition of Valley-derived OPs used in dormant spraying is an example. Chlorpyrifos, diazinon, and parathion and their oxons were detected in air, rain, and snow samples collected in December-March in the Sierra Nevada mountains southeast of the San Joaquin Valley orchard regions (*48*). Air concentrations at pg/m^3 levels were determined at elevations of 533 and 1920 m in the mountains. Rainwater concentrations were at ppt (ng/l) levels at the mountain sites. Clearly, long-range transport of OPs can occur, but the indications are that removal processes (exchange with soil and plant surfaces, wet deposition, and chemical breakdown *via* oxon formation followed by hydrolysis) operating in the Valley itself may limit the buildup of OPs at non-target sites outside the Valley.

Recently a number of pesticides and PCBs were detected in snow and some surface waters in the Sierra Nevada mountains, apparently resulting from wet deposition of residues in clouds (*49*). The significance of these aerially transported residues is under investigation (*50*).

Conclusions

Fog represents an atmospheric medium in which pesticides may be degraded or undergo vapor-liquid distribution, and subsequent deposition. It probably does not alone present a direct health hazard to humans, but it will add to total exposure from ingestion of pesticides in food crops and drinking water, and residential exposure. Fogwater residues may have a major health impact on animals which reside in or around pesticide-treated areas, from deposition to their feathers or fur. More work is needed to define risks associated with pesticides, as well as other toxicants in foggy atmospheres.

Literature Cited

(1) Lewis, R.G. and R.E. Lee, Jr. 1976. Air pollution from pesticides: Sources, occurrence and dispersion. In: ***Air Pollution from Pesticides and Agricultural Processes.*** R.E. Lee, Jr. (Ed), CRC Press, Cleveland, OH. pp 5-50.
(2) Bidleman, T.F. 1988. Atmospheric processes. ***Environ. Sci. Technol.*** **22**:361-367.
(3) Woodrow, J.E., D.G. Crosby and J.N. Seiber. 1983. Vapor-phase photochemistry of pesticides. ***Res. Rev.*** **85**:111-125.
(4) Seiber, J.N. and J.E. Woodrow. 1995. Origin and fate of pesticides in air. Reprinted from ACS Conference Proceedings Series ***Eighth International Congress of Pesticide Chemistry: Options 2000*** (N.N Ragsdale, P.C. Kearney and J.R. Plimmer, Eds), pp 157-172.
(5) Yvon-Lewis, S.A. and J.H. Butler. 1997. The potential effect of oceanic biological degradation on the lifetime of atmospheric CH_3Br. ***Geophys. Res. Lett.*** **24**:1227-1230.

(6) Rice, C.P. 1996. Pesticides in fogwater. *Pesticide Outlook* 7:31-36.
(7) Goolsby, D.A., E.M. Thurman, M.L. Pommes and W.A. Battaglin. 1994. Temporal and geographic distribution of herbicides in precipitation in the midwest and northeast United States, 1990-91. In *Proceedings of the Fourth National Pesticide Conference*, Richmond, VA., November 1-3, 1993.
(8) Majewski, M.S. and P.D. Capel. 1995. In: *Pesticides in the Atmosphere*, Ann Arbor Press, Chelsea, MI, 214 pp.
(9) Kurtz, D.A. (Ed). 1990. *Long Range Transport of Pesticides*, Lewis Publishers, Chelsea, MI.
(10) Glotfelty, D.E., G.H. Williams, H.P. Freeman and M.M. Leech. 1990. Regional atmospheric transport and deposition of pesticides in Maryland. In: *Long Range Transport of Pesticides* (Kurtz, D.A., Ed), Lewis Publishers, Chelsea, MI. Pp 199-221.
(11) Eisenreich, J.S., B.B. Looney and J.D. Thornton. 1981. Airborne organic contaminants in the Great Lakes ecosystem. *Environ. Sci. Technol.* **15**:30-38.
(12) Glotfelty, D.E., J.N. Seiber, and L.A. Liljedahl. 1987. Pesticides in fog. *Nature* **325**:602-605.
(13) Turner, B., S. Powell, N. Miller and J. Melvin. 1989. *A Field Study of Fog and Dry Deposition as Sources of Inadvertant Pesticide Residues on Row Crops*. Environmental Hazards Assessment Program. State of California, Department of Food and Agriculture. Sacramento, CA. EH-89-11, 42p + appendices.
(14) Wilson, B.W., M.J. Hooper, E.E. Littrell, P.J. Detrich, M.E. Hansen, C.P. Weisskopf and J.N. Seiber. 1991. Orchard dormant sprays and exposure of red-tailed hawks to organophosphates. *Bull. Environm. Contamin. Toxicol.* **47**:717-724.
(15) Chernyak, S., C.P. Rice and L.L. McConnell. 1996. Evidence of currently-used pesticides in air, ice, fog, seawater, and surface microlayer in the Bering and Chikchai seas. *Marine Pollut. Bull.* **32**:410-419.
(16) California Department of Pesticide Regulation. 1995. Pesticide Use Report, 1993 Annual, Sacramento, CA, June.
(17) Glotfelty, D.E., C.J. Schomburg, M.M. McChesney, J.C. Sagebiel and J.N. Seiber. 1990. Studies of the distribution, drift, and volatilization of diazinon resulting from spray application to a dormant peach orchard. *Chemosphere* **21**:1301-1314.
(18) Seiber, J.N., B.W. Wilson and M.M. McChesney. 1993. Air and fog deposition residues of four organophosphate insecticides used on dormant orchards in the San Joaquin Valley, California. *Environ. Sci. Technol.***27**:2236-2243.
(19) Jacob, D.J., R.F. Wang and R.C. Flagan. 1984. Fogwater collector design and characterization. *Environ. Sci. Technol.* **31**:827-833.
(20) Collett, Jr., J.L., B.C. Daube, Jr., J.W. Munger and M.R. Hoffmann. 1990. A comparison of two cloudwater/fogwater collectors: The rotating arm collector and the Caltech active strand cloudwater collector. *Atmos. Environ.* **24A**:1684-1692.
(21) Daube, Jr., B.C., R.C. Flagan and M.R. Hoffmann. 1987. Active cloudwater collector. United States Patent No. 4,697,462.
(22) Glotfelty, D.E., J.N. Seiber, and L.A. Liljedahl. 1987. Pesticides and other organics in fog. In *Measurement of Toxic Air Pollutants*, Proceedings of the 1986 EPA/APCA Symposium. Air Pollution Control Association/EPA, Raleigh, NC. April 27-30, 1986. pp. 168-175.

(23) Mallant, R.K.A.M. and G.P.A. Kos. 1990. An optical device for the detection of clouds and fog. *Aerosol Sci. Technol.* **13**:196-202.
(24) Bidleman, T.F. and C.E. Olney. 1975. Long-range transport of toxaphene insecticide in the atmosphere of the western North Atlantic. *Nature* **257**:475.
(25) Thomas, T.C. and J.N. Seiber. 1974. Chromosorb 102, an efficient medium for trapping pesticides from air. *Bull. Environ. Contamin. Toxicol.* **12**:17-25.
(26) Wehner, T.A., J.E. Woodrow, Y.-H. Kim and J.N. Seiber. 1984. Multiresidue analysis of trace organic pesticides in air. In: *Identification and Analysis of Organic Pollutants in Air*, L.H. Keith (Ed.), Butterworth Publishers, Woburn, MA. pp. 273-290.
(27) Seiber, J.N., D.E. Glotfelty, A.D. Lucas, M.M. McChesney, J.C. Sagebiel, and T.A. Wehner. 1990. A multiresidue method by high performance liquid chromatography-based fractionation and gas chromatographic determination of trace levels of pesticides in air and water. *Arch. Environ. Contamin. Toxicol.* **19**:583-592.
(28) Spear, R.C., W.J. Popendorf, J.T. Leffingwell and D. Jenkins. 1975. Parathion residues on citrus foliage. Decay and composition as related to worker hazard. *J. Agric. Food Chem.* **23**:808-810.
(29) Woodrow, J.E., J.N. Seiber, D.G. Crosby, K.W. Moilanen, C.J. Soderquist and C. Mourer. 1977. Airborne and surface residues of parathion and its conversion products in a treated plum orchard environment. *Arch. Environm. Contam. Toxicol.* **6**:175-191.
(30) Henderson, J.D., J.T. Yamamoto, D.M. Fry, J.N. Seiber and B.W. Wilson. 1994. Oral and dermal toxicity of organophosphate pesticides in the domestic pigeon (Columba livia). *Bull. Environ. Contam. Toxicol.* **52**:633-640.
(31) Suntio, L.R., W.Y. Shiu, D. Mackay, J.N. Seiber, and D. Glotfelty. 1988. Critical review of Henry's Law constants for pesticides. *Rev. Environ. Contam. Toxicol.* **103**:1-59.
(32) Glotfelty, D.E., M.S. Majewski and J.N. Seiber. 1990. Distribution of several organophosphorus insecticides and their oxygen analogues in a foggy atmosphere. *Environ. Sci. Technol.* **24**:353-357.
(33) Schomburg, Charlotte J., Dwight E. Glotfelty and James N. Seiber. 1991. Pesticide occurrence and distribution in fog collected near Monterey, California. *Environ. Sci. Technol.* **25**:155-160.
(34) Sagebiel, J.C. and J.N. Seiber. 1993. Studies on the occurrence and distribution of wood smoke marker compounds in foggy atmospheres. *Environ. Toxicol. Chem.* **12**:813-822.
(35) Harder, H.W., E. Christenson, J.R. Mathews and T.F. Bidleman. 1980. *Estuaries* **3**:142-147.
(36) Ligocki, M.P., C. Luenberger and J.F. Pankow. 1985. *Atmos. Environ.* **19**:1609-1617.
(37) Pankow, J.F., L.M. Isabelle and W.E. Asher. 1984. Trace organic compounds in rain. 1. Sampler design and analysis by adsorption/thermal desorption (ATD). *Environ. Sci. Technol.* **18**:310-318.
(38) Gill, P.S., T.E. Graedel and C.J. Weschler. 1983. Organic films on atmospheric

aerosol particles, fog droplets, cloud droplets, raindrops, and snowflakes. *Revs. Geophys. Space Physics* **21**:903-920.
(39) Chiou, C.T., R.L. Malcolm, T.I. Brinton and D.E. Kile. 1986. Water solubility enhancement of some organic pollutants and pesticides by dissolved humic and fuluic acids. *Environ. Sci. Technol.* **20**:502-508.
(40) Capel, P.D., R. Gunde, F. Zürcher and W. Giger. 1990. Carbon speciation and surface tension of fog. *Environ. Sci. Technol.* **24**:722-727.
(41) Goss, K-U. 1994. Predicting the enrichment of organic compounds in fog caused by adsorption on the water surface. *Atmos. Environ.* **28**:3513-3517.
(42) Perona, M.J. 1992. The solubility of hydrophobic compounds in aqueous droplets. *Atmos. Environ.* **26**:2549-2533.
(43) Valsaraj, K.T. 1988. On the physico-chemical aspects of partitioning of non-polar hydrophobic organics at the air-water interface. *Chemosphere* **17**:875-887.
(44) Valsaraj, K.T., G.J. Thoma, D.D. Reible and L. Thibodeaux. 1993. On the enrichment of hydrophobic organic compounds in fog droplets. *Atmos. Environ.* **27**:203-210.
(45) Hoff, J.T., D. Mackay, R. Gillham and W.Y. Shiu. 1993. Partitioning of organic chemicals at the air-water interface in environmental systems. *Environ. Sci. Technol.* **27**:2174-2180.
(46) Oudiz, D. and A.K. Klein. 1998. *Evaluation of Ethyl Parathion as a Toxic Air Contaminant.* California Department of Food and Agriculture. EH-88-5. June. 179 pp.
(47) Hooper, M.H., P.J. Detrich, C.P. Weisskopf and B.W. Wilson. 1989. Organophosphorous insecticide exposure in hawks inhabiting orchards during winter dormant spraying. *Bull. Environ. Contamin. Toxicol.* **42**:651-659.
(48) Zabik, J.M. and J.N. Seiber. 1993. Atmospheric transport of organophosphate pesticides from California's Central Valley to the Sierra Nevada mountains. *J. Environ. Qual.* **22**:80-90.
(49) McConnell, L.L., J.S. LeNoir, S.K. Datta and J.N. Seiber. 1998. Wet deposition of current-use pesticides in the Sierra Nevada mountain range. *Environ. Toxicol. Chem.* **17**:1908-1916.
(50) Seiber, J.N. and J.E. Woodrow. 1998. Air transport of pesticides. *Revs. Toxicol.* **2**:287-294.

Chapter 22

The Role of Dissolved Organic Matter in Pesticide Transport through Soil

J. Letey[1], C. F. Williams[1], W. J. Farmer[1], S. D. Nelson[1], M. Agassi[2], and M. Ben-Hur[3]

[1]Department of Environmental Science, Soil and Water Unit, University of California, Riverside, CA 92521
[2]Soil Erosion Research Station, Rupin Institute Post, 60960 Israel
[3]Agricultural Research Organization, Volcani Center, P. O. Box 6, Bet Dagan, Israel

Soil derived dissolved organic matter (DOM) can complex with pesticides and facilitate their transport through soil. Soil column leaching experiments showed that, in the absence of preferential flow, the herbicide napropmaide was present in the initial leachate. A stable, non-retarded, complex formed between napropamide and DOM that was capable of transport through soil columns. The presence of a napropamide – DOM complex in the leachate was confirmed using an equilibrium dialysis technique. Batch equilibrium adsorption techniques qualitatively predicted the transport of napropamide in soil. The formation of the napropamide – DOM complex was dependent on a drying event where napropamide was applied to the soil and allowed to dry prior to water application.

Applications of agricultural chemicals over the past century have been identified as a major contributor of organic pesticides in drinking wells and groundwater supplies in the USA (1,2). Potential environmental degradation due to pesticides is high since they are designed to be released directly into the environment. However, the environmental impacts of most pesticides are minimized due to soil processes that reduce their mobility while natural biogeochemical processes detoxify them. Any transport process that results in increased movement of pesticides deep into the profile where the rate of the biogeochemical degradation is reduced increases the threat of groundwater degradation. It was speculated that the rapid flow of relatively small amounts of pesticides may be more perilous to groundwater than the slower-moving major mass due to greater degradation of the slow moving mass (3).

Considerable research has focused on agricultural pesticide use to better understand and predict their environmental impacts and transport in soils (4-6). Various approaches have been used to model the movement of organic chemicals through soil based primarily on the assumption that pesticides move by flowing water and are retarded by adsorptive and dispersive forces caused by movement through the soil matrix (7). However, field observations have revealed that a fraction of some surface-applied pesticides move deeper into the soil profile than would be expected from predictions using these transport models (8-12). It has been observed that 19, 16, and 9% of the applied atrazine [6-chloro-N-ethyl-N'-methylethyl)-1,3,5-triazine-2,4-diamine, CAS: 1912-24-9], napropamide [2(α-naphthoxy)-N,N-diethyl propionamide, CAS: 15299-99-7], and prometryn [2,4-bis(isopropylamino)-6-(methylthio)-s-triazine, CAS: 7287-19-6] moved below depths at which models based on convection, dispersion equations predicted (12). Much of the unpredictable transport of such compounds has been attributed to the rapid flow of the solute through soil macropores (13). A review of the experimental literature has suggested that there is still a lack of knowledge about the mechanisms responsible for rapid chemical transport (10).

Recently faster than predicted transport has been attributed to water flow via preferential pathways that exhibit increased flow velocities compared to the average bulk soil. It is evident from the literature that the presence of preferential flow channels results in pesticide transport to deeper levels than present theory predicts. However, if preferential flow is considered, the distribution of pesticides in the soil profile will not be accurately predicted if the pesticide is transported on a carrier to which the pesticide is adsorbed. Fine mineral or organic particles suspended in and transported by the flowing water and/or dissolved organic matter could serve as carriers.

Agricultural chemicals are adsorbed to soil, to some extent, according to the compound's physiochemical properties. Chemical adsorptive properties are typically determined by simple batch equilibrium procedures performed in the laboratory. Sorption coefficients derived from these procedures are used to determine herbicide retention in soil, to estimate its mobility, and to predict potential groundwater contamination (14). Herbicide distribution between the soil and solution is commonly evaluated using the Freundlich equation

$$C_S = K_f C_L^N \qquad [1]$$

where C_S is the amount of organic sorbed per mass of soil, C_L is the equilibrium solution phase organic concentration, K_f is the adsorption coefficient, and N accounts for the degree of nonlinearity in the sorption isotherm. Miller et al. (1985) demonstrated that K_f was inversely proportional to the aqueous phase solubility of compounds of similar chemical grouping. For many soil-pesticide combinations the sorption isotherm is linear with N = 1 such that Equation [1] is reduced to:

$$C_S = K_D C_L \qquad [2]$$

where K_D is the linear distribution coefficient for a soil.

Soil organic matter is the dominating factor controlling pesticide adsorption to soil (15). Soil organic matter is complex and has been described as a three-dimensional carbon-based network that allows for solute inclusion into intramolecular hydrophobic sites having various sizes and shapes (16). The complexity of this framework has given rise to several proposed mechanisms by which organic compounds associate with organic matter. Two possible mechanisms that bind hydrophobic organic chemicals to soil organic matter are sorption and partitioning (17). Sorption theory postulates that the hydrophobic organics are attached to specific binding sites on soil organic matter. The partition theory involves hydrophobic compounds partitioning into hydrophobic microsites within the organic matter where they are exposed to more favorable thermodynamic conditions than in the aqueous phase. These mechanisms are not mutually exclusive and it is probable that both mechanisms occur simultaneously in soils.

Because of the strong association of organic compounds to soil organic matter, the distribution coefficient K_D is commonly normalized on the basis of the fraction of organic C in soil organic matter according to Equation [3]

$$K_{OC} = \frac{K_D}{f_{OC}} \quad [3]$$

where K_{OC} is the organic carbon distribution coefficient and f_{OC} is the fraction of soil organic C on a mass basis.

Soil organic matter is composed of very complex, non-repeating, polymer like organic molecules ranging from easily decomposed simple sugars to recalcitrant humic substances. Dissolved organic matter (DOM) is the fraction of soil organic matter soluble in water over the range of normal soil pH. Soil derived dissolved organic matter is comprised mostly of the lower molecular weight soil organic fraction (i.e. sugars, amino acids, proteins, etc.), fulvic acid, and some humic acids (18). It should be noted that considerable disagreement as to the definition of 'dissolved' organic matter exists. Some definitions are based on light refraction while others are based on size. For the purpose of the research conducted here an operational definition of any organic carbon that passed through a column is considered 'dissolved' but definitions may vary in the literature that is cited here.

The presence of DOM has been shown to increase the solution phase concentration of organics in natural waters through the formation of stable complexes between the organic and DOM. Solution phase concentration of DDT [1,1,1,-Trichloro-2,2-bis(p-chlorophenyl)ethane, CAS: 50-29-3] at equilibrium with river sediments was 3.75 times higher when DOM was present in the aqueous phase (19). It has also been found that 4 to 6% of soluble DDT was bound to humic acid in natural river waters (20). A number of polychlorinated biphenyls (PCB) congeners as well as other hydrophobic organics were shown to have increased solution phase concentrations in the presence of DOM (21-23). The presence of DOM in natural waters could increase the solution phase concentration of DDT and PCBs 2.2 to 7 times (23). In addition it was found that the higher the water solubility of the PCB congener the less of an effect DOM had on increasing solution phase concentration (23). A linear relationship between increased solution phase concentration and DOM concentration was shown with DDT and PCBs (24). In the same study it was shown that the linearity of the relationship was independent of DOM type but that the magnitude of increased solution phase concentration was greater for humic acids and weakest for fulvic acids. The presence of an organic – DOM complex will increase C_L in both Equations [1] and [2] reducing the immobile phase concentration allowing transport deeper within the profile. Therefore, model predictions based on sorption theory that fails to account for the presence of organic solutes complexed with DOM will underestimate the extent of transport in soil systems.

The presence of an organic – DOM complex can be confirmed in a number of ways. Three common measurement techniques are; 1) decreases in K_D, 2) reduced extraction efficiencies in liquid-liquid partitioning, and 3) maintaining concentration gradients across a dialysis membrane.

Experiments to determine K_D usually involve bringing untreated soil in contact with water containing the organic. The mixture is allowed to equilibrate, the resulting solution phase concentration measured, and the reduction in solution phase concentration is assumed to have been caused by the organic being sorbed to the solid phase. The presence of specific types of DOM in solution have been shown to reduce the K_D of organics to soil solids (4, 25). A reduction of K_D will result in a higher solution phase concentration for a given mass sorbed providing an indirect measure of the complex formed.

Measurements using liquid-liquid partitioning determine if the organic contaminant at a specific concentration has a lower chemical potential in the extracting phase than in the aqueous phase. An interaction between the contaminant and DOM that lowers the aqueous phase chemical potential of the contaminant will result in a lower extraction efficiency. It has been shown that DOM as well as a DOM - pH interaction reduced the extraction efficiency of several chlorinated compounds (26). Results showed that at pH ≤ 7.0, extraction efficiencies were reduced with increasing DOM concentration. After extraction at low pH, base was used to increase the pH of all samples followed by a second extraction. It was found that, for solutions already extracted at low pH, an additional 1 to 45% of applied chemical was extracted by hexane at pH 12 (26). When cholesterol and PCB's were dissolved in distilled and river waters, the presence of DOM reduced the extraction efficiency of dichloromethane in liquid-liquid partitioning experiments (27). Dichloromethane was also used to show that DOM could increase the solubility of C_{12} to C_{26} n-alkanes (28). Hexane extraction efficiency of DDT and DDE [1,1-Dichloro-2,2-bis(p-chlorophenyl)ethylene, CAS: 72-55-9] was also reduced in the presence of DOM (29).

To include the effects of organics complexed with DOM in transport models, the formation of a complex must be predicted and measured. Equilibrium dialysis is a common technique used for detecting organic – DOM complexes. Equilibrium dialysis employs a membrane with size dependant permeability. The sample is placed on one side of the membrane and allowed to come to equilibrium with compound-free solution on the other side. Any increase in concentration on the sample side of the membrane is due to lowered chemical potential, increased solution phase concentration, and the presence of a stable organic – DOM complex. An equilibrium dialysis technique found that ^{14}C labeled napropamide formed a complex capable of overcoming the diffusion gradient across a membrane (30). The membrane chosen had a molecular weight cutoff of 1 000 Daltons allowing free napropamide (molecular weight, 271 Daltons) to pass through the membrane but preventing any DOM or complex that had a molecular weight >1 000 Daltons from passing through. Nine percent of the napropamide was complexed by soil derived humic acid inside the membrane. Others have used the equilibrium dialysis technique and gel iso-electrofocusing to show that napropamide formed a stable complex with soil derived humic acid. It was also shown that the napropamide – DOM complex was formed with two distinctly different humic acids (31). The equilibrium dialysis technique was also used with atrazine as well as napropamide to measure increased solution phase concentration (32).

The degree to which DOM will complex with organic chemicals depends on the nature and source of the DOM. It was found that napropamide had a higher affinity for dissolved humic acids than dissolved fulvic acids (30). Napropamide was found to have a greater association with soil-derived than peat-derived humic acids (33). Sewage-borne particulate matter has been shown to be a more effective adsorbent to organic compounds than river-borne particulate matter (34).

Studies involving the use of sewage wastes suggest that sewage-derived organic matter may enhance herbicide transport. Batch equilibrium studies suggest that sewage sludge may facilitate atrazine transport as decreased adsorption occurred when sewage sludge was present (4). Enhanced transport of DDT was observed in laboratory columns in the presence of suspended solids from sewage effluent (35). Increased mobility of organic pollutants occurred in cropland irrigated with sewage effluent and it was postulated that the cause was the result of DOM acting as a carrier in solution (36). Additionally it was suggested that sewage-derived DOM may have been the cause for enhanced atrazine transport in soil irrigated with secondary effluent (37).

In addition to enhanced movement due to the addition of organic wastes, naturally occurring DOM may also act as a carrier in facilitated transport. Water-dispersible colloids (including organics) fractionated from soil samples with diverse physicochemical characteristics were transported by water through intact soil columns in the laboratory (38, 39). The solution phase concentration and therefore the mobility of DDT (19) and some PCB congeners (40) increased with an increase in DOM content in sediment systems. The mobility of DDT (41) and of toxaphene (CAS: 8001-35-2) (42) in soil was enhanced by the addition of urea and anhydrous ammonia, which raised soil pH, thereby solublizing soil organic matter. Dissolved organic carbon can readily move through soil columns in the absence of preferential flow pathways elucidating DOM as a potential carrier mechanism for herbicide transport (43). Formation of a herbicide-DOM complex may lead to enhanced chemical transport if the complex has a lower adsorption affinity to the soil. Recent efforts have been made to include the effects of macromolecules on the modeling of the transport of organic pollutants in soils (44).

We initiated a laboratory research program to investigate facilitated transport of pesticides through soils. The research was initially directed toward determining if mineral particulates, when mobilized, could facilitate the transport of a pesticide adsorbed to the particulate. The first experiments were designed to create conditions whereby mineral particulates would be transported through soil columns. Napropamide was applied to the surface of dry soil columns and then water was flowed through the columns. The effluent was periodically collected from the bottom of the column and analyzed for napropamide and mineral particulate concentration. Concentrations of both mineral particulates and napropamide were highest in the initial effluent sample collected and progressively decreased in subsequent samples (unpublished data). A positive correlation between mineral particulate concentration and napropamide concentration suggested that, indeed, the mineral particulates could have facilitated napropamide transport. However, in subsequent experiments designed to prevent mineral particulate transport, napropamide was still found in the initial effluent sample and decreased in successive samples. The latter observation did not negate the possibility of facilitated transport by mineral particulates, but indicated that another mechanism was also operative.

Based on information reported in the above literature review and other factors, it was postulated that the observed napropamide transport through our soil columns was facilitated by the formation of a napropamide – DOM complex. The research program was redirected toward verifying and understanding the role of DOM in pesticide transport through soil. This paper summarizes some of the more important findings of this research.

Procedures

Many of the procedural details are reported elsewhere (25, 45 and additional manuscripts submitted for publication). Brief summaries of the procedures are presented below. Napropamide was selected as a model compound to investigate DOM complexation and facilitated transport in soil. It represents a group of polar nonionic herbicides and has an aqueous solubility of 74 mg L^{-1}; vapor pressure of 1.7 X 10^{-7} mm Hg; K$_{oc}$ of 700 L kg^{-1}; and degradation half-life of 70 days (46). It is a selective herbicide used for controlling several grass and broadleaf weeds in various crops.

Analytical grade compound was obtained from Chem Service, Inc., West Chester, PA. Sigma Chemical, St. Louis, MO supplied ^{14}C labeled napropamide. Radiolabled napropamide had a radiochemical (^{14}C-α-napthoxy) purity of 99% (specific activity 9.8 Ci mol^{-1}) napropamide with approximately one percent being other ^{14}C labeled chemicals. High pressure liquid chromatography followed by radio chemical detection showed that all of the ^{14}C impurities were associated with a single retention time corresponding to a compound having a lower water solubility than napropamide

Soils. Soils used in the study were Hanford sandy loam (Typic Xerothent), Domino sandy clay loam (Xerollic Calciorthid), and Tujunga loamy sand (Typic Xeropsamment). These soils were collected from the 0-15 cm layer at various locations in southern California. Studies were also conducted using the surface 18 cm of an Airport silt loam (Typic Natrustalf) collected from two adjacent sites in Davis County, Utah. One soil (SS) had received annual amendments of 137 Mg (dry wt) sewage sludge ha^{-1} for 3 consecutive years. The soil from the second site did not

Table I. Some characteristics of the soils studied.

Soil	Sand	Silt	Clay	CEC (cmol(+) kg^{-1} soil)	Organic Carbon[¥] %	pH
	----%----					
Hanford sandy loam	67.1	25.8	7.1	4.5	0.43	6.1
Domino sandy clay loam	50.5	24.3	25.2	15.4	0.52	8.0
Tujunga loamy sand	82.4	13.9	3.7	2.3	0.71	7.9
SS[‡], (Airport silt loam)	38	32	30	nm[§]	2.72	7.6
NoSS[†], (Airport silt loam)	45	32.5	22.5	nm[§]	2.45	7.0

[†] Airport silt loam
[‡] Sewage sludge added at rate of 137 Mg ha^{-1} y^{-1} for 3 years. Soil collected 2 years following last sludge application.
[¥] Walkley-Black procedure used (Nelson and Sommers, 1982).
[§] Not measured

receive sewage sludge application (NoSS). Soils were collected 2 years following the last sludge application allowing for humification of the applied organic matter. All soils were air dried, sieved to 2.0 mm, and some of their physical properties measured by standard techniques (Table I).

Transport Studies. The Hanford, Domino, and, Tujunga soils were packed into columns to a soil depth of 12 cm (45). Napropamide in acetone was dripped on the soil surface to create an equivalent napropamide application rate of 8 kg ha^{-1}, and the acetone was then allowed to completely evaporate. The columns were then irrigated and the leachate collected for ^{14}C activity and DOM analysis. Napropamide concentrations in the leachates, as determined by ^{14}C activity were confirmed using gas chromatography (GC).

Sewage sludge (SS) and NoSS soils were packed in columns to a soil depth of 30 cm (25). Napropamide in acetone was dripped on the soil surface to create an equivalent napropamide application rate of 9.5 kg ha^{-1}, and the acetone allowed to completely evaporate. Columns were flooded with a constant head and leachate was collected. Leachate samples were analyzed for ^{14}C activity and DOM. Napropamide concentrations in the leachate were determined by liquid scintillation and DOM was measured using ultra-violet promoted persulfate oxidation followed by infrared detection. At the end of all the transport experiments the soils were allowed to drain and each column was sectioned and analyzed for ^{14}C activity.

Sorption Studies. Several sorption studies were conducted (25). The approach used was designed to obtain data that represented pesticide application and movement in soils. In the field pesticides are typically sprayed on the soil surface and dried before rain or irrigation can transport the chemicals into the profile. After water is applied, some of the pesticide is desorbed forming a soil extract. This process will be referred to as an extraction event. As the extract moves down the profile, there is opportunity for the pesticide to be adsorbed from the extract onto the soil surfaces. This process will be referred to as adsorption from the extract.

Extraction Event. Extraction event studies were only done on the SS and NoSS soils. They were conducted by applying napropamide in acetone to both soils and allowing the acetone to dry completely in a dark hood. The acetone treatment was used to simulate field application of napropamide in an emulsifiable concentrate. Napropamide solutions of 50, 25, 10, and 5 mg L^{-1} were prepared in pesticide grade acetone at a ^{14}C-labeled to non-labeled ratio of 1:24 (w/w). Five milliliters of the napropamide solution were applied to 50 g soil in aluminum pans for final herbicide treatment concentrations of 5.0, 2.5, 1.0, and 0.5 mg kg^{-1} soil. The napropamide solution was applied to the soil in a dropwise fashion using an automatic pipette while the soil was stirred continuously for uniform herbicide distribution. The acetone was then allowed to completely evaporate at room temperature in a dark hood.

Soil pretreated with herbicide was placed into a 50-mL Teflon centrifuge tube and filled with napropamide free water. Napropamide free water was prepared by adding sufficient NaCl and CaCl$_2$ such that resulting solution had an EC of 1 dS m^{-1} and a SAR of 2. The final soil to water ratio was 1:2 (w/w). Treatments were replicated 3 times for each soil. Tubes were capped, shaken for 2 h, centrifuged and supernatant extracted and stored in glass vials. The extract solution was analyzed for ^{14}C activity by liquid scintillation and the amount of napropamide retained by the soil determined.

Traditional Adsorption Procedures. In the traditional batch adsorption procedures, untreated soil was immersed in solutions with different concentrations of the chemical to be adsorbed. After equilibrium the concentration of the chemical of interest was measured, and the amount of adsorption was calculated from the difference between initial and final concentrations. This basic procedure was used with three different types of solutions. In one case, referred to as adsorption from the extract, the extract solutions from the above described procedures for the extraction event were used. In another case napropamide was dissolved in water to create the solution. In the third case napropamide free water was applied to untreated soil and then extracted. Napropamide was added to this solution to create different concentrations and then these solutions were used for adsorption measurements with untreated soil. For the latter two procedures, the initial solution concentrations were 50, 25, 10, and 5 mg L^{-1}. For all cases, the solutions were mixed with untreated (herbicide – free) soil in centrifuge tubes with a soil to extract ratio of 1:2 (w/w). Centrifuge tubes were sealed, shaken, centrifuged, the supernatant extracted, and the amount of napropamide adsorbed determined.

Partitioning Between Hexane and Water. Napropamide will partition between hexane and water until the chemical potentials in the two phases are equal. 'Free' napropamide would be partitioned with more than 99% in the hexane phase. Therefore, it was postulated that less than 99% partitioning into hexane would be evidence that a portion of the napropamide was complexed with DOM which would lower its chemical potential.

These studies were performed with extract from the SS and NoSS soils. Napropamide in acetone was added to the soils, dried, and then extracted with water using a 1:2 (w/w) soil to water ratio. In a separate case, napropamide was dissolved in water and the aqueous solution was added to soil. The wet soil was allowed to sit without evaporating for 48 hr. In each case, napropamide was applied at 5 µg g^{-1} soil. The extracts from these two procedures were then separately combined with hexane at a 1:1 (v/v) ratio in separatory funnels. The funnels were rigorously shaken and the two phases allowed to separate. Napropamide concentration was determined in each phase by liquid scintillation and each treatment was replicated 5 times.

Equilibrium Dialysis Membrane Measurements. The presence of a napropamide-DOM complex was evaluated on 27 samples using a dialysis equilibrium technique. The first, third, and last effluent samples from each column in the transport studies were placed inside dialysis tubing with a molecular weight cutoff of 500 Daltons. Dialysis tubing was then placed in centrifuge tubes and bathed in napropamide free water. Tubes were shaken and the outside solution analyzed for ^{14}C activity and replaced with fresh deionized water every eight hours. After 24 hours no ^{14}C activity was measured in the outside solution and at that time samples from inside were analyzed for ^{14}C activity. Any activity remaining inside the dialysis tubing was concluded to be napropamide complexed to DOM.

Results and Discussion

The leachate collected from the columns of all three California soils exhibited initially high napropamide concentrations, as determined by ^{14}C analysis, that decreased with cumulative leachate (Figure 1). The amount of napropamide leached through the soils in the first 2.4 cm of cumulative leachate ranged from 1.5-1.8% of the total applied. This leachate represents a volume of less than 60% of one void volume. Napropamide concentration in the effluent was independently confirmed using gas chromatography (GC). A linear relationship between napropamide measured as ^{14}C activity and that from the GC had a R^2 of 0.98 and a slope of 0.825 such that in all cases ^{14}C activity underestimated the concentration of napropamide as measured by GC.

A trend similar to that of napropamide occurred with DOM with high initial DOM concentrations which decreased with increasing effluent volume (data not shown). Dissolved organic matter concentrations ranged from a high of 745 to a low of 10 mg C L^{-1}.

The distribution of napropamide within the soil at the end of the experiment is shown in Figure 2. For all three soils the napropamide concentration dropped to below detection (approximately 100 ng kg^{-1}) at some point in the profile. Therefore, herbicide which moved entirely through the soil column must have existed in a non-reactive form that did not adsorb to the soil. Napropamide distribution within the soil columns is as expected on the basis of clay content and organic matter with the napropamide being retained nearer the surface for the Domino sandy clay loam. Napropamide recovery in all the soils averaged 95% of the total applied, which gave an average mass balance of 97% for soil and leachates of all the treatments.

Dialysis tubing was used to determine if soil leachate from the soil columns contained a stable napropamide – DOM complex. The equilibrium dialysis technique resulted in 17 to 56% of the total activity in the effluent samples remaining inside the dialysis tubing which provides evidence that a fraction of the applied napropamide formed a stable complex with DOM which could contribute to facilitated transport. It was also observed that in all cases the percentage of activity that remained inside the dialysis tubing increased for the effluents that came out at latter times.

The results from the column transport studies on these three soils clearly identified facilitated transport of a small fraction of the applied napropamide. Similar observations in the field were formerly attributed only to preferential flow. The GC analysis verified that the ^{14}C was associated with napropamide and not other chemical impurities. The experiment was conducted to eliminate preferential flow. The results provided indirect evidence that napropamide – DOM complexes were responsible for the facilitated transport. Detailed studies on the SS and NoSS soils were conducted to verify this hypothesis. These studies included adsorption, solvent partitioning, leaching and equilibrium dialysis procedures.

Extraction event sorption isotherms for the SS and NoSS soils are plotted in Figure 3. A greater amount of napropamide was retained on the pretreated SS soil (K_D = 8.7) than pretreated NoSS soil (K_D = 6.5). Sewage sludge-derived organic matter apparently has a higher affinity for napropamide than natural soil organic matter as the organic matter contents in NoSS (2.45%) and SS (2.72%) soils were similar and yet the difference in the sorption capacity of the SS soil was greater. In contrast to the extraction event results, when extracts from pretreated soils were used, more adsorption occurred on the NoSS than the SS soil. The K_D values were 2.7 for SS and 4.4 for NoSS (Figure 4). The lower adsorption of napropamide in SS may be

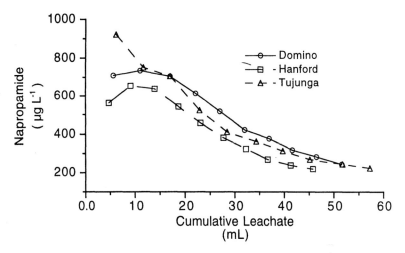

Figure 1. Napropamide concentration in the effluent from the Domino, Hanford, and Tujunga soil columns. Columns were air dry before irrigation began.

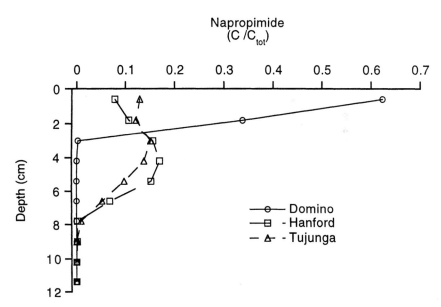

Figure 2. Final napropamide distribution in the Domino, Hanford, and Tujunga soil columns after leaching. Napropamide concentration is reported as the fraction of napropamide applied.

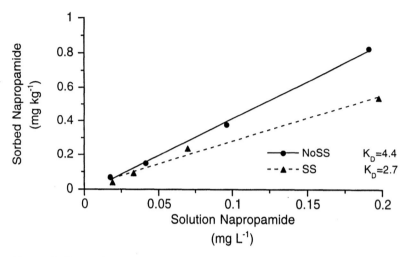

Figure 3. Extraction event isotherms for sewage sludge treated (SS) and native (NoSS) soils. Soils were pretreated with naproamide, dried, equilibrated with water and then napropamide concentration measured in supernatant.

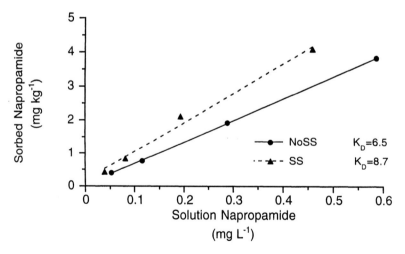

Figure 4. Adsorption isotherms from SS and NoSS soil extract. Napropamide extract from treated soils equilibrated with untreated soil and then napropamide concentration measured in supernatant.

associated with a napropamide-DOM complex in solution or may also be attributed to a co-solute effect by the DOM present in the extract solution.

Further studies found that the reduction in K_D was due to a napropamide – DOM complex. Napropamide adsorption from water resulted in a K_D of 11.9 for the SS soil and 7.4 for the NoSS soil. Adsorption from a solution where napropamide was added to the extract from an untreated soil resulted in a K_D of 11.7 for the SS soil and 6.9 for the NoSS soil. Statistical analysis showed that there was no difference between adsorption from the two different solutions. The presence of DOM in the traditional extract solution did not reduce the sorption of napropamide compared to the traditional water solution and therefore the reduction in K_D is not the result of DOM acting as a co-solute. Rather a drying event, as with the extraction event study was criticle to the formation of the complex.

The partitioning of napropamide between hexane and water also provided evidence of a complex between napropamide and DOM and that drying napropamide on the soil resulted in the formation of a complex capable of reducing partitioning into hexane. When the napropamide was applied to soil and allowed to dry before extraction, 82% of the napropamide was partitioned into hexane from the NoSS and 75% from the SS soil. When napropamide solution was applied to the soils and not allowed to dry before extraction 99% of the napropamide was partitioned into hexane from the NoSS and 98% from the SS soil. These results are consistent with other observations in our studies that drying the napropamide in layered soil columns was important in the formation of the napropamide – DOM complex which led to facilitated transport (manuscript submitted for publication). Therefore, combining napropamide and DOM in solution without drying is not very effective in creating a complex between the two.

Results from the column transport study with the SS and NoSS soils support the theory of facilitated napropamide transport through soil. Napropamide concentrations were highest in the initial leachate and decreased with cumulative leachate (Figure 5). Initial napropamide concentration in the leachate was higher for the SS soil than NoSS. The pattern of DOM concentration in the leachate was similar to napropamide. The highest DOM concentration was found in the initial leachate and rapidly decreased with cumulative leachate (data not shown).

Equilibrium dialysis was used to confirm the presence of a napropamide – DOM complex in the effluent. All effluent samples containing napropamide had activity that remained inside the dialysis tubing. The concentration remaining inside the dialysis tubing was normalized based on the total napropamide released into solution at the soil surface, which was estimated from the solution concentration in the extraction event. A strong positive linear relationship was found between normalized napropamide and DOM concentration remaining inside the dialysis bag for both soils that is indicative of an association between the two.

These results are consistent with the adsorption data and the results of napropamide partitioning between hexane and water. After napropamide was applied to soil, dried, and extracted, adsorption from the extract was less than adsorption from an aqueous napropamide solution. Furthermore the reduction in adsorption was greater in the SS soil than the NoSS soil indicating a stronger complex with DOM from the SS soil. The reduction in adsorption could be attributed to a fraction of the napropamide being complexed in a form that was not adsorbed. In this form, it would be transported through the column as a nonreactive solute and appear in the first leachate. Also partitioning of napropamide into hexane from this extract was reduced suggesting a stable complex. Both the adsorption and partitioning data indicated that there was a higher percentage of complexation in the SS than the NoSS soil and this is consistent with a higher concentration of napropamide in the SS than the NoSS leachate (Figure 5).

The distribution of napropamide in the soil columns following leaching is shown in Figure 6. The majority of napropamide was transported less than 23 cm. Napropamide retention was greater in SS compared to the NoSS soil. The center of mass was transported further in the NoSS soil than in the SS soil. This is consistent

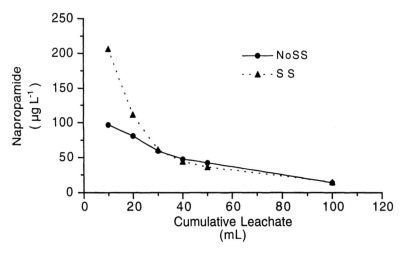

Figure 5. Napropamide concentration in the effluent from SS and NoSS soil columns. Columns were air dry before irrigation began (Adapted with permission from reference 25. Copyright American Society of Agronomy 1998.)

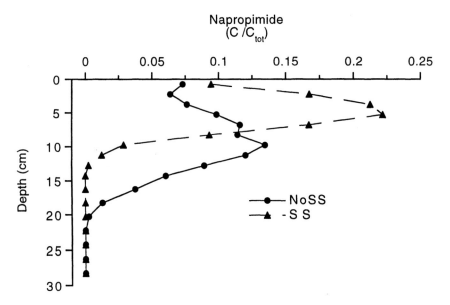

Figure 6. Final napropamide distribution in the SS and NoSS soil columns after leaching. Napropamide concentration is reported as the fraction of napropamide applied (Adapted with permission from reference 25. Copyright American Society of Agronomy 1998.)

with the extraction event adsorption results where the K_D values were significantly lower for NoSS than SS indicating a stronger herbicide affinity to SS than NoSS soil. Therefore, the batch extraction event predicted the transport behavior of the major chemical mass by the soil organic matter and the adsorption from the extract predicted the transport behavior facilitated by DOM.

Conclusion

Batch equilibrium sorption studies were used to qualitatively predict facilitated transport by using an extraction event to simulate the process that extracts the napropamide from a dried treated soil into the solution phase, and adsorption from that extract to simulate adsorption to the solid phase deeper in the profile. An interaction that decreased the chemical potential of napropamide in the aqueous phase was measured by partitioning of napropamide between hexane and water. A drying event was necessary in the formation of a stable DOM – napropamide complex. Allowing the napropamide to dry on the soil led to lower extraction efficiency as well as lower K_D values in the extract adsorption. Facilitated transport of napropamide by DOM was observed in five different soils. In all cases less than 2% of the applied napropamide was transported via facilitation and the major mass of applied napropamide was transported consistent with models based on sorption/desorption theory. The presence of a complex in the leachate was confirmed through equilibrium dialysis.

Acknowledgements. This research was supported by the University of California Kearney Foundation of Soil Science.

Literature Cited.

1. U. S. Environmental Protection Agency (USEPA). Pesticides in ground water: Background document. Office of Groundwater Protection. WH5506, Washington, D.C. **1986**.
2. Zsolnay, A. *Chemosphere*. **1992**, 24:663-669.
3. Utermann, J., E. J. Kladivko, and W. A. Jury. *J. Environ. Qual.* **1990**, 19:707-714.
4. Barriuso, E., U. Baer, and R. Calvet. *J. Environ. Qual.* **1992**, 21:359-367.
5. Clay, S. A. and W. C. Koskinen. *Weed Sci.* **1990**, 38:262-266.
6. Madhun, Y. A., J. L. Young, and V. H. Freed. *J. Envron. Qual.* **1986**, 15:64-68.
7. van Genuchten, M. T. and J. C. Parker. *Soil Sci. Soc. Am. J.* **1984**, 48:703-708.
8. Kladivko, E. J., G. E. van Scoyor, E. J. Monke, K. M. Oates, and W. Posk. *J. Environ. Qual.,* **1991**, 20:264-270.
9. Shipitalo, M. J., M. M. Edwards, W. A. Dick, and L. B. Owens. *Soil Sci. Soc. Am. J.* **1990**, 54:1530-1536.
10. Flury, M. *J. Environ. Qual.* **1996**, 25:25-45.
11. Jury, W. A., H. Elabd, L. D. Clendening, and M. Resketo. Evaluation of pesticide transport screening models under field conditions. In: *Evaluation of Pesticide and Groundwater*. Editors, W. Y. Garner et al. ACS Symposium, **1991**, Series 315, American Chemical Society, Washington, D.C. pp. 384-395.
12. Ghodrati, M., and W. A. Jury. *J. Contam. Hydrol.* **1992**, 11:101-125.
13. Grochulska, J., and E. J. Kladivko. *J. Environ. Qual.* **1994**, 23:498-507.
14. Barriuso, E., D. A. Laird, W. C. Koskinen, and R. H. Dowdy *Soil Sci. Soc. Am. J.* **1994**, 58:1632-1638.
15. Stevenson, F. J. *J. Environ. Qual.* **1972**, 1:333-343.
16. Shulten, H. R., and M. Schnitzer. *Naturwiss.* **1995**, 82:487-498.
17. Chiou, C. T. Roles of organic matter, minerals, and moisture in sorption of nonionic compounds and pesticides by soil. *In* Humic Substances in Soil and Crop Sciences. **1990**, ASA and SSSA, 677 South Segoe Road, Madison, WI, 53711, USA. pp. 111-161.

18. Paul. E. A., and F. E. Clark. *Soil Microbiology and Biochemistry*. **1989**, Academic Press, Inc, New York.
19. Caron, G., I. H. Suffet, and T. Belton. *Chemosphere* **1985**, 14:993-1000.
20. Kulovaara, M. *Chemosphere.* **1993**, 27:233-2340.
21. Wershaw, R. L., P. J. Burcar, and M. C. Goldberg. *Envron. Sci. Technol.* **1969**, 3:271-273.
22. Jota, M. A. T., and J. P. Hassett. *Envron. Toxicol Chem.* **1991**, 10:483-491.
23. Chiou, C. T, R. L. Malcolm, T. I. Brinton, and D. E. Kile. *Environ. Sci. Technol.* **1986**, 20:502-508.
24. Chiou, C.T., D.E. Kile, T.I. Brinton, R.L. Malcolm, and J.A. Leenheer. *Environ. Sci. Techol.* **1987**, 21(12):1231-1234.
25. Nelson, S. D., J. Letey, W. J. Farmer, C. F. Williams, and M. Ben-Hur. *J. Environ. Qual.* **1998** 27(5): 1194- 1200.
26. Maguire, R. J., S. P. Batchelor, and C. A. Sullivan. *Envron. Toxicol Chem.* **1995**, 14:389-393.
27. Hassett, J. P. and M. A. Anderson. *Emvron. Sci. Technol.* **1979**, 13:1526-1529.
28. Maguire, R. J., R. J. Tkacz, and S. P. Batchelor. *Environ. Toxicol. Chem.* **1993**, 12:805-811.
29. Driscoll, M.S., J.P. Hassett, C.L. Fish, and S. Litten. *Environ. Sci. Technol.* **1991**, 25:1432-1439.
30. Lee, D.-Y., and W. J. Farmer. *J. Environ. Qual.* **1989**, 18:468-474.
31. Liu, R., C. E. Clapp, M. H. B. Hayes, and U. Mingelgrin. Stability of complexes formed by the herbicide napropamide and soluble humic acids, In *Humic substances and organic matter in soil and water environments: characterization, transformations and interactions.* C. E. Clapp et al. (eds.) IHSS, Birmingham, UK. **1996**, pp. 305-315.
32. Clapp, C.E., U. Mingelgren, R. Liu, H. Zhang, and M.H.B. Hayes. *J. Environ. Qual.* **1997**, 26:1277-1281.
33. Clapp, C. E., R. Liu, , M. H. B. Hayes, and U. Mingelgrin (eds). Stability of complexes formed by the herbicide napropamide and soluble humic acids,. In *Humic substances and organic matter in soil and water environments: characterization, transformations and interactions.* **1996**, pp.305-315IHSS, Birmingham, UK.
34. Hassett, J. P. and M. A. Anderson. *Water Res.* **1982**, 16:681-686.
35. Vinten, A. J. A., B. Yaron, and P. H. Nye. *J. Agric. Food. Chem.* **1983**, 31:662-664.
36. Muszkat, L., D. Raucher, M. Magaritz, D. Ronen, and A.J. Amiel. *Groundwater.* **1993**, 31(4):556-565.
37. Graber, E. R, Z. Gerstl, E. Fischer, and U. Mingelgrin. *Soil Sci. Soc. Am. J.* **1995**, 59:1513-1519.
38. Seta, A. K, and A. D. Karathanasis. *Soil. Sci. Soc. Am. J.* **1997**, 61:604-611.
39. Seta, A. K, and A. D. Karathanasis. *Soil. Sci. Soc. Am. J.* **1997**, 61:612-617.
40. Gschwend, P. M., and S. Wu. *Environ. Sci. Technol.* **1985**, 19:90-96.
41. Ballard, T. M. *Soil Sci. Soc. Amer. Proc.* **1971**, 35:145-147.
42. Smith, S., and G. H. Willis. *Environ. Toxicol. Chem.* **1985**, 4:425-434.
43. Dunnivant, F. M., D. M. Jardine, D. L. Taylor, and J. F. McCarthy. *Soil Sci. Soc. Am. J.* **1992**, 56:437-444.
44. Enfield, C. G., G. Bengtsson, and R. Lindquist. *Environ. Sci. Technol.* **1989**, 23:1278-1286.
45. Williams, C. F., J. Letey, W. J. Farmer, S. D. Nelson, and M. Ben-Hur. Facilitated transport of napropamide by dissolved organic matter through soil columns. *Soil Sci. Soc. Am. J.* **1999** (in press).
46. Wauchope, R. D., T. M. Buttler, A. G. Hornsby, P. W. M. Augustijn Beckers, and J. P. Burt. *Rev. Environ. Contain. Toxicol.* **1992**, 123:1-155.

AUTHOR INDEX

Author Index

Agassi, M., 347
Anderson, R. D., 95
Arora, K., 272
Baker, David B., 46
Baker, J. L., 20, 272
Battaglin, W. A., 248
Bengston, R. L., 159
Ben-Hur, M., 347
Brightman, S. K., 126
Bucks, D. A., 232
Capel, Paul D., 172
Chirnside, A. E. M., 80
Crosby, D. G., 65
Dubrovsky, Neil M., 306
Durst, R. A., 135
Farmer, W. J., 347
Forney, D. R., 20, 126, 135
Foster, Gregory D., 115
Fouss, J. L., 159
Gronberg, Jo Ann M., 306
Hall Jr., L. W., 95
Hallberg, G. R., 217
Hatfield, J. L., 201, 232
Havens, P. L., 288
Hochstedler, M. E., 217
Horton, M. L., 232
Humiston, M. C., 126
Johnson, W. E., 95
Kanwar, R. S., 185
Karlen, D. L., 185
Kratzer, Charles R., 306
Kuivila, Kathryn M., 306
Larabee-Zierath, D., 217
Lee, J. M., 65, 135

Letey, J., 347
Lippa, Katrice A., 115
Meek, D. W., 159
Mickelson, S. K., 272
Misra, A. K., 272
Moorman, T. B., 185
Nelson, S. D., 347
Panshin, Sandra Y., 306
Peter, C. J., 20
Pfeiffer, R. L., 248
Poletika, N. N., 288
Pruegger, J. H., 201
Rankin, C., 20
Rice, C. P., 95
Richards, R. Peter, 46
Ritter, W. F., 80
Robb, C. K., 288
Ross, Lisa, 2, 65
Sauer, T. J., 201
Scarborough, R. W., 80
Scoggin, K. D., 146, 248
Seiber, James N., 323
Smith, R. D., 288
Southwick, L. M., 159
Spittler, Terry, 2, 20, 126, 135
Steenhuis, T. S., 80
Steffin, D., 20
Steinheimer, Thomas, 2, 146, 248
Strahan, J., 20
Williams, C. F., 347
Willis, G. H., 159
Woodrow, James E., 323
Zhang, Hua, 172

SUBJECT INDEX

Subject Index

A

Acetanilide herbicides
  annual concentrations and losses from watershed B and B-R, 27t, 28t
  calculating pesticide loading at German Branch, 108
  concentrations relative to storm hydrograph at German Branch, 106f
  concern for aquatic and toxicological risk, 21–22
  corn and soybean farmers, 22
  enzyme immunoassay, 130–131
  gas chromatography with electron capture detection (GC–ECD), 129–130
  monthly time weighted mean concentration at German Branch, 107f
  multi-residue scheme coupling solid phase extraction and GC–ECD, 127
  receipt to analyses intervals and storage temperatures, 131t
  storage stability, 131–132
  See also Watershed monitoring in sustainable agriculture studies
Acetochlor
  detection in all sample types, 222–223
  detection in rainwater at four sites, 226f
  usage in agricultural watershed study, 100t
  See also Pesticides in air and precipitation
Agricultural operations
  Chesapeake Bay, 8–9
  east coast of United States, 7
Agricultural pesticides
  degradation in soil environment, 3
  nonpoint-source runoff, 95–96
Agricultural runoff, scale of studies, 96
Agricultural watershed study
  acetanilide herbicides and dissolved organic carbon (DOC) concentrations relative to storm hydrograph at German Branch, 106f
  atrazine and metolachlor concentrations in bulk atmospheric deposition samples within German Branch watershed, 111f
  deethylatrazine/atrazine ratio (DAR), 108
  dissolved inorganic nitrogen determination, 102
  dissolved nitrogen, 108–109
  DOC analysis, 102
  ecological risk assessment of atrazine, 108
  enzyme linked immunosorbent assay (ELISA) for 2,4-D, 102
  estimated pesticide usage in watershed, 100t
  gas chromatography/mass spectrometry for pesticides in study, 101–102
  German Branch discharge and rainfall in watershed during study, 103f
  German Branch watershed characteristics, 99t
  high pesticide concentrations in German Branch, 104
  hydrogeomorphology of watersheds of

Chesapeake Bay, 96
land use in German Branch watershed, 98*f*
monthly flow weighted mean concentrations (FWMC) equation, 108
monthly time weighted mean concentrations (TWMC) estimating pesticide exposure to aquatic life, 104, 108
monthly TWMC for triazine and acetanilide herbicides at German Branch, 107*f*
nonpoint-source runoff, 95–96
pesticide levels in sediment and sediment pore water at German Branch, 110*f*
pesticide loading at station B for triazines and acetanilide herbicides, 108
pesticides in rainfall, 109
pesticides in surface water, 102, 104, 108
pesticide use on Maryland's eastern shore as percentage of statewide use, 97*f*
rainfall and hydrology, 102
sample analysis, 101–102
scale of agricultural runoff studies, 96
sediment and pore water, 109
semi-volatile pesticide extraction from sediment and water, 101
site geography and land use, 96, 99
stream hydrology and sample collection method, 99–101
surface water–DOC, 109
triazine and 2,4-D herbicide concentrations relative to storm hydrograph at German Branch, 105*f*
water quality criteria and guidelines, 102, 104
water sample analysis for chlorinated phenoxy herbicide 2,4-D, 104
Agrochemical movement
agricultural operations, 7
Chesapeake Bay and eastern shore, 8–11
Midwestern plains, 11–12
minimizing off-field displacement, 12
new agricultural ethic, 14–15
pesticide detection in surface and ground water, 14
presenters and program titles, 15*t*, 16*t*, 17*t*
San Joaquin Valley of California, 12–14
water quality, 5–7
watershed concept, 7–8
Agrochemical runoff patterns, scale effects in assessment, 7–8
Agrochemicals
mobilization in snowmelt, 150–151
new agricultural ethic, 14–15
potential to damage nontarget environmental settings, 6–7
use causing concern for soil and ground water contamination, 135–136
Airborne pesticides
description, 323
processes for removal, 323–324
removal, 14
See also Pesticide transport and fate in fog
Alachlor
comparing calculated and measured enrichment factors, 341*t*
concern for aquatic and toxicological risk, 21–22
detection in ground water, 88
detection in ground water for May to September, 85*t*
effects of tillage on mean annual losses of herbicides in subsurface drainage from plots near Nashua, Iowa, 192*t*
field experiment for leaching, 81
maximum concentration at Chesapeake Farms, 34*t*
maximum runoff concentrations, 39–40

preemergence application after corn
  planting, 83–84
rainfall simulation experiment, 83–84,
  91–93
riparian forest buffer system/natural
  rainfall, 280–281
soil pore water concentrations in
  suction and passive lysimeter, 41t
tillage effects on herbicide loss, 193
usage in agricultural watershed study,
  100t
See also Ground water contamination
  on sandy soils; Vegetated filter strips
  (VFS); Watershed fluxes of
  pesticides; Watershed monitoring in
  sustainable agriculture studies
Atlantic Coast
agricultural operations, 7
See also Agricultural watershed study
Atmospheric transport
human risk assessment, 12
See also Pesticides in air and
  precipitation
Atrazine
annual chemographs for Maumee
  River and Rock Creek, 48, 51f
annual herbicide concentrations and
  losses from watershed A and A2, 26t
annual herbicide concentrations and
  losses from watershed B and B-R,
  27t, 28t
annual losses in drainage waters as
  percentage of applied herbicide, 191f
annual losses in subsurface drainage in
  corn and corn-soybean production
  systems, 190t
application rates on two plot sizes,
  160t
buffer strips-tile outlet terrace/natural
  rainfall, 281–282
comparing calculated and measured
  enrichment factors, 341t
comparing chemographs for suspended
  solids, atrazine, and nitrate, 60, 61f

concentration exceedency curves, 48,
  54f, 55
concentration in bulk atmospheric
  deposition samples within German
  Branch watershed, 111f
concentration in runoff as function of
  soil concentration, 163, 165, 167
concentration in runoff on two plot
  sizes, 161–162
concentration in soil and runoff for
  two plot sizes, 164t
concentration in surface water runoff
  from system A, 27f, 36
concentrations and water flow rate
  from subsurface drainage, 187f
deethylatrazine/atrazine ratio (DAR),
  108
detection in all sample types, 222–223
detection in ground water, 88
detection in ground water for May to
  September, 86t
detection in rainwater at four sites,
  225f
detection without application during
  study, 42–43
determinations in snowmelt runoff,
  149–150
distribution between fogwater and
  interstitial air in California, 333t
distribution of annual maximum
  concentration, 48, 52f
distribution within lowest quartile of
  values during May through August,
  48, 53f
ecological risk assessment, 108
effect of different microbial
  populations, 242
effects of tillage on mean annual losses
  of herbicides in subsurface drainage
  from plots near Nashua, Iowa, 192t
example chemical in GLEAMS model,
  175–176
fate based on interval from application
  to storm event, 90t

fate of ^{14}C-atrazine in manured and untreated Floyd soil, 196, 198t
field experiment for leaching, 81
losses at field and watershed scales, 186, 187f
losses in runoff and subsurface drainage, 188–189
magnitude of herbicide loss, 189–190
maximum concentration at Chesapeake Farms, 34t
modeling study using GLEAMS model, 89–91
monitoring data from tributaries, 10
oats strip/natural rainfall, 279
percent total load versus percent total storm discharge for suspended solids, atrazine, and nitrate, 60, 62f
pesticides in rainfall, 109
potential loss in runoff as percentage of amount applied for Fillmore County based on GLEAMS, 179f
potential loss in runoff as percentage of amount applied for Forestville Creek springshed, 181f
predicted loss in runoff to surface water as function of soil type and slope, 177f
predictions using transport models, 347–348
preemergence application after corn planting, 83–84
properties, 160t
rainfall simulation experiment, 83–84
riparian forest buffer system/natural rainfall, 280–281
soil pore water concentrations in suction and passive lysimeter, 41t
statistical distribution of predicted potential loss in runoff to surface-water, 178t
tillage effects on herbicide loss, 193
time dependent desorption-phase distribution coefficients from four watersheds, 155t
usage in agricultural watershed study, 100t
vegetated filter strips (VFS)/simulated rainfall, 275–277
VFS/natural rainfall, 280
water quality concerns, 21
See also GLEAMS (ground water loading effects of agricultural management systems) model; Ground water contamination on sandy soils; Pesticides in air and precipitation; Runoff; Snowmelt runoff; Vegetated filter strips (VFS); Watershed fluxes of pesticides; Watershed monitoring in sustainable agriculture studies

B

Bensulfuron methyl
chemical properties and environmental characteristics, 253t
structure, 252f
See also Herbicides in surface streams and ground water
Buffer strips. *See* Vegetated filter strips (VFS)

C

California's Central Valley
watershed regions, 8
See also Pesticide transport and fate in fog; Rice herbicide discharges
Ceriodaphnia dubia
concentration showing toxicity, 318, 320
toxicity testing, 289, 307
See also Organophosphorus insecticides
Chesapeake Bay
agricultural operations, 7–9
assessment of agrochemical movement, 8–11
breeding ground for migratory marine species and waterfowl, 8

collection of delicate ecosystems, 115
contribution of western shore
tributaries, 10
monitoring data for metolachlor and
atrazine from tributaries, 10
studies on larger areas, 10
watershed diversity, 9
See also Agricultural watershed study;
Watershed fluxes of pesticides
Chesapeake Farms. *See* Pesticide runoff
and leaching from farming systems;
Watershed monitoring in sustainable
agriculture studies
Chloride
percent of total load account for by
1% of time with highest loading
rates, 60t
See also Watershed size
Chlorimuron ethyl
annual herbicide concentrations and
losses from watershed B and B-R,
27t, 28t
annual herbicide concentrations and
losses from watershed C, 29t
chemical properties and environmental
characteristics, 253t
maximum concentration at Chesapeake
Farms, 34t
maximum runoff concentrations, 39–40
preemergence and postemergence
treatment, 39
preemergence and postemergence
uses, 22
soil pore water concentrations in
suction and passive lysimeter, 41t
structure, 251f
See also Herbicides in surface streams
and ground water
Chloroacetamides, concern for aquatic
and toxicological risk, 21–22
Chloroacetanilides, results on
reconnaissance duplicates, 266, 268t
Chlorothalonil, usage in agricultural
watershed study, 100t
Chlorpyrifos

aqueous phase enrichment factors for
distribution between air and
fogwater, 335t
chemical applications, 293, 295t, 296t,
297t
chemical concentrations, 293, 298–300
comparing calculated and measured
enrichment factors, 341t
daily average air concentrations during
intense dormant spraying, 326, 327f
distribution between fogwater and
interstitial air in California, 333t
monthly pesticide application,
concentration in samples, and
discharge for San Joaquin River,
317f
oxon/thion ratios in fogwater using
active strand fogwater sampler, tree
drip, and nonfog air, 337t
temporal variation in occurrence and
concentration, 316
turf strip/simulated rainfall, 277–278
usage in agricultural watershed study,
100t
worst case exposure analysis for
Orestimba Creek contributions, 303
See also Organophosphorus
insecticides; Pesticide transport and
fate in fog; Vegetated filter strips
(VFS)
Chlorsulfuron
chemical properties and environmental
characteristics, 253t
structure, 251f
See also Herbicides in surface streams
and ground water
Collectors, fogwater, 328, 330f, 331f
Colusa basin drain
peak molinate and thiobencarb
concentrations, 75f
watershed monitoring, 74
See also Rice herbicide discharges
Conferences, Management Systems
Evaluation Areas (MSEA), 241–242
Conservation tillage, reducing mass of
water and sediment carriers, 273

Corn-soybean production systems. *See* Herbicide transport in subsurface drainage water

Cyanazine
  annual herbicide concentrations and losses from watershed A and A2, 26*t*
  annual herbicide concentrations and losses from watershed B and B-R, 27*t*, 28*t*
  buffer strips-tile outlet terrace/natural rainfall, 281–282
  concentration in surface water runoff from system A, 37*f*
  detection in all sample types, 222–223
  detection without application during study, 42–43
  effects of tillage on mean annual losses of herbicides in subsurface drainage from plots near Nashua, Iowa, 192*t*
  maximum concentration at Chesapeake Farms, 34*t*
  monthly pesticide application, concentration in samples, and discharge for San Joaquin River, 317*f*
  preemergence application after corn planting, 83–84
  rainfall simulation experiment, 83–84, 91–93
  soil pore water concentrations in suction and passive lysimeter, 41*t*
  temporal variation in occurrence and concentration, 316
  tillage effects on herbicide loss, 193
  usage in agricultural watershed study, 100*t*
  vegetated filter strips (VFS)/simulated rainfall, 276–277
  VFS/natural rainfall, 280
  water quality concerns, 21
  See also Pesticides in air and precipitation; Vegetated filter strips (VFS); Watershed fluxes of pesticides; Watershed monitoring in sustainable agriculture studies

D

2,4-D
  concentrations relative to storm hydrograph at German Branch, 105*f*
  2,4-D in rainwater at four sites, 226*f*
  detection in air and precipitation, 222–223
  grassed waterway/simulated rainfall, 274–275
  turf strip/simulated rainfall, 277–278
  usage in agricultural watershed study, 100*t*
  See also Pesticides in air and precipitation; Vegetated filter strips (VFS)

DDT. *See* 1,1,1-Trichloro-2,2-bis(*p*-chlorophenyl)ethane (DDT)

Deethylatrazine
  deethylatrazine/atrazine ratio (DAR), 108
  usage in agricultural watershed study, 100*t*

Degradation
  airborne pesticide removal, 14
  airborne pesticide removal process, 323–324
  See also Pesticide transport and fate in fog

Deisopropylatrazine, usage in agricultural watershed study, 100*t*

Delaware, infiltration transport mechanism through sandy soil, 9

Delmarva Peninsula
  agricultural operations, 7
  See also Ground water contamination on sandy soils

Deposition in precipitation, human risk assessment, 12

Diazinon
  aqueous phase enrichment factors for distribution between air and fogwater, 335*t*
  chemical applications, 293, 295*t*, 296*t*, 297*t*

chemical concentrations, 293, 298–300
comparing calculated and measured enrichment factors, 341*t*
daily average air concentrations during intense dormant spraying, 326, 327*f*
distribution between fogwater and interstitial air in California, 333*t*
monthly pesticide application, concentration in samples, and discharge for San Joaquin River, 317*f*
oxon/thion ratios in fogwater using active strand fogwater sampler, tree drip, and nonfog air, 337*t*
temporal variation in occurrence and concentration, 316
toxicity to *Ceriodaphnia dubia*, 318, 320
transport by storm runoff, 316, 318–320
usage in agricultural watershed study, 100*t*
worst case exposure analysis for Orestimba Creek contributions, 303
See also Organophosphorus insecticides; Pesticide transport; Pesticide transport and fate in fog; Watershed fluxes of pesticides
Dicamba
turf strip/simulated rainfall, 277–278
See also Vegetated filter strips (VFS)
Dichlorprop
detection in all sample types, 222–223
See also Pesticides in air and precipitation
Discharges. *See* Rice herbicide discharges
Dissolved organic carbon (DOC)
analysis method, 102
concentrations relative to storm hydrograph at German Branch, 106*f*
levels in surface water, 109
Dissolved organic matter (DOM) in pesticide transport through soil

adsorption isotherms from sewage sludge treated (SS) and native (NoSS) extracts, 354, 356*f*
adsorption to soil by physiochemical properties, 348
characteristics of soils, 352*t*
confirming presence of organic–DOM complex, 349
degree of DOM complexation with organic chemicals, 350
DOM acting as carrier in facilitated transport, 350–351
equilibrium dialysis membrane measurements, 353–354
evidence of complex between napropamide and DOM, 357
evidence supporting facilitated napropamide transport through soil, 357, 359
experimental procedures, 351–354
extraction event isotherms for SS and NoSS soils, 354, 356*f*
extraction event studies, 352–353
facilitated transport of fraction of applied napropamide, 354
final napropamide distribution in Domino, Hanford, and Tujunga soils after leaching, 355*f*
final napropamide distribution in SS and NoSS soil columns after leaching, 358*f*
Freundlich equation for herbicide distribution between soil and solution, 348
including effects of organic–DOM complex in transport models, 350
leachate from columns of three California soils, 354
liquid-liquid partitioning measurements, 349–350
minimizing environmental impacts of pesticides, 347
modeling movement of organic chemicals through soil, 347–348

napropamide concentration in effluent from Domino, Hanford, and Tujunga soils, 355f
napropamide concentration in effluent from SS and NoSS soil columns, 358f
organic carbon distribution coefficient, 348–349
partitioning between hexane and water, 353
presence of DOM increasing solution phase concentration of organics, 349
sewage-derived organic matter enhancing transport, 350
soil organic matter controlling pesticide adsorption to soil, 348
soils under study, 351–352
sorption studies, 352–354
traditional adsorption procedures, 353
transport studies, 352
Dry deposition
airborne pesticide removal, 14
airborne pesticide removal process, 323–324
See also Pesticide transport and fate in fog

E

Eastern shore
coastal plain sections affording agrochemical transport, 9
See also Chesapeake Bay
Education programs, Management Systems Evaluation Areas (MSEA), 237–241
Enantiomers, organic pesticides, 4
Endosulfan (α,β), usage in agricultural watershed study, 100t
Enrichment factors. See Pesticide transport and fate in fog
Environmental fate
understanding for pesticide usage, 2–3
See also Pesticide transport and fate in fog

Environmentalists, new agricultural ethic, 14–15
Enzyme linked immunosorbent assays (ELISA), complement to multi-residue scheme, 127–128
EPTC, monthly pesticide application, concentration in samples, and discharge for San Joaquin River, 317f
Esfenvalerate, gas chromatography with electron capture detection (GC–ECD), 130
Estimating pesticide losses in runoff. See GLEAMS (ground water loading effects of agricultural management systems) model

F

Fall Line Toxics Monitoring Program (FLTMP)
focus, 116
primary goal, 115–116
See also Watershed fluxes of pesticides
Farmers, new agricultural ethic, 14–15
Farming systems
evaluation of impact on water quality, 237, 239
See also Pesticide runoff and leaching from farming systems
Fate. See Pesticide transport and fate in fog
Field-scale process-based model. See GLEAMS (ground water loading effects of agricultural management systems) model
FILIA. See Flow injection liposomal immunoassay (FILIA)
Fillmore County, Minnesota. See GLEAMS (ground water loading effects of agricultural management systems) model
Fish protection, California's Central Valley, 8
Flow injection liposomal immunoassay (FILIA)

analysis of real environmental samples, 136
apparatus, 136, 138
detection of imazethapyr in water samples, 139, 141
development for imazethapyr, 128
dose-response relationship for imazethapyr from spiked water samples, 139, 140f
experimental materials, 136
FILIA assay, 139
FILIA data summary for standard curve plus field samples, 142t
FILIA method, 138
FILIA run of actual field sample, 143f
imazethapyr, 135, 137f
preparation of capillary column, 138–139
production of liposomes, 138
purification of monoclonal antibody, 138
schematic diagram of FILIA system, 137f
solvent extraction method for FILIA, 129
typical analytical run, 140f
water sample analysis, 139
See also Imazethapyr; Watershed monitoring in sustainable agriculture studies
Flumetsulam
chemical properties and environmental characteristics, 253t
structure, 250f
See also Herbicides in surface streams and ground water
Flux-gradient theory
bulk Richardson number, 205
diabatic stability correction functions for momentum and sensible heat flux, 204
gradient profile for metolachlor, 203
theoretical considerations, 203–205
See also Volatilization

Fog. *See* Pesticide transport and fate in fog
Freundlich equation, herbicide distribution between soil and solution, 348
Fungicides
production and consumption, 4t
registered chemicals, 4–5

G

GENEEC (generic estimated environmental concentration), model predicting surface water runoff, 34–35
Geographic information system (GIS). *See* GLEAMS (ground water loading effects of agricultural management systems) model
German Branch watershed characteristics, 99t
German Branch discharge and rainfall in watershed during study, 103f
See also Agricultural watershed study
GLEAMS (ground water loading effects of agricultural management systems) model
combined use of GLEAMS and geographic information system (GIS) for predictions of potential losses, 182
combining model with GIS for interpreting data generated in surface and ground water monitoring programs, 182
corn crop and atrazine and permethrin, 175–176
data preparation and model calculations, 175–176
description of study area, 173–174
estimating potential loss in pesticide runoff, 173, 176
field-size area design, 176

general GLEAMS model results, 176–177
hydrology component, 175
limitations in combining process-based models with GIS technology, 182
location and topography of Fillmore County, Minnesota, 174f
mapping GLEAMS model results, 178
modeling study, 81–83
pesticide component, 175
potential for herbicide leaching into water table, 93
potential loss in runoff for atrazine as percentage of amount applied, 179f
potential loss in runoff for atrazine as percentage of amount applied for Forestville Creek springshed, 181f
potential loss in runoff for permethrin as percentage of amount applied, 180f
predicted loss of atrazine and permethrin in runoff to surface water as function of soil type and slope, 177f
process-based simulation model, 172–173
simple field validation of GLEAMS combined with GIS, 178, 181–182
soil half-life calculation, 83
sources of environmental data, 174–175
spatial distribution of soils in Fillmore County, 176
statistical distribution of predicted potential loss of atrazine and permethrin in runoff to surface-water for Fillmore County, 178t
study results, 89–91
See also Ground water contamination on sandy soils
Golf courses, standards of performance and maintenance, 10–11
Ground water contamination
management of irrigated corn and soybeans to minimize, 239–240
pesticide detection, 14
See also Herbicides in surface streams and ground water
Ground water contamination on sandy soils
agronomic factors of calibration trials, 82
alachlor detections in ground water for May to September, 85t
analytical methods, 84
atrazine detections in ground water for May to September, 86t
atrazine fate from application to storm event, 90t
climatic data for calibrating model, 82
field experiment, 81
field experiment results, 84, 88–89
frequency of detection in relation to half-life of herbicide, 93
GLEAMS soil half-life calculation, 83
modeling study results, 89–91
modeling study using GLEAMS model, 81–83
pesticide transport research, 80–81
rainfall simulation experiment, 83–84, 91–93
sensitivity analysis of plant uptake variable, 82–83
simazine detections in ground water for May to September, 87t
Ground water loading effects of agricultural management systems. See GLEAMS (ground water loading effects of agricultural management systems) model

H

Halosulfuron methyl
chemical properties and environmental characteristics, 253t
structure, 252f

See also Herbicides in surface streams and ground water
Henry's law constant, pesticide distribution between vapor and liquid phases, 329, 334
Herbicides
field losses by volatilization, 11
fish kills, 8
production and consumption, 4t
registered chemicals, 4–5
See also Ground water contamination on sandy soils; Rice herbicide discharges; Runoff; Snowmelt runoff
Herbicides in surface streams and ground water
advantages of using single-ion criterion mode of mass spectral analysis of low analyte concentration, 263
analytical equipment, 259
analytical method, 258–259, 261
analytical separation for sixteen target analytes, 259, 260f
attributes and advantages of ideal quantitative method for pesticide residue determination, 266, 269
capability of method to detect and quantify imidazolinone (IMI) and sulfonylurea (SU) families, 266
chemical properties and environmental characteristics of sixteen target analytes, 253t
chromatogram of sample extract from Iowa River near Rowan showing imazethapyr and nicosulfuron, 262f
chromatogram of sample extract showing detects in Iowa River, 264f
comparison of two selected-ion monitoring (SIM) approaches to identifying nicosulfuron and imazethapyr, 265f
comparison of two SIM approaches for identification, 263, 266
detection of target herbicides, 261, 263

DuPont-multianalyte method for monitoring water for herbicide residues, 269
intensely cultivated agricultural region of midwest, 249, 254
method validation, 261
molecular structures, common chemical names, and formulated product name for sulfonamide (SA), SU, and IMI herbicides, 250f, 251f, 252f
newer herbicide classes for conservation tillage grain production, 248–249
objectives of reconnaissance survey, 254
reassessing original results after eleven months, 263
reconnaissance duplicates for triazines and chloroacetanilides, 266, 268t
reconnaissance results for SU's, IMI's, and SA, 267t
reconnaissance survey, 254, 258
sampling schedule, procedure, and field quality assurance, 258
sampling site locations showing drainage relationship to Mississippi River, 255f
standards, instrument tuning, and calibration, 259, 261
study plan and sampling sites, 254
surface-water sampling data, 256t
well water sampling, 257t
Herbicide transport in subsurface drainage water
amounts of atrazine and metolachlor lost from Nashua plots in subsurface drainage, 195f
annual losses of atrazine in drainage waters as percentage of applied herbicide in relation to drainage water or annual precipitation, 191f
annual losses of atrazine in subsurface

drainage waters in different corn and corn-soybean production systems, 190*t*
application rates and herbicide losses, 194
comparing atrazine losses at different scales by monitoring drainage at field and watershed scales, 186, 187*f*
contradictory studies on effects of manure on herbicide persistence, 196
deethylatrazine/atrazine (DAR) ratios in river water, 189
effects of banding and broadcast application methods on mean annual losses of atrazine and metolachlor in subsurface drainage water from corn plots, 195*f*
effects of tillage on mean annual losses of herbicides in subsurface drainage from plots near Nashua, Iowa, 192*t*
fate of ^{14}C-atrazine in manured and untreated Floyd soil from Nashua soil, 196, 198*t*
little linkage between nitrate and herbicide in subsurface drainage, 186, 188
losses of atrazine in runoff and subsurface drainage, 188–189
magnitude of herbicide loss, 189–190
magnitude of herbicide losses in subsurface drainage waters from corn and soybean cropping systems, 185–186
manure effects on herbicide loss, 194, 196–198
mean annual herbicide losses in subsurface drainage water from continuous corn plots treated with manure and broadcast herbicides, 197*f*
methodology and scale, 188–189
soils and herbicide behavior, 190, 192–193
tillage effects on herbicide loss, 193

use of relatively small field sites to estimate subsurface drainage contribution, 189
Hexazinone. *See* Watershed fluxes of pesticides
Human risk assessment, atmospheric transport and pesticide deposition in precipitation, 12

I

Imazapyr
chemical properties and environmental characteristics, 253*t*
structure, 250*f*
See also Herbicides in surface streams and ground water
Imazaquin
chemical properties and environmental characteristics, 253*t*
structure, 250*f*
See also Herbicides in surface streams and ground water
Imazethapyr
chemical properties and environmental characteristics, 253*t*
chromatogram of sample extract from Iowa River, 261, 262*f*
comparison of two selected-ion monitoring (SIM) approaches for identification, 263, 265*f*
detection in water samples, 139, 141
development of flow injection liposomal immunoassay (FILIA), 128
dose-response curve from spiked water samples, 139, 140*f*
FILIA method, 138
imidazolinone class, 135
immunoassay and capillary electrophoresis methods, 32–33
low-use-rate herbicide, 141
maximum concentration at Chesapeake Farms, 34*t*

receipt to analyses intervals and
storage temperatures, 131t
stability, 132
structure, 137f, 250f
See also Flow injection liposomal
immunoassay (FILIA); Herbicides in
surface streams and ground water;
Watershed monitoring in sustainable
agriculture studies
Imidazolinone herbicides. *See*
Herbicides in surface streams and
ground water
Infiltration
pesticides leaching to water table, 7
transport mechanism through
Delaware sandy soil, 9
Insecticides
production and consumption, 4t
registered chemicals, 4–5
Iowa
evaluating impact of farming systems
on water quality, 237, 239
herbicide usage, 11
Iowa Ground Water Monitoring
(IGWM) program, 254
See also Herbicide transport in
subsurface drainage water;
Pesticides in air and precipitation;
Snowmelt runoff

L

Lake Erie basin
scale effect observations, 159–160
scale effects in assessment, 7–8
Leaching
impacting nonpoint source (NPS)
processes, 6
pesticides to water table, 7
See also Pesticide runoff and leaching
from farming systems
Long-term agriculture studies. *See*
Watershed monitoring in sustainable
agriculture studies

M

Malathion
comparing calculated and measured
enrichment factors, 341t
distribution between fogwater and
interstitial air in California, 333t
usage in agricultural watershed study,
100t
See also Watershed fluxes of pesticides
Management Systems Evaluation Areas
(MSEA)
alternative management systems for
enhancing water of aquifer
underlying claypan soils (Missouri),
239
buried valley aquifer management
systems evaluation area (Ohio), 240
conferences and workshops, 241–242
cropping systems evaluation working
group, 236
data base development and
management technical
subcommittee, 235
description of cropping and tillage
practices across projects, 238t
evaluation of impact of farming
systems on water quality (Iowa),
237, 239
expectation of projects to enhance
water quality, 240–241
formation, 233
future efforts, 245
general objectives, 233
hydrologic components for each
project, 243t
location of research sites in midwest
United States, 238f
management of irrigated corn and
soybeans to minimize ground water
contamination (Nebraska), 239–240
Midwest Initiative on Water Quality
(Minnesota), 239
overall accomplishments, 244–245
plot and field studies, 241

process modeling working group, 236–237
protecting water quality and assessing impact of chemicals and practices on ecosystem, 23
quality assurance/quality control technical subcommittee, 235–236
research studies, 242, 244
responsibility of steering committee, 234–235
socioeconomic evaluations working group, 237
structure, 233–237
study of best management practices (BMP) and water quality, 146–147
technical subcommittees and working groups, 234
technology transfer technical subcommittee, 236
watershed scale studies in Iowa and Missouri, 241
Manure
contradictory studies on effects on herbicide persistence, 196
effects on herbicide loss, 194, 196–198
mean annual herbicide losses in subsurface drainage water from continuous corn plots treated with manure and broadcast herbicides, 197f
See also Herbicide transport in subsurface drainage water
Maryland. See Pesticide runoff and leaching from farming systems
Mechanical incorporation, pesticide losses, 273
Mecoprop
turf strip/simulated rainfall, 277–278
See also Vegetated filter strips (VFS)
Merced River. See Pesticide transport
Methidathion
aqueous phase enrichment factors for distribution between air and fogwater, 335t
chemical applications, 293, 295t, 296t, 297t
chemical concentrations, 293, 298–300
daily average air concentrations during intense dormant spraying, 326, 327f
distribution between fogwater and interstitial air, 333t
oxon/thion ratios in fogwater using active strand fogwater sampler, tree drip, and nonfog air, 337t
worst case exposure analysis for Orestimba Creek contributions, 303
See also Organophosphorus insecticides; Pesticide transport and fate in fog
Metolachlor
annual herbicide concentrations and losses from watershed B and B-R, 27t, 28t
application rates on two plot sizes, 160t
buffer strips-tile outlet terrace/natural rainfall, 281–282
choice preemergence herbicide among Midwest farmers, 202
concentration exceedency curves, 48, 54f, 55
concentration in bulk atmospheric deposition samples within German Branch watershed, 111f
concentration in runoff as function of soil concentration, 163, 165, 167
concentration in runoff on two plot sizes, 162–163
concentration in soil and runoff for two plot sizes, 164t
concern for aquatic and toxicological risk, 21–22
detection in rainfall samples, 202, 222–223
detection in rainwater at four sites, 225f
detection rates, 42
detection without application during

study, 42–43
determination in snowmelt runoff, 149–150
distribution of annual maximum concentration, 48, 52f
distribution within lowest quartile of values during May through August, 48, 53f
magnitude of herbicide loss, 189–190
maximum concentration at Chesapeake Farms, 34t
maximum runoff concentrations, 39–40
monitoring data from tributaries, 10
pesticides in rainfall, 109
preemergence application after corn planting, 83–84
properties, 160t
rainfall simulation experiment, 83–84, 91–93
soil pore water concentrations in suction and passive lysimeter, 41t
tillage effects on herbicide loss, 193
time dependent desorption-phase distribution coefficients from four watersheds, 155t
usage in agricultural watershed study, 100t
vegetated filter strips (VFS)/simulated rainfall, 276–277
VFS/natural rainfall, 280
VFS/natural-simulated rainfall, 278–279
volatilization flux estimates, 210, 214
volatilization profiles, 11
See also Pesticides in air and precipitation; Runoff; Snowmelt runoff; Vegetated filter strips (VFS); Volatilization; Watershed fluxes of pesticides; Watershed monitoring in sustainable agriculture studies

Metribuzin
annual herbicide concentrations and losses from watershed B and B-R, 27t, 28t
lacking detection in lysimeter samples, 42
maximum concentration at Chesapeake Farms, 34t
maximum runoff concentrations, 39–40
soil pore water concentrations in suction and passive lysimeter, 41t
triazine herbicide for soybeans, 21
VFS/natural-simulated rainfall, 278–279
See also Vegetated filter strips (VFS); Watershed monitoring in sustainable agriculture studies

Metsulfuron methyl
chemical properties and environmental characteristics, 253t
structure, 250f
See also Herbicides in surface streams and ground water

Midwestern plains
atmospheric transport and deposition in precipitation, 12
intensity-of-usage patterns, 11
minimizing off-field agrochemical displacement, 12
patterns of herbicide usage, 12
volatilization as loss mechanism, 11

Midwest Water Quality Initiative
project description, 239
steering committee, 234–235
See also Management Systems Evaluation Areas (MSEA)

Minnesota
Midwest Initiative on Water Quality, 239
See also GLEAMS (ground water loading effects of agricultural management systems) model

Missouri, alternative management systems for enhancing water of aquifer underlying claypan soils, 239

Mobility, impacting nonpoint source (NPS) processes, 6

Models. *See* Dissolved organic matter

(DOM) in pesticide transport through soil; GLEAMS (ground water loading effects of agricultural management systems) model; SCI–GROW (screening concentration in ground water)

Molinate
 action levels and water quality goals, 69, 72
 dissipation and persistence, 68–69
 dissipation rates in rice field water, 70f
 history of use restrictions, 73t
 peak concentrations in Colusa basin drain, 75f
 peak concentrations in Sacramento River, 76f
 physical and chemical properties, 71t
 regulatory goals, 73t
 See also Rice herbicide discharges

N

Napropamide
 predictions using transport models, 347–348
 See also Dissolved organic matter (DOM) in pesticide transport through soil
National Stream Quality Accounting Network (NASQAN), 254
National Water Quality Assessment (NAWQA)
 protocols for surface and ground water, 258
 study plan and sampling sites, 254
Natural rainfall. *See* Vegetated filter strips (VFS)
Nebraska, management of irrigated corn and soybeans to minimize ground water contamination, 239–240
Nematocides, production and consumption, 4t
Nicosulfuron
 annual herbicide concentrations and losses from watershed A and A2, 26t
 annual herbicide concentrations and losses from watershed B and B-R, 27t, 28t
 annual herbicide concentrations and losses from watershed C, 29t
 annual herbicide concentrations and losses from watershed D, 29t
 chemical properties and environmental characteristics, 253t
 chromatogram of sample extract from Iowa River, 261, 262f
 comparison of two selected-ion monitoring (SIM) approaches for identification, 263, 265f
 concentration in surface water runoff from system A, 36, 37f
 maximum concentration at Chesapeake Farms, 34t
 postemergence spray, 22
 soil pore water concentrations in suction and passive lysimeter, 41t
 structure, 250f
 See also Herbicides in surface streams and ground water
Nitrate
 comparing chemographs for suspended solids, atrazine, and nitrate, 60, 61f
 cumulative loading curves, 59f
 effects of watershed size on concentrations exceedency curves, 57f
 percent of total load account for by 1% of time with highest loading rates, 60t
 percent total load versus percent total storm discharge for suspended solids, atrazine, and nitrate, 60, 62f
 See also Snowmelt runoff; Watershed size
Nitrite. *See* Watershed size
Nonpoint sources (NPS) pollution
 characteristics, 5–6
 in-field and off-site best management practices (BMP), 272–273
 multiplicity of operations contributing to agrochemical movement, 9–10

new agricultural ethic, 14–15
potential to damage nontarget environmental settings, 6–7
runoff and leachate as primary processes, 6
runoff of agricultural pesticides, 95–96
soil composition and landscape topography, 6
See also Vegetated filter strips (VFS)
Nontarget natural resources, issues of impact from pesticide usage, 2–3
Nutrients
concentration patterns, 55
loading patterns, 55
See also Watershed size

O

Ohio, buried valley aquifer management systems evaluation area, 240
Orestimba Creek. *See* Organophosphorus insecticides; Pesticide transport
Organic matter. *See* Dissolved organic matter (DOM) in pesticide transport through soil
Organic pesticides, variety and chemical complexity, 4
Organophosphorus insecticides
agricultural fields with insecticide applications draining into Orestimba Creek or falling within spray drift zone, 294f
chemical analysis methods, 293
chemical applications, 293
chemical concentrations, 293, 298–300
chemical mass loading downstream from sampling location, 302f
chemical monitoring data, 299f
comparison of weekly grab samples, 298
crops and methods in study area, 289, 292
field losses by volatilization, 11
first chemical applications to fields draining into Orestimba Creek or falling within spray drift zone, 295t
instrumentation, 292
interpretations regarding sources of chemical contamination, 298
map of central California showing study area in southern Stanislaus, 290f
materials and methods, 289, 292–293
monitoring information for chlorpyrifos, diazinon, and methidathion, 289
rainfall data from weather station and reported discharge, 301f
relationship between season and concentration/mass loading, 303
sample handling and shipping, 293
second chemical applications to fields draining into Orestimba Creek or falling within spray drift zone, 296t
storage stability, 132
stream hydrology and chemical loading into San Joaquin River, 300, 303
study area, 289, 292
third chemical applications to fields draining into Orestimba Creek or falling within spray drift zone, 297t
toxicity testing with invertebrate *Ceriodaphnia dubia*, 289
uncertainties and future research needs, 304
water features and measurement stations in study area, 291f
water sample collection and analysis, 292
worst case exposure analysis for Orestimba Creek contributions, 303
See also Pesticide transport and fate in fog

P

Paraoxon

aqueous phase enrichment factors for distribution between air and fogwater, 335t
comparing calculated and measured enrichment factors, 341t
See also Pesticide transport and fate in fog

Parathion
aqueous phase enrichment factors for distribution between air and fogwater, 335t
comparing calculated and measured enrichment factors, 341t
daily average air concentrations during intense dormant spraying, 326, 327f
distribution between fogwater and interstitial air, 333t
oxon/thion ratios in fogwater using active strand fogwater sampler, tree drip, and nonfog air, 337t
See also Pesticide transport and fate in fog

Partition theory, binding hydrophobic chemical to soil organic matter, 348

Pendimethalin
application rates on two plot sizes, 160t
comparing calculated and measured enrichment factors, 341t
distribution between fogwater and interstitial air, 333t
properties, 160t
usage in agricultural watershed study, 100t
yield in runoff, 167
See also Runoff

Pentachlorophenol
detection in all sample types, 222–223
See also Pesticides in air and precipitation

Permethrin
example chemical in GLEAMS model, 175–176
potential loss in runoff as percentage of amount applied for Fillmore County based on GLEAMS, 180f

predicted loss in runoff to surface water as function of soil type and slope, 177f
statistical distribution of predicted potential loss in runoff to surface-water, 178t
See also GLEAMS (ground water loading effects of agricultural management systems) model

Persistence, impacting nonpoint source (NPS) processes, 6

Pesticide loss
concept of mixing zone at soil surface, 273
conservation tillage, 273
herbicide banding in row crops, 273
mechanical incorporation, 273
See also GLEAMS (ground water loading effects of agricultural management systems) model

Pesticide runoff and leaching from farming systems
annual herbicide applications and surface water runoff losses from four watersheds, 34t
annual herbicide concentrations and losses for watershed A and A2, 26t
annual herbicide concentrations and losses for watershed B and B-R, 27t, 28t
annual herbicide concentrations and losses for watershed C, 29t
annual herbicide concentrations and losses for watershed D, 29t
Chesapeake Farms description, 23–24
concentrations of atrazine, cyanazine, and nicosulfuron in surface water runoff from system A, 36, 37f
concentrations of thifensulfuron methyl and tribenuron methyl in surface water runoff, 38f
cropping system A, 24
cropping system B, 26, 28
cropping system C, 28
cropping system D, 28, 30
description of small plots, 31

description of watersheds, 30–31
differences in maximum concentrations among systems, 42
drinking water standard as risk endpoint, 35
factors controlling concentrations and runoff of herbicides, 40
herbicide detection without application during study, 42–43
herbicide incorporation versus surface application, 35–36
herbicides in study, 21–23
immunoassay and capillary electrophoresis methods, 32–33
leaching losses, 40–43
materials and methods, 23–33
maximum concentrations of detected herbicides at Chesapeake Farms, 34t
models predicting surface water runoff and leaching, 34–35
runoff losses, 33–40
sample analysis methods, 32–33
sample handling and processing, 31–32
sampling equipment, 31
soil pore water concentrations in suction and passive lysimeter in systems, 41t
soil pore water (leachate) sample collection, 32
sustainable agriculture project at Chesapeake Farms, 23–24, 25f
water quality attributes of shorter residual sulfonylureas, 36, 39
Pesticide transport
annual diazinon application occurrence, 318
Chesapeake Bay and eastern shore, 9
comparing concentrations of pesticides to established water-quality criteria, 313
concentration showing toxicity to *Ceriodaphnia dubia*, 318, 320
correspondence between application time and occurrence, 320
correspondence between time of application and occurrence, 316
diazinon concentration in samples and stream flow for winter storm runoff in tributary and San Joaquin River, 319f
examining spatial and temporal variability of dissolved pesticide occurrence and concentrations, 307
factors affecting spatial and temporal occurrence of pesticides in surface water, 320
factors in transport of diazinon, 316
frequency of pesticide detection for each subbasin site and San Joaquin River site, 312f
main source of diazinon to San Joaquin River, 318
Merced River, 314
methods, 307, 309
monthly pesticide application, pesticide concentration in samples, and discharge for San Joaquin River, 317f
number of detected pesticides per sample at each subbasin site, 315f
Orestimba Creek, 313–314
overall occurrence, 309, 313
quality assurance of pesticide data, 309
reduction by vegetated filter strips (VFS) for hypothetical conditions, 284t
Salt Slough, 314
sampling frequency, 309
spatial correspondence between pesticide application and occurrence, 314, 316
spatial distribution of pesticides in surface waters, 313
spatial variation in pesticide occurrence and concentrations, 313–316
study area map including basin boundaries and sample site locations, 308f
summary of pesticide analyses for four

sites, 310t, 311t
temporal variation in pesticide occurrence and concentrations, 316
transport of diazinon by storm runoff, 316, 318–320
See also Dissolved organic matter (DOM) in pesticide transport through soil; Vegetated filter strips (VFS)
Pesticide transport and fate in fog
air residues, 324, 326
air-water distribution of pesticides, 329
airborne pesticide removal processes, 323–324
analysis of extracts of fogwater, particle filters, and vapor traps, 328
aqueous phase enrichment factors (EFs) for pesticide distribution between fogwater and air, 335t
arguments in favor of enrichment hypothesis, 339
Caltech active strand cloudwater collector, 330f
Caltech rotating arm collector, 330f
comparison of calculated and measured EFs, 341t
correlations including interface-air and interface-water coefficients, 339
daily average air concentrations of chlorpyrifos, diazinon, methidathion, and parathion, 326, 327f
determining uptake of pesticides in fog in wintertime, 334
distribution of parathion in foggy atmosphere near Parlier, CA, 335f
distribution of pesticides in air, fogwater, and plant surfaces at Kearney Agricultural Research Center at Parlier, CA, 336
distribution of pesticides in foggy Central Valley atmospheres, 334
EFs of pesticides and wood smoke markers versus octanol-water partition coefficient and water solubility, 340f
enrichment factor (EF), 334
EFs of chlorpyrifos, diazinon, parathion, and methidathion, 338
enrichment of hydrophobic versus hydrophilic pesticides, 336
enrichment phenomenon of chemical solutes in suspended aqueous phase of foggy atmosphere, 336
enrichment reproducibility in fog sampling, 338
evidence of interface involvement in enrichment process, 338–339
fog sampling and simultaneous collection of interstitial air, 328
fogwater collector methodology, 328
fogwater concentrations, 328
fogwater deposition to non-target food crops for indirect human exposure, 342
fogwater entrainment and concentration of various chemicals, 334, 336
Henry's law constant, 329, 334
high-volume dichotomous impactor for sampling interstitial pesticide vapors in fog, 332f
hypotheses explaining organic enrichment in fogwater, 338
indicator of pesticide distribution between vapor and liquid, 329, 334
occurrence of fog episodes in Central Valley, 324, 326
oxon/thion ratios in fogwater using active strand fogwater sampler, tree drip, and non-fog air, 337t
pesticides and alteration products in fog in California, 333t
pesticide use in California, 324
potential for regional or longer-range transport and deposition of Valley organophosphates, 343
residue averages of four organophosphates and their oxons in day and night air samples, 326, 327t
risks in humans and wildlife, 342–343
samples containing oxygen analogues of organophosphates, 329

significance of occurrence in foggy
  atmospheres and fogwater, 342–343
susceptibility of orchard dormant
  sprays entering air, 326
top 25 pesticides in California, 325t
types of fogwater collectors, 330f,
  331f
USDA high volume rotating screen
  atmospheric fog extractor, 331f
Pesticide usage
dissipation in field over time, 4
growth, 3
partitioning among hydrogeologic
  compartments, 3
potential impacts on nontarget natural
  resources, 2–3
production statistics, 4t
Pesticides
approaches to estimate contamination
  of surface and ground water, 23
component of American farming
  systems, 185
environmental impact minimization by
  soil processes, 347
leaching to water table by infiltration
  and runoff, 7
new agricultural ethic, 14–15
See also Watershed fluxes of pesticides
Pesticides in air and precipitation
acetochlor in rainwater at four sites,
  226f
air sample collection and measurement,
  218, 221
air sample results, 223, 229
air sampler data for all analytes from
  each site, 228t
atrazine in rainwater at four sites, 225f
comparison of atrazine in bulk and wet
  precipitation, 224f
2,4-D in rainwater at four sites, 226f
detections from dry deposition sampler
  data from each site, 230t
dry sample results, 229
highest pesticide detections, 222–223
maps of sampling sites in Johnson
  County, Iowa, 219f

materials and methods, 218–222
metolachlor in rainwater at four sites,
  225f
pesticides and related compounds in
  study, 220t
precipitation and dry deposition sample
  collection, 221–222
precipitation results, 223
quality assurance and quality control,
  222
rain and bulk precipitation data for
  chlorinated insecticides, 227t
rainfall data from Iowa Institute and
  Hydraulic Research, 224f
Phosphorus. *See* Watershed size
Plot sizes. *See* Runoff
Pollution
water quality, 5–7
See also Nonpoint sources (NPS)
  pollution
Poultry farms, Chesapeake Bay and
  eastern shore, 8–9
Precipitation
human risk assessment, 12
See also Pesticide transport and fate in
  fog; Pesticides in air and
  precipitation
Presidential Initiative to Enhance Water
  Quality
central objectives, 146
principles, 232
See also Management Systems
  Evaluation Areas (MSEA)
Primisulfuron methyl
chemical properties and environmental
  characteristics, 253t
structure, 251f
See also Herbicides in surface streams
  and ground water
Prometon. *See* Watershed fluxes of
  pesticides
Prometryn, predictions using transport
  models, 347–348
Prosulfuron
chemical properties and environmental
  characteristics, 253t

structure, 252f
See also Herbicides in surface streams and ground water
Pyrethroids
  receipt to analyses intervals and storage temperatures, 131t
  storage stability, 132

Q

Quality, water. *See* Water quality

R

Rainfall, natural. *See* Vegetated filter strips (VFS)
Rainfall, simulated. *See* Vegetated filter strips (VFS)
Rainfall simulation
  research method, 83–84
  results, 91–93
  See also Ground water contamination on sandy soils
RaPID assays (rapid pesticide immuno detection assays), reanalysis of triazines and acetanilides, 130–131
Reconnaissance survey. *See* Herbicides in surface streams and ground water
Regulation
  action levels and water quality goals for rice herbicides in California, 69, 72
  limiting pesticide input to environment, 72
  need for, 68
  regulatory goals, 73t
  See also Rice herbicide discharges
Research programs, Management Systems Evaluation Areas (MSEA), 237–241
Research studies, Management Systems Evaluation Areas (MSEA), 242–244
Rice herbicide discharges
  action levels and water quality goals, 69, 72
  dissipation and persistence of rice herbicides, 68–69
  dissipation rates for molinate and thiobencarb in rice field water, 70f
  history of use restrictions of molinate and thiobencarb, 73t
  hydrology of California's Central Valley, 65, 67
  influence of environmental factors, 74, 77
  limiting pesticide input to environment, 72
  map of Sacramento River hydrologic basin, 66f
  need for regulation, 68
  peak molinate and thiobencarb concentrations in Colusa basin drain, 75f
  peak molinate and thiobencarb concentrations in Sacramento River, 76f
  physical and chemical properties of molinate and thiobencarb, 71t
  reactions of hydroxyl radicals with thiobencarb, 71f
  regulatory goals, 73t
  rice herbicides in Sacramento River watershed, 67–68
  rice production and herbicide use, 67
  use of major rice herbicides in California in 1982, 1988, and 1995, 70t
  watershed monitoring of Colusa basin drain and Sacramento River, 72, 74
Ridge-tillage system
  snowmelt runoff hydrographs, 153–154
  terraced, snowmelt runoff, 154
Risk assessment, human, atmospheric transport and pesticide deposition in precipitation, 12
Root Zone Water Quality Model (RZWQM)

data base development and management, 235
process modeling, 236–237

Runoff
atrazine and metolachlor concentration in soil and runoff for two plot sizes, 164t
atrazine concentration, 162t
comparative variable width notched box plots of sediment yield/runoff for two plot sizes, 168f
concentrations of atrazine and metolachlor as functions of soil concentrations, 163, 165, 167
concentrations of atrazine and metolachlor on two plot sizes, 161–163
herbicide application rates on two plot sizes, 160t
herbicide concentrations in surface water, 11
herbicide properties, 160t
impacting nonpoint source (NPS) processes, 6
losses with respect to plot size, 159–160, 169
metolachlor concentration, 164t
pesticide input from irrigation, 13
pesticides leaching to water table, 7
scale of agricultural studies, 96
sediment yield, 167–168
trifluralin and pendimethalin yield, 167
See also Pesticide runoff and leaching from farming systems; Snowmelt runoff

Runoff estimates. *See* GLEAMS (ground water loading effects of agricultural management systems) model

S

Sacramento River
hydrology, 65, 67
map of hydrologic basin, 66f
peak molinate and thiobencarb concentrations, 76f
rice herbicides, 67–68
watershed monitoring, 74
See also Rice herbicide discharges

Salt Slough. *See* Pesticide transport

Sandy soil
infiltration transport mechanism for pesticides, 9
See also Ground water contamination on sandy soils

San Joaquin Basin, 12–13

San Joaquin River. *See* Organophosphorus insecticides; Pesticide transport

San Joaquin Valley
airborne pesticide measurement, 14
drainage basins, 12–13
factors in transport to surface water, 13–14
major crop acreage, 13
pesticides input from irrigation runoff, 13
pesticide transport through soil profiles, 14
removal of airborne pesticides, 14
temporal and spatial pesticide distribution, 13

SCI–GROW (screening concentration in ground water)
agreement with observations, 42
model predicting leaching, 34–35

Simazine
detection in ground water, 88–89
detection in ground water for May to September, 87t
distribution between fogwater and interstitial air in California, 333t
field experiment for leaching, 81
preemergence application after corn planting, 83–84
rainfall simulation experiment, 83–84, 91–93
usage in agricultural watershed study, 100t

See also Ground water contamination on sandy soils; Watershed fluxes of pesticides
Simulated rainfall. *See* Vegetated filter strips (VFS)
Simulation models. *See* GLEAMS (ground water loading effects of agricultural management systems) model
Snowmelt runoff
  aerial photographs showing field boundaries and soil associations for four watersheds, 147*f*
  average depth of thaw equation, 156
  chemical determinations, 149–150
  comparison of four watersheds in terms of topographic characteristics, erosion properties, and agronomic practices, 148*t*
  conventional tillage of watersheds 1 and 2, 151–152
  desorption characteristics, 155–156
  environmental conditions, 150
  field drainage at watershed outlet, 151
  hydrograph from watershed 1 with fitted nitrate and herbicides concentration data, 151*f*
  hydrograph from watershed 2 with fitted nitrate and herbicides concentration data, 152*f*
  hydrograph from watershed 3 with fitted nitrate and herbicides concentration data, 153*f*
  hydrograph from watershed 4 with fitted nitrate and herbicides concentration data, 154*f*
  Management System Evaluation Area (MSEA) program, 146–147
  mobilization of agrochemicals, 150–151
  representative daily cycle of solar radiation, air temperature at soil surface, and soil temperature, 148–149*f*
  ridge-tillage of watershed 3, 153–154
  site description at Deep Loess Research Station, 147*f*, 148
  surface hydrology, 150
  surface water sampling, 149
  temperature measurement, 149
  temporal variation in concentration of agrochemicals, 151–156
  terraced ridge-tillage of watershed 4, 154
  time dependent desorption-phase distribution coefficients for atrazine and metolachlor release from four watersheds, 155*t*
  typical Loess hills surface topography showing relationship to sampling location for surface runoff, 149*f*
Soil composition
  affecting nonpoint source (NPS) processes, 6
  See also Dissolved organic matter (DOM) in pesticide transport through soil
Soil organic matter
  description, 348
  partition theory, 348
  sorption theory, 348
  See also Dissolved organic matter (DOM) in pesticide transport through soil
Sorption theory, binding hydrophobic chemical to soil organic matter, 348
Spray drift zone. *See* Organophosphorus insecticides
Spring season pattern. *See* Snowmelt runoff
Storm hydrographs
  comparing various storm characteristics, 48, 50*f*
  effects of watershed size on unit area hydrographs, 47–48, 49*f*
Subsurface drainage water. *See* Herbicide transport in subsurface drainage water
Sulfometuron methyl
  chemical properties and environmental

characteristics, 253t
structure, 251f
See also Herbicides in surface streams and ground water
Sulfonamide herbicides. *See* Herbicides in surface streams and ground water
Sulfonylureas
analysis by immunoassays, 33
annual concentrations and losses from watershed B and B-R, 27t, 28t
annual concentrations and losses from watershed C, 29t
annual concentrations and losses from watershed D, 29t
impacting quantities of pesticides, 22
leaching losses, 40–43
little water quality monitoring data, 22
See also Herbicides in surface streams and ground water
Surface streams. *See* Herbicides in surface streams and ground water
Surface water, pesticide detection, 14
Suspended solids
comparing chemographs for suspended solids, atrazine, and nitrate, 60, 61f
cumulative loading curves, 59f
effects of watershed size on concentrations exceedency curves, 57f
percent of total load account for by 1% of time with highest loading rates, 60t
percent total load versus percent total storm discharge for suspended solids, atrazine, and nitrate, 60, 62f
scale effects in assessment, 7–8
yield in runoff, 167–168
See also Runoff; Watershed size
Sustainable agriculture studies. *See* Watershed monitoring in sustainable agriculture studies

T

Tefluthrin, gas chromatography with electron capture detection (GC–ECD), 130
Terraced ridge-tillage system, snowmelt runoff hydrographs, 154
Thifensulfuron methyl
annual concentrations and losses from watershed B, 27t
annual concentrations and losses from watershed C, 29t
chemical properties and environmental characteristics, 253t
concentration in surface water runoff, 38f
lacking detection in lysimeter samples, 42
maximum concentration at Chesapeake Farms, 34t
minimal threat to surface and ground water quality, 22
soil pore water concentrations in suction and passive lysimeter, 41t
structure, 250f
See also Herbicides in surface streams and ground water
Thiobencarb
action levels and water quality goals, 69, 72
dissipation and persistence, 68–69
dissipation rates in rice field water, 70f
history of use restrictions, 73t
peak concentrations in Colusa basin drain, 75f
peak concentrations in Sacramento River, 76f
physical and chemical properties, 71t
reactions of hydroxyl radicals with, 71f
regulatory goals, 73t
See also Rice herbicide discharges

Tillage
  effects on herbicide loss, 193
  reducing mass of water and sediment carriers, 273
  snowmelt runoff hydrographs, 151–152
Topography, affecting nonpoint source (NPS) processes, 6
Transport. *See* Dissolved organic matter (DOM) in pesticide transport through soil; Pesticide transport; Pesticide transport and fate in fog
Triasulfuron
  chemical properties and environmental characteristics, 253*t*
  structure, 251*f*
  See also Herbicides in surface streams and ground water
Triazines
  annual concentrations and losses from watershed A and A2, 26*t*
  annual concentrations and losses from watershed B and B-R, 27*t*, 28*t*
  calculating pesticide loading at German Branch, 108
  concentrations relative to storm hydrograph at German Branch, 105*f*
  corn and soybean farmers, 22
  enzyme immunoassay, 130–131
  gas chromatography with electron capture detection (GC–ECD), 129–130
  herbicides in U.S. crop protection, 21
  monthly time weighted mean concentration at German Branch, 107*f*
  multi-residue scheme coupling solid phase extraction (SPE) and GC–ECD, 127
  receipt to analyses intervals and storage temperatures, 131*t*
  results on reconnaissance duplicates, 266, 268*t*
  storage stability, 131–132
  See also Watershed monitoring in sustainable agriculture studies
Tribenuron methyl
  annual concentrations and losses from watershed C, 29*t*
  annual concentrations and losses from watershed D, 29*t*
  concentration in surface water runoff, 38*f*
  lacking detection in lysimeter samples, 42
  maximum concentration at Chesapeake Farms, 34*t*
  minimal threat to surface and ground water quality, 22
  soil pore water concentrations in suction and passive lysimeter, 41*t*
  water quality attributes, 36, 39
1,1,1-Trichloro-2,2-bis(*p*-chlorophenyl)ethane (DDT)
  dissolved organic matter (DOM) increasing solution phase concentration, 349
  mobility enhancement by DOM, 350–351
  See also Dissolved organic matter (DOM) in pesticide transport through soil
Trifensulfuron methyl
  annual concentrations and losses from watershed D, 29*t*
  water quality attributes, 36, 39
Trifluralin
  application rates on two plot sizes, 160*t*
  grassed waterway/simulated rainfall, 275
  properties, 160*t*
  usage in agricultural watershed study, 100*t*
  yield in runoff, 167
  See also Runoff; Vegetated filter strips (VFS)
Triflusulfuron methyl
  chemical properties and environmental characteristics, 253*t*

structure, 251f
See also Herbicides in surface streams and ground water
Tulare Basin, San Joaquin Valley, 12–13
Turf, contributing mobile agrochemicals, 10–11

V

Vegetated filter strips (VFS)
annual reduction in runoff, metolachlor, and metribuzin by VFS as affected by tillage/cropping, 279t
application of rainfall, 282
buffer strip efficiency in herbicide removal, 277t
buffer strip efficiency in herbicide retention with and without vegetation, 277t
buffer strips in conjunction with in-field best management practices (BMP), 283
buffer strips-tile outlet terrace/natural rainfall/atrazine, metolachlor, and cyanazine, 281–282
concentrations and reduction in atrazine and alachlor transport as runoff passes through grass and forest buffer strips, 281t
field runoff source area to buffer strip area ratio, 282–283
function, 273–274
grassed waterway/simulated rainfall/2,4-D, 274–275
grassed waterway/simulated rainfall/trifluralin, 275
methods estimating VFS effects on pesticide transport, 282
oats strip/natural rainfall/atrazine, 279
percent herbicide loss from terraces after four runoff events, 282t
percent reduction in atrazine transport using oat strips, 279t
percent reduction of atrazine and sediment with 5:1 and 10:1 area ratio, 276t
percent reductions of dicamba, 2,4-D, mecoprop, and chlorpyrifos in runoff water, 278t
pesticide sorption potential for soil and sediment, 283
presence of dense, close-grown vegetation, 274
reduction in runoff and trifluralin in grassed waterway, 275t
reduction of water, sediment, and 2,4-D in grassed waterway, 275t
riparian forest buffer system/natural rainfall/atrazine and alachlor, 280–281
studies for pesticide transport reduction, 274–282
turf strip/simulated rainfall/dicamba, 2,4-D, mecoprop, and chlorpyrifos, 277–278
VFS/natural rainfall/atrazine, metolachlor, and cyanazine, 280
VFS/natural-simulated rainfall/metolachlor and metribuzin, 278–279
VFS pesticide transport reduction for hypothetical conditions, 284t
VFS/simulated rainfall/atrazine, 275–276
VFS/simulated rainfall/atrazine, metolachlor, and cyanazine, 276–277
Volatilization
ambient micrometeorological conditions, 207
bulk Richardson number, 205
continuous metolachlor flux estimates for 11 days, 213f
daily average metolachlor flux estimates, 209f
detection of metolachlor in rainfall

samples, 202
diabatic stability correction functions for momentum and sensible heat flux, 204
environmental factors, 201–202
flux-gradient theory using aerodynamic profile for metolachlor, 203–205
gradient profile, 203
loss mechanism of pesticide chemicals, 11
materials and methods, 205–207
metolachlor concentration profiles, 208, 210
metolachlor extraction method, 207
metolachlor flux estimates, 210, 214
metolachlor flux estimates for 24-hour period under clear sky conditions, 212f
metolachlor flux estimates for 24-hour period under patchy-cloud conditions, 213f
metolachlor vapor concentration profiles during first two hours after application, 211f
metolachlor vapor concentration profiles seven days after application in 1995, 212f
micrometeorological instruments measuring surface energy balance components, 206
pesticide sampling mast, 206–207
preliminary results from 1992, 208
preliminary study using flux-gradient technique, 205–206
relative humidity affecting dieldrin, 202
series of field scale volatilization studies, 205
theoretical considerations, 203–205

W

Water quality
alternative management systems for enhancing water of aquifer underlying claypan soils, 239
characteristics of pollution from nonpoint sources (NPS), 5–6
concentrations in agricultural watershed study, 102, 104
criteria and guidelines, 100t
evaluating impact of farming systems, 237, 239
expectations of projects to enhance, 240–241
goals for molinate and thiobencarb, 69, 72, 73t
potential to damage nontarget environmental settings, 6–7
Presidential Initiative to Enhanced Water Quality, 146, 232
See also Management Systems Evaluation Areas (MSEA); Organophosphorus insecticides; Rice herbicide discharges
Watershed fluxes of pesticides
basin-specific concentrations and fluxes at river fall lines, 122–123
comparison of estimated pesticide fluxes above fall line of Susquehanna River for synoptic and intensive studies, 123f
concentrations in Choptank and Nanticoke Rivers, 119–120
concentrations of pesticides in river fall line samples, 119, 121t
estimated pesticide fluxes for each river basin, 120, 122t
estimates of fluxes above river fall lines, 120–121
Fall Line Toxics Monitoring Program (FLTMP), 115
flux estimates, 119
focus of FLTMP, 116
Goulden large-sample extractor (GLSE) procedure, 117
hydrologic conditions for spring and fall synoptic sampling, 119
map of Chesapeake Bay showing watershed locations of monitored tributaries, 117f
materials and methods, 116–119

motives determining basin specific
 fluxes, 121–122
pesticide analysis in GLSE extracts,
 118
pesticides in water samples from
 tributary fall lines, 118t
primary goal of FLTMP, 115–116
quality assurance samples, 118
river fall line sampling, 116
sample processing and analysis, 116–
 118
spring and autumn river discharges and
 basin areas above fall lines of
 monitored tributaries, 120t
Watershed monitoring in sustainable
 agriculture studies
analytical requirements and strategy,
 126–128
EIA (enzyme immunoassay) of
 triazines and acetanilides, 130–131
EIA false positive threshold at
 protocol level of quantitations
 (LOQ), 133t
EIA reruns and verification of GC–
 ECD data, 132–133
EIA sensitivity and cross-reactivity,
 133t
enzyme linked immunosorbent assays
 (ELISA) complementing multi-
 residue scheme, 127–128
experimental, 128–131
flow injection liposomal immunoassay
 (FILIA) development, 128
GC–ECD determination of
 esfenvalerate and tefluthrin, 130
GC–ECD determination of triazines
 and acetanilides, 129–130
goal to analyze pesticides in series of
 samples with varying matrices and
 concentration extremes, 126
imazethapyr, 132
importance of attaining maximum
 sensitivity for pesticides, 127
multi-residue scheme coupling solid
 phase extraction and GC–ECD, 127
pesticides applied, 127t
receipt to analyses intervals and
 storage temperatures, 131t
sample preparation, 128
solid extraction for FILIA, 129
solid phase extraction method, 128–
 129
solvent extraction for GC–ECD, 129
storage stability at 0°C and -20°C, 131
storage stability of triazines and
 acetanilides, 132t
synthetic pyrethroids and
 organophosphates, 132
triazines and acetanilides, 131–132
validation, 132
Watershed size
analytical program, 47
annual unit area hydrographs for
 Maumee River and Rock Creek, 49f
comparison of chemographs for
 suspended sediments, atrazine, and
 nitrate, 60, 61f
concentration exceedency curves for
 atrazine and metolachlor, 48, 54f, 55
concentration exceedency curves for
 various nutrients, 56f
distribution of annual maximum
 concentrations of atrazine and
 metolachlor, 48, 52f
distribution of atrazine and
 metolachlor concentrations within
 lowest quartile of values during May
 through August, 48, 53f
effects on annual herbicide
 chemographs, 48, 51f
effects on concentration exceedency
 curves for suspended solids and
 nitrate, 57f
effects on peak flow/storm volume,
 storm runoff duration, and peak flow
 to basin area, 48, 50f
effects on percent of loads versus
 percent of time for suspended solids
 and nitrate, 55, 59f
effects on unit area hydrographs, 49f

factors contributing to differences in runoff characteristics among watersheds, 46–47
factors giving rise to concentration and flow differences in two streams, 60
hypothesized causes of scale effects, 60, 63
illustrating asynchrony between chemical transport and water discharge during storm runoff event, 60, 63
increasing amounts of storage of pre-storm water in stream channels, 63
methods, 47
monitoring program for agricultural watersheds in Lake Erie Basin, 48t
nutrient and sediment concentrations patterns, 55
nutrient and sediment loading patterns, 55
percent of total load exported during 1% of time with highest loading rates, 55, 60t
percent of total load versus percent of total storm discharge for suspended sediments, atrazine, and nitrate, 60, 62f
percent total load versus percent of time for various parameters, 55, 58f
pesticide concentration patterns, 48, 55
storm hydrographs, 47–48

Watersheds
Chesapeake Bay and eastern shore, 9
comparing water quality, 8
concept for ranges of spatial scales, 7–8
factors contributing to differences in runoff characteristics, 46–47
hydrogeomorphology of areas of Chesapeake Bay, 96
monitoring program for agricultural watersheds in Lake Erie basin, 48t
See also Agricultural watershed study; Pesticide runoff and leaching from farming systems; Rice herbicide discharges

Wet deposition
airborne pesticide removal, 14
airborne pesticide removal process, 323
See also Pesticide transport and fate in fog

Wildlife protection, California's Central Valley, 8

Workshops, Management Systems Evaluation Areas (MSEA), 241–242

RETURN TO: CHEMISTRY LIBRARY
100 Hildebrand Hall • 642-3753

LOAN PERIOD	1	2	3 MONTH
4		5	6

ALL BOOKS MAY BE RECALLED AFTER 7 DAYS.
~~Renewable by telephone.~~

DUE AS STAMPED BELOW.

NON-CIRCULATING
UNTIL: 84′1,

DEC 19 2006

FORM NO. DD 10 UNIVERSITY OF CALIFORNIA, BERKELEY
3M 3-00 Berkeley, California 94720–6000